PERIODIC TABLE OF THE ELEMENTS

Key

1 — Atomic number	
Hydrogen — Name	
H — Symbol	
1.008 — Atomic weight	

Legend: Metals · Metalloids · Nonmetals

Group

Period	1 / 1A	2 / 2A	3 / 3B	4 / 4B	5 / 5B	6 / 6B	7 / 7B	8 / 8B	9 / 8B	10 / 8B	11 / 1B	12 / 2B	13 / 3A	14 / 4A	15 / 5A	16 / 6A	17 / 7A	18 / 8A
1	1 Hydrogen **H** 1.008																	2 Helium **He** 4.003
2	3 Lithium **Li** 6.94	4 Beryllium **Be** 9.01											5 Boron **B** 10.81	6 Carbon **C** 12.01	7 Nitrogen **N** 14.01	8 Oxygen **O** 16.00	9 Fluorine **F** 19.00	10 Neon **Ne** 20.18
3	11 Sodium **Na** 22.99	12 Magnesium **Mg** 24.31											13 Aluminum **Al** 26.98	14 Silicon **Si** 28.09	15 Phosphorus **P** 30.97	16 Sulfur **S** 32.07	17 Chlorine **Cl** 35.45	18 Argon **Ar** 39.95
4	19 Potassium **K** 39.10	20 Calcium **Ca** 40.08	21 Scandium **Sc** 44.96	22 Titanium **Ti** 47.87	23 Vanadium **V** 50.94	24 Chromium **Cr** 52.00	25 Manganese **Mn** 54.94	26 Iron **Fe** 55.85	27 Cobalt **Co** 58.93	28 Nickel **Ni** 58.69	29 Copper **Cu** 63.55	30 Zinc **Zn** 65.38	31 Gallium **Ga** 69.72	32 Germanium **Ge** 72.64	33 Arsenic **As** 74.92	34 Selenium **Se** 78.96	35 Bromine **Br** 79.90	36 Krypton **Kr** 83.80
5	37 Rubidium **Rb** 85.47	38 Strontium **Sr** 87.62	39 Yttrium **Y** 88.91	40 Zirconium **Zr** 91.22	41 Niobium **Nb** 92.91	42 Molybdenum **Mo** 95.96	43 Technetium **Tc** [98]	44 Ruthenium **Ru** 101.07	45 Rhodium **Rh** 102.91	46 Palladium **Pd** 106.42	47 Silver **Ag** 107.87	48 Cadmium **Cd** 112.41	49 Indium **In** 114.82	50 Tin **Sn** 118.71	51 Antimony **Sb** 121.76	52 Tellurium **Te** 127.60	53 Iodine **I** 126.90	54 Xenon **Xe** 131.29
6	55 Cesium **Cs** 132.91	56 Barium **Ba** 137.33	71 Lutetium **Lu** 174.97	72 Hafnium **Hf** 178.49	73 Tantalum **Ta** 180.95	74 Tungsten **W** 183.84	75 Rhenium **Re** 186.21	76 Osmium **Os** 190.23	77 Iridium **Ir** 192.22	78 Platinum **Pt** 195.08	79 Gold **Au** 196.97	80 Mercury **Hg** 200.59	81 Thallium **Tl** 204.38	82 Lead **Pb** 207.2	83 Bismuth **Bi** 208.98	84 Polonium **Po** [209]	85 Astatine **At** [210]	86 Radon **Rn** [222]
7	87 Francium **Fr** [223]	88 Radium **Ra** [226]	103 Lawrencium **Lr** [262]	104 Rutherfordium **Rf** [265]	105 Dubnium **Db** [268]	106 Seaborgium **Sg** [271]	107 Bohrium **Bh** [272]	108 Hassium **Hs** [277]	109 Meitnerium **Mt** [276]	110 Darmstadtium **Ds** [281]	111 Roentgenium **Rg** [280]	112 Copernicium **Cn** [285]						

Lanthanides

57 Lanthanum **La** 138.91	58 Cerium **Ce** 140.12	59 Praseodymium **Pr** 140.91	60 Neodymium **Nd** 144.24	61 Promethium **Pm** [145]	62 Samarium **Sm** 150.36	63 Europium **Eu** 151.96	64 Gadolinium **Gd** 157.25	65 Terbium **Tb** 158.93	66 Dysprosium **Dy** 162.50	67 Holmium **Ho** 164.93	68 Erbium **Er** 167.26	69 Thulium **Tm** 168.93	70 Ytterbium **Yb** 173.05

Actinides

89 Actinium **Ac** [227]	90 Thorium **Th** 232.04	91 Protactinium **Pa** 231.04	92 Uranium **U** 238.03	93 Neptunium **Np** [237]	94 Plutonium **Pu** [244]	95 Americium **Am** [243]	96 Curium **Cm** [247]	97 Berkelium **Bk** [247]	98 Californium **Cf** [251]	99 Einsteinium **Es** [252]	100 Fermium **Fm** [257]	101 Mendelevium **Md** [258]	102 Nobelium **No** [259]

CHEMISTRY

IN THE COMMUNITY

Sixth Edition
Volume 2 Units 5-7

AMERICAN CHEMICAL SOCIETY

Chief Editor: Angela Powers
Revision Team: Laurie Langdon, Thomas Pentecost, Cece Schwennsen, Michael Mury
ACS: LaTrease Garrison, Mary Kirchhoff, Marta Gmurczyk, Karen Kaleuati, Terri Taylor, Emily Bones, Cornithia Harris
Chemistry at Work: Christen Brownlee
ACS Committee on Chemical Safety: Harry J. Elston
Ancillary Materials: Susan Cooper, Michael Dianovsky, Sara Marchlewicz, Stephanie Ryan
Editorial Advisory Board: Steven Long (Chair), Henry Heikkinen, Cathy Middlecamp, Barbara Sitzman, Michael Tinnesand, Don Wink
Teacher Reviewers and Pilot Testers: Bonnie Bloom, Patricia Deibert, Pamela Diaz, Regis Goode, Jennifer Kieffer-Gerckens, John Novak, Deborah Pusateri, Barbara Sitzman, Linda Tilton

W. H. FREEMAN/BFW

Executive Editor: Ann Heath
Assistant Editor: Dora Figueiredo
Development Editor: Don Gecewicz
Media Editor: Dave Quinn
Executive Marketing Manager: Cindi Weiss
Photo Editor: Bianca Moscatelli
Photo Researcher: Jacqui Wong
Project Editor: Vivien Weiss
Design Manager: Blake Logan
Text Designer: Rae Grant Design
Cover Image: Per Eriksson/The Image Bank/Getty Images, illustration by ACS
Illustrations: Network Graphics
Illustrations Coordinator: Janice Donnola
Production Manager: Susan Wein
Composition: MPS Content Services, A Macmillan Company; Rae Grant Design
Printing and Binding: Quad Graphics

This material is based upon work supported by the National Science Foundation under Grant No. SED-88115424 and Grant No. MDR-8470104. Any opinions, findings, and conclusions or recommendations expressed in this publication are those of the authors and do not necessarily reflect the views of the National Science Foundation. Any mention of trade names does not imply endorsement by the National Science Foundation.

Library of Congress Control Number: 2011928734

ISBN-13: 978-0-692-96437-8

CHEMISTRY
IN THE COMMUNITY
Sixth Edition

CHEMCOM

A PROJECT OF THE
AMERICAN CHEMICAL SOCIETY

ACS
Chemistry for Life®

AMERICAN CHEMICAL SOCIETY

Important Notice

Chemistry in the Community (ChemCom) is intended for use by high school students in the classroom laboratory under the direct supervision of a qualified chemistry teacher. The experiments described in this book involve substances that may be harmful if they are misused or if the procedures described are not followed. Read cautions carefully and follow all directions. Do not use or combine any substances or materials not specifically called for in carrying out investigations. Other substances are mentioned for educational purposes only and should not be used by students unless the instructions specifically so indicate.

The materials, safety information, and procedures contained in this book are believed to be reliable. This information and these procedures should serve only as a starting point for good laboratory practices, and they do not purport to specify minimal legal standards or to represent the policy of the American Chemical Society. No warranty, guarantee, or representation is made by the American Chemical Society as to the accuracy or specificity of the information contained herein, and the American Chemical Society assumes no responsibility in connection therewith. The added safety information is intended to provide basic guidelines for safe practices. It cannot be assumed that all necessary warnings and precautionary measures are contained in the document or that other additional information and measures may not be required.

Safety and Laboratory Activity

In *ChemCom*, you will frequently complete laboratory investigations. While no human activity is completely risk free, if you use common sense, as well as chemical sense, and follow the rules of laboratory safety, you should encounter no problems. Chemical sense is just an extension of common sense. Sensible laboratory conduct won't happen by memorizing a list of rules, any more than a perfect score on a written driver's test ensures an excellent driving record. The true "driver's test" of chemical sense is your actual conduct in the laboratory.

You will find Rules of Laboratory Conduct on pages 8–10 in Unit 0.

BRIEF CONTENTS

CONTENTS

UNIT 4

WATER: EXPLORING SOLUTIONS   388

x   Contents

UNIT **5**

INDUSTRY: APPLYING CHEMICAL
      REACTIONS   502

UNIT **6**

ATOMS: NUCLEAR INTERACTIONS   584

INDUSTRY: APPLYING CHEMICAL REACTIONS

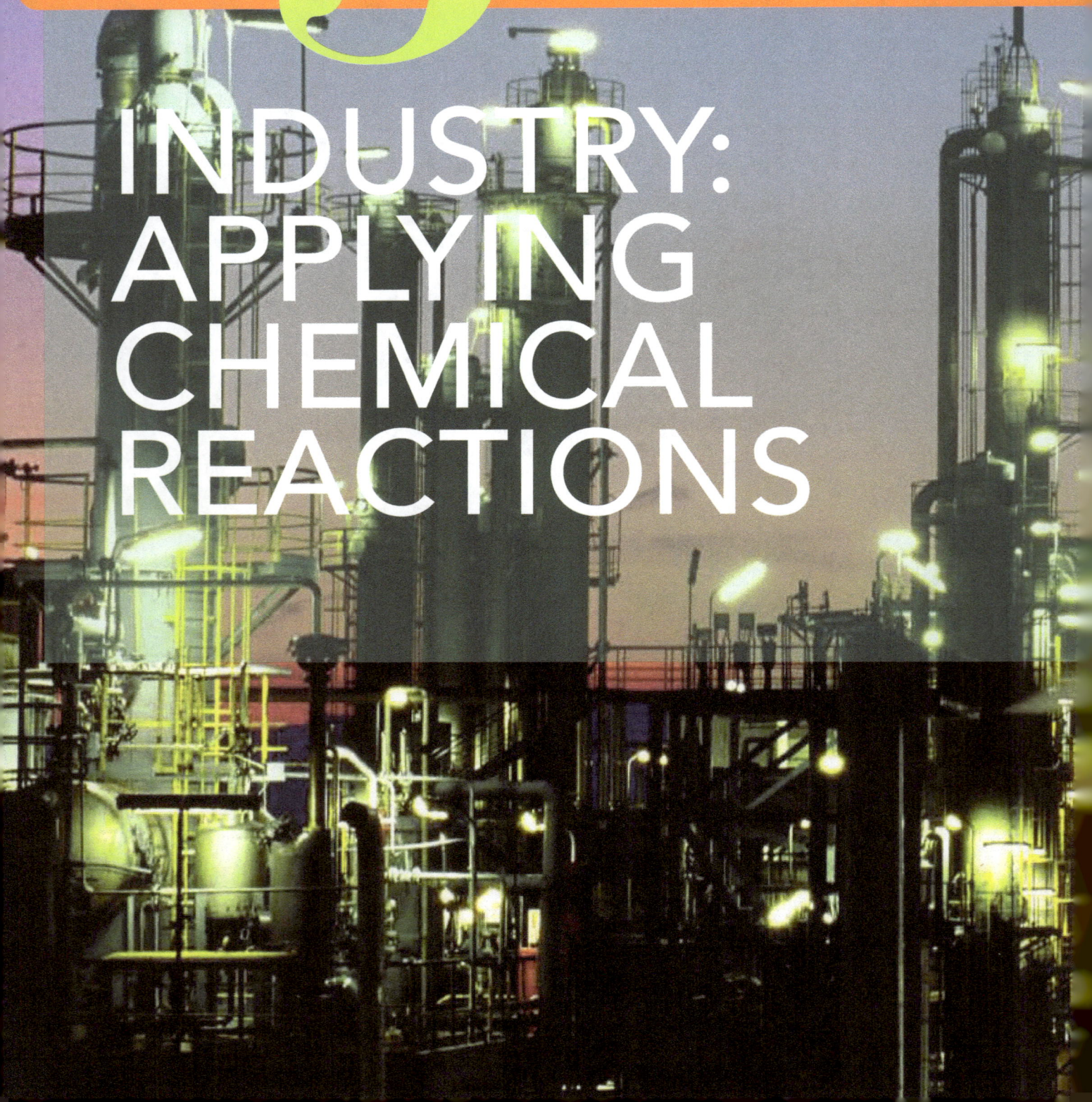

What role does nitrogen chemistry play in agriculture?

What chemical principles can be used in the responsible manufacture of ammonia?

How is chemical energy transformed into electrical energy?

What challenges must be met to optimize production and use of batteries?

?

Two large companies are interested in developing manufacturing facilities—and much-needed job opportunities—in Riverwood. What issues should citizens consider before inviting either company to locate in Riverwood?

Turn the page to learn about some chemical industries and the information the community needs to make a good decision.

More Job Options for Riverwood

A Riverwood News Commentary
By Greta Lederman   RIVERWOOD RESIDENT

Several weeks ago, the town council announced that two industrial firms are interested in converting the empty Riverwood Corportion building into a chemical manufacturing plant in Riverwood. Because of a lack of employment opportunities in Riverwood, many of your friends and neighbors have expressed relief at this news from EKS Nitrogen Products Company and WYE Battery Technology Corporation.

When the Riverwood Corporation declared bankruptcy last year, many residents lost their jobs. Although some have found new jobs, many have not. The unemployment rate in Riverwood has hovered near 15% for most of the past year.

Can anything be done to boost our local economy? Yes. We can allow one of these eager companies to build a chemical plant in our town. New jobs would be created for chemists, chemical technicians, and office personnel. Each company claims it would need to hire about 200 new employees.

We need more job opportunities; no one denies that. But some people may wonder whether a chemical plant is the best option for Riverwood. Aren't chemical plants dangerous? Should our community wait for a different opportunity (whenever and whatever that might be)?

I admit I was concerned about the issue of safety; thus I started doing my own research. I've found that both companies have strong safety and environmental records. They both advocate green technologies. The fact that both companies confirm plans to adapt the former River-

> **"** *We have a unique opportunity to choose between two companies that are competing to locate here. . . .* **"**

wood Corporation site instead of building an entirely new plant illustrates their commitment to sustainability.

EKS and WYE are attracted by our well-educated workforce and by access to abundant resources and electrical energy, among other factors. Both are well suited to our community.

We have a unique opportunity to choose between two companies that are competing to locate here. Usually it's the reverse—two or more cities compete for one company.

Learn more about each company: Visit the Web site of each company and attend the discussions scheduled around town.

If you still have doubts, notice how this news has affected your friends and neighbors. A new sense of hope has emerged over recent weeks. Let's turn weeks of hope into years of Riverwood prosperity.

Two companies are interested in redeveloping the old Riverwood Corporation building and site. Are the benefits to Riverwood worth the potential risks?

New Jobs May Be Too Costly

A Riverwood News Commentary

By Pak Jin-Woo   RIVERWOOD RESIDENT

Although no one can deny that our Riverwood community needs new jobs, it's foolish to invite either EKS Nitrogen Products Company or WYE Battery Technology Corporation to locate here to manufacture, respectively, ammonia or lithium-ion batteries, without first considering all the consequences.

The promise of 200 new jobs is alluring. However, all of these jobs would be in only one company. What will happen if there's a decline in the market for ammonia or lithium-ion batteries?

Wouldn't it be more prudent to distribute those 200 jobs among several different companies? Let's learn from our recent experience with the city's former main employer, the now bankrupt Riverwood Corporation.

Although long-term economic health is important, we must question the potential safety and environmental risks that each of these two chemical plants would pose to our community.

For instance, ammonia manufacturing requires very high pressures and temperatures. Although accidents are uncommon, the potential consequences of an explosion or spill are great. Several illnesses and even deaths of workers have been documented at ammonia plants.

Battery production also likely involves risks to its workers. Now, most lithium-ion batteries are produced outside the United States, and statistics on work-related injuries and deaths are difficult to find. However, a few large fires have been reported, both in battery production plants and battery recycling facilities. Also, do you remember large recalls of lithium-ion computer batteries in the last several years? Laptops were bursting into flames due to the batteries shorting out and heating up. Imagine what would happen if this caused a fire within the WYE battery plant.

Perhaps the greatest long-term concern is the potential for environmental harm. Ammonia-based contamination of the Snake River could, for example, cause another fish kill crisis. Along the same lines, manufacturing batteries comes with its own concerns regarding toxic metal waste. We should also consider what happens to batteries once their useful life is over. Will they be recycled? Or will they simply be left in a landfill somewhere, where their metals can leach into that community's drinking water?

I realize that not all of these scenarios are likely to happen. However, we must be willing to accept the consequences should something negative happen. With what level of risk are you and your friends willing to live?

Riverwood could participate in an exciting new industry if WYE Battery Technology Corporation locates here. Their proposed facility would be used to build lithium-ion batteries (shown on left) for electric and hybrid vehicles.

As you can infer from the newspaper commentaries you just read, two companies want to establish a chemical plant in Riverwood. As both commentators acknowledge, either company would offer at least 200 new job opportunities to the Riverwood community. However, job creation is far from the only factor to consider.

Later in this unit, you will help to decide whether to invite EKS or WYE (or neither company) to Riverwood. The chemistry you learn in this unit will prepare you to make informed decisions about the risks and benefits of operating such plants. Keep in mind the concerns in the two commentaries as you learn about the chemistry involved in manufacturing ammonia and batteries.

SECTION A PROVIDING NUTRIENTS FOR AGRICULTURE

What role does nitrogen chemistry play in agriculture?

In the first four units, you learned that chemistry is concerned with the composition and properties of matter, changes in matter, and the energy involved in those changes. In this unit, you will explore how chemical industries use chemical knowledge and reactions to produce a wide range of useful material goods on a large scale (see Figure 5.1).

In particular, this unit offers you the opportunity to evaluate the chemical operations of EKS Nitrogen Products Company and WYE Battery Technology Corporation. This knowledge will help you later as you debate whether or not a new chemical plant should locate in Riverwood, and, if so, which one.

Figure 5.1 *Agriculture and the chemical industry are involved in growing, preserving, and packaging the food products available in your local grocery store.*

GOALS

- List the main elements found in fertilizer.
- Describe the effect of each ingredient in a typical fertilizer on plant growth.
- Describe the nitrogen cycle, specifically referring to processes by which nitrogen gas is fixed (converted into nitrogen-containing compounds).
- Predict relative electronegativity trends among several elements.
- Determine whether covalently bonded atoms exhibit positive or negative oxidation states, based on their electronegativity values.

concept check 1

1. You studied the carbon cycle in Unit 2 and the water cycle in Unit 4.
   a. Does the cycling of carbon throughout the environment involve mainly chemical changes or physical changes? Explain.
   b. Does the cycling of water throughout the environment involve mainly chemical changes or physical changes? Explain.
2. Why might you perform a "confirming test" in the laboratory? Does a confirming test tell you how much of a substance is in your sample? Explain.
3. List three or more nutrients that plants need to live.

■ INVESTIGATING MATTER
A.1 FERTILIZER COMPONENTS

As you saw in the opening commentaries, the EKS Nitrogen Products Company wants to convert the old Riverwood Corporation building into a chemical plant. This plant would produce fertilizers, a product you may use at home. (See Figure 5.2.) What is in fertilizer that makes it useful for plant nourishment and growth? How can you decide whether a particular fertilizer is best for a specific application, such as on houseplants, lawns, or cornfields? One way is to find out if the fertilizer contains the proper ingredients. Complete fertilizers, such as those manufactured by EKS, contain the three main elements that growing plants need—nitrogen, phosphorus, and potassium—as well as trace ions and filler material.

Figure 5.2 *Some common forms of household fertilizer.*

Figure 5.3 shows the label of a typical commercial fertilizer bag. The numbers on the label indicate the percent values of key ingredients contained in the fertilizer: nitrogen, N; phosphorus, P (expressed as percent P_2O_5); and potassium, K (expressed as percent K_2O). The proportion of each component varies according to crop needs. Most lawn grasses need nitrogen: A 20–10–10 fertilizer is a good choice for lawns. Phosphorus is especially useful in promoting fruit and vegetable growth, so some gardeners prefer a 10–30–10 mixture over a balanced (10–10–10) fertilizer.

Many plant nutrients provided in fertilizers are in the form of cations and anions. Cations are likely to include potassium (K^+), ammonium (NH_4^+), and iron(II) (Fe^{2+}) or iron(III) (Fe^{3+}); anions include nitrate (NO_3^-), phosphate (PO_4^{3-}), and sulfate (SO_4^{2-}).

In this laboratory investigation, you will use confirming tests to determine whether a fertilizer solution contains specified nutrients.

Reporting P and K as P_2O_5 and K_2O (as in Figure 5.3) is traditional and originates from early research on fertilizers. When chemists burned fertilized plants to ash and analyzed the ash, they obtained the masses of P_2O_5 and K_2O.

You used confirming tests in the water testing investigation in Unit 4—see page 452.

Ideal for roses, azaleas, camellias, ferns, fuchsias, begonias, and other acid or shade loving plants.

GUARANTEED ANALYSIS

Total Nitrogen (N)... 4.00%
Available Phosphoric Acid (P_2O_5) 5.00%
Potash (K_2O)... 2.00%
Iron (Fe).. 1.00%

Derived from processed organic materials: Ammonium Nitrate, Ammonium Phosphate, Sulfate of Potash, and Iron Sulfate.

DIRECTIONS
Apply evenly by hand around base of plant out to drip line and water in well. A 1/8" deep layer around plant three times annually. For potted plants apply 2 tablespoons full per 6" pot.
LAWN & DICHONDRA — Apply by hand or spreader 2-1/2 lbs. per 100 sq. ft. (10 x 10). A 1 lb. coffee can holds approximately 2-1/2 lbs. of 4-5-2.

Figure 5.3 *This label indicates that this fertilizer contains 4% nitrogen, 5% P_2O_5, 2% K_2O, and 1% iron.*

Preparing to Investigate

You will test a fertilizer solution for six particular ions (three anions and three cations). In Part I, you will perform tests on known solutions of those ions to become familiar with each confirming test. In Part II, you will decide which ions are present in an unknown fertilizer solution. Read *Gathering Evidence* and prepare a suitable data table to record your observations.

Gathering Evidence

Part I. Ion Tests

1. Before you begin, put on your goggles, and wear them properly throughout the investigation.

2. Prepare a warm-water bath by adding ~30 mL of water to a 100-mL beaker. Place the beaker on a hot plate. The water must be warm, but it should not boil. Control the hot plate accordingly. You will use this bath in Step 7d.

> Is it possible to have a solution that contains only NO_3^- ions? What else must be in the solution for it to be electrically neutral?

3. Obtain a set of Beral pipets containing solutions of each of six known ions: nitrate (NO_3^-), phosphate (PO_4^{3-}), sulfate (SO_4^{2-}), ammonium (NH_4^+), iron(III) (Fe^{3+}), and potassium (K^+). Observe and record the colors of all six solutions.

$BaCl_2$ Test

Several of the ions you are studying in this activity can be identified first by their reaction with barium cations (Ba^{2+}) and then by their behavior in the presence of an acid.

> A *blank* was explained and used in Unit 4, page 452.

4. a. Add 2 to 3 drops of each test solution to six separate, clean test tubes. Add 2 to 3 drops of distilled water to a seventh clean test tube as a blank.

   b. Test each solution individually by adding 1 to 2 drops of 0.1 M barium chloride ($BaCl_2$) solution (see Figure 5.4). Record your observations.

   c. Add 3 drops of 6 M hydrochloric acid (HCl) to each of the seven test tubes. (*Caution: 6 M HCl is corrosive. If any splashes on your skin, wash it off thoroughly with water and inform your teacher. Do not inhale any HCl fumes.*) Record your observations.

   d. Dispose of the solutions as instructed, then clean and rinse the test tubes.

Figure 5.4 *Adding drops to a test tube.*

Brown-Ring Test

In the presence of nitrate ions (NO_3^-), mixing iron(II) ions (Fe^{2+}) with sulfuric acid (H_2SO_4) produces a distinctive result—a brown ring. You can use this *brown-ring test* to detect nitrate ions in a solution.

5. a. Add eight drops of sodium nitrate ($NaNO_3$) solution to a small, clean test tube. Place eight drops of distilled water in a second small, clean test tube, which will serve as a blank.

    b. Add about one milliliter of iron(II) sulfate ($FeSO_4$) solution to each of the two test tubes. Gently swirl to mix the contents of each tube.

    c. Obtain about one milliliter of concentrated sulfuric acid (H_2SO_4) for each tube. Your teacher will carefully pour H_2SO_4 along the inside of each test tube wall so the acid forms a second layer under the undisturbed solution already in the tube (see Figure 5.5). (**Caution:** *Concentrated H_2SO_4 is a very strong, corrosive acid. If any touches your skin, immediately wash affected areas with abundant running tap water and inform your teacher.*)

    d. Allow the test tubes to stand, unstirred, for 1 to 2 minutes.

    e. Observe any change that occurs at the interface of the two liquid layers. Record your observations.

6. Dispose of the solutions as instructed, then clean and rinse the test tubes.

NaOH and Litmus Tests

Now you will investigate which of the three cations (NH_4^+, Fe^{3+}, K^+) you can identify by observing their characteristic behavior in the presence of a strong base.

7. a. Add four drops of NH_4^+ cation test solution to a clean test tube. Repeat for the other two test solutions. Add four drops of distilled water to a fourth clean test tube, which will serve as a blank.

    b. Moisten four strips of red litmus paper with distilled water. Place them on a watch glass.

    c. Add 10 drops of 3 M sodium hydroxide (NaOH) directly to the solution in one of the four test tubes. (**Caution:** *3 M NaOH is corrosive. If any splashes on your skin, wash it off thoroughly with water and inform your teacher. Do not allow any NaOH solution to contact the test tube lip or inner wall.*) Immediately stick one of the four moistened red litmus-paper strips from Step 7b across the top of the test tube (see Figure 5.6). The strip must not contact the solution.

    d. Warm the test tube gently in the hot water bath for 1 min. Wait 30 s and then record your observations.

    e. Repeat Steps 7c and 7d for each of the other three test tubes.

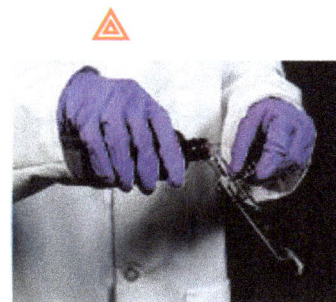

Figure 5.5 *To minimize mixing, concentrated sulfuric acid is allowed to run down the inside surface of the test tube. Possessing a larger density than water, concentrated sulfuric acid forms a layer beneath the water.*

Figure 5.6 *Place each strip of moistened pH paper across the top of a test tube.*

Flame Test

You can identify many metal ions by the characteristic color they emit when heated in a burner flame. For example, it is common to use a flame test to identify potassium ions.

8. a. Obtain a platinum or nichrome wire inserted into glass tubing or a cork stopper.

   b. Set up and light a Bunsen burner. Adjust the flame to produce a light blue, steady inner cone and a luminous, pale blue outer cone.

   c. To clean the wire, place ~10 drops of 2 M hydrochloric acid (HCl) in a small test tube. (**Caution:** *2 M HCl is corrosive. If any splashes on your skin, wash it off thoroughly with water and inform your teacher. Do not inhale any HCl fumes.)* Dip the wire into the hydrochloric acid and then heat the wire tip in the burner flame. Position the wire at the intersection of the two parts of the flame, not in the center cone. As the wire heats bright red, the burner flame may become colored, as illustrated in Figure 5.7. The characteristic colors are due to metallic cations held on the wire's surface.

Figure 5.7 *Completing a flame test. Metallic ions emit characteristic colors when heated.*

> The lack of color change in the flame indicates that the wire is clean.

   d. Continue dipping the wire into the acid solution and inserting the wire into the flame until there is little or no change in flame color as the wire heats to redness.

   e. Place 7 drops of the solution containing potassium ions into a clean test tube. Dip the cool, cleaned wire into this solution. Then insert the wire into the flame. Note any change in flame color, the intensity of the color, and the estimated time (in seconds) that the color was visible.

> The colored glass may make the characteristic potassium flame color easier to see against the color of the burner flame.

   f. Repeat the potassium-ion flame test, this time observing the burner flame through cobalt or didymium glass. Again, note the color, intensity, and duration of the color. Your partner can hold the wire in the flame while you observe through the colored glass. Then exchange roles. Record all observations.

KSCN Test

When potassium thiocyanate (KSCN) is added to an aqueous solution containing iron(III) ions (Fe^{3+}), a deep red color appears due to formation of $[FeSCN]^{2+}$ cations. Appearance of this characteristic color confirms the presence of iron(III) in a solution.

9. a. Place three drops of iron(III)-containing solution into a clean test tube. Place three drops of water into a second clean test tube.

   b. Add 1 drop of 0.1 M potassium thiocyanate (KSCN) solution to each test tube. Record your observations.

   c. Dispose of the solutions as instructed, then clean and rinse the test tubes.

Part II. Tests on Unknown Fertilizer Solution

1. Obtain a Beral pipet containing an unknown fertilizer solution. Record its code number. Your unknown solution contains one of the cations and one of the anions that you tested in Part I. Observe and record the color (if any) of the unknown solution.

2. Complete the same laboratory tests as in Part I on the fertilizer solution to identify the two unknown ions it contains. Record all observations and conclusions. Repeat a particular test, if you wish, to confirm your initial observations.

3. Dispose of all solutions used in this investigation as directed by your teacher.

4. Wash your hands thoroughly before leaving the laboratory.

Interpreting Evidence

1. Consider the test performed using $BaCl_2$ solution.

   a. Which of the six known ions are detected using this test?

   b. Write the net ionic equation for the reaction between $BaCl_2$ solution and the solution containing phosphate ions.

2. What would you observe when NaOH is combined with a solution containing Fe^{3+}?

3. Write the net ionic equation for the reaction between solutions containing NaOH and NH_4^+.

4. Describe what you would see if you added KSCN solution to a solution containing Fe^{3+} ions.

Making Claims

5. Identify the ions that were present in your unknown fertilizer sample, citing evidence from your investigation.

6. Is barium nitrate, $Ba(NO_3)_2$, soluble or insoluble in water? Cite specific evidence from your investigation to support your answer.

7. Based on your observations, which of the six ions studied in this investigation are always soluble?

Reflecting on the Investigation

8. Look up the components of common commercial fertilizers and provide the name and formula for two of the ingredients that contain one or both of the ions in your fertilizer sample. For example, sodium chloride (NaCl) would supply sodium ions (Na^+) and chloride ions (Cl^-) to a solution, while potassium carbonate (K_2CO_3) would furnish potassium ions (K^+) and carbonate ions (CO_3^{2-}).

9. Using evidence collected in your investigation, describe a test you could complete to verify whether a fertilizer sample contains phosphate ions (PO_4^{3-}).

10. Explain why the kind of data gathered in this investigation about a specific fertilizer solution is inadequate for you to judge whether the fertilizer is suitable for a particular use.

11. Consider the following results from an investigation of an unknown solution: The sample produced no visible flame color, although some sparks were produced. The solution formed a precipitate with NaOH that was red and jelly-like. When a drop of barium chloride was added to the solution, the resulting mixture was cloudy. The brown-ring test produced no visible brown-ring. The solution turned deep red when a drop of KSCN was added. Identify the ions in the solution.

A.2 FERTILIZER AND THE NITROGEN CYCLE

Liquefied ammonia can also be directly applied to soil as a fertilizer.

Every year, EKS manufactures about 3 million tons of ammonia and more than 1.5 million tons of nitric acid worldwide. Most is used to manufacture fertilizers sold to farmers and gardeners.

The purpose of all fertilizers is to add nutrients to soil so growing plants have adequate supplies. The raw materials that growing crops use are mainly carbon dioxide from the atmosphere and water and nutrients from the soil. Plant roots absorb water and nutrients such as phosphate, magnesium, potassium, and nitrate ions. You tested for the presence of many of these nutrients in a fertilizer in Investigating Matter A.1. How do plants use nutrients to grow and thrive?

Phosphate becomes part of the energy-storage molecule ATP (adenosine triphosphate) and the nucleic acids RNA and DNA. Magnesium ions are a key component of chlorophyll, which is required for photosynthesis. Potassium ions—found in the fluids and cells of most living things—help maintain a growing plant's ability to convert carbohydrates from one form to another, and to synthesize proteins. Nitrate ions aid vigorous plant growth and supply nitrogen atoms that are incorporated into proteins. Figure 5.8 illustrates how plants use some nutrients.

Potassium (K)
• Promotes cell-wall thickening, protecting plants against cold temperatures, pests, and disease
• Regulates water use and nutrient movement in cells
• Activates plant enzymes involved in processes such as photosynthesis and respiration

Nitrogen (N)
• Promotes new green growth
• Component of proteins, DNA, RNA, ATP, and chlorophyll

Phosphorus (P)
• Promotes production of blooms, flowers, seeds, and root development
• Component of DNA, RNA, and ATP

Adenosine triphosphate (ATP)—provides energy for cellular reactions

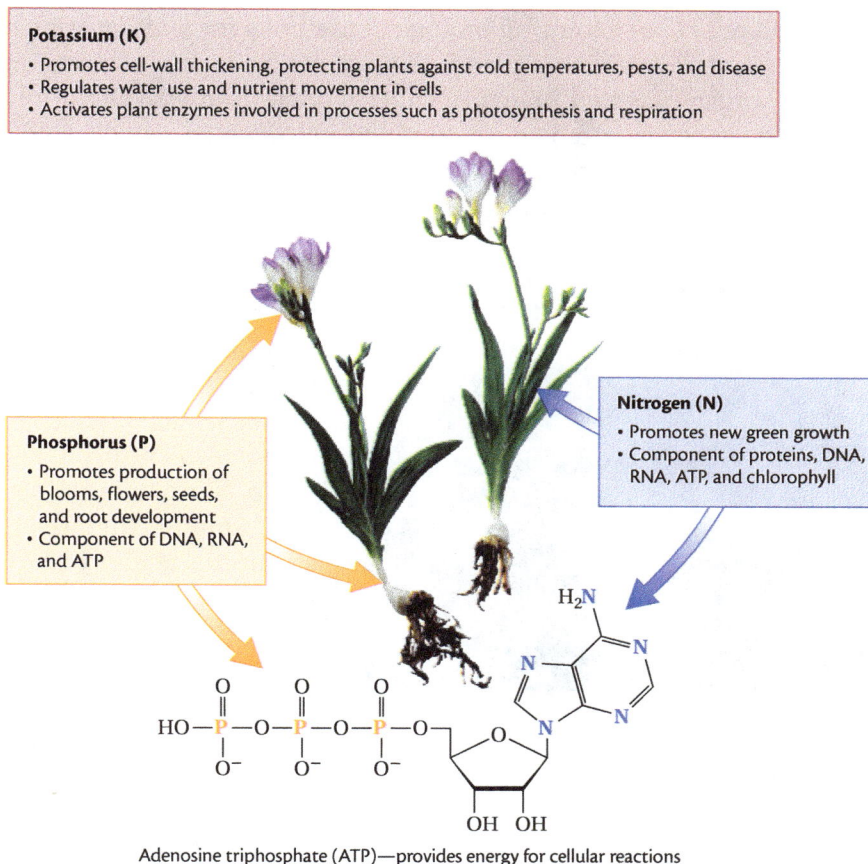

Figure 5.8 *All plants depend on particular nutrients participating as reactants in vital chemical reactions.*

Proteins are a major constituent of all living organisms, including plants. In plants, proteins usually account for 5 to 20% of the plant's mass. Nitrogen makes up about 16% of the mass of those protein molecules. Although nitrogen gas (N_2) is abundant in the atmosphere, it is so chemically stable (non-reactive) that plants cannot use it directly. However, nitrogen gas can be **fixed**—that is, combined with other elements to produce nitrogen-containing compounds that plants can use chemically. Some plants called legumes, such as clover and alfalfa, have nitrogen-fixing bacteria in their roots. Lightning (see Figure 5.9) is the other natural method of nitrogen fixation, but it accounts for a much smaller fraction of the total nitrogen fixed. Combustion can also fix atmospheric nitrogen by causing it to combine with other elements, especially oxygen.

Proteins are large molecules made up of various combinations of 20 different amino acids.

Figure 5.9 *Energy supplied by lightning facilitates reactions between atmospheric nitrogen and oxygen. This is one way to fix nitrogen, that is, to incorporate it into nitrogen-based compounds.*

Scientists are exploring biological methods for making atmospheric nitrogen more available to plants. These methods include engineering some microorganisms and plants to contain genes that will direct the production of nitrogen-fixing enzymes. Having such enzymes would allow plants or their bacteria to produce their own nitrogen-based fertilizers, just as the bacteria associated with legumes do.

When ammonia (NH_3) and ammonium ions (NH_4^+) enter soil from decaying matter and from other sources (see Figure 5.10), soil bacteria oxidize them to nitrate ions. Plants reduce nitrate ions to nitrite ions (NO_2^-), then to ammonia. They then use ammonia directly to synthesize amino acids. Unlike humans or other animals, many plants are able to synthesize all their needed amino acids by using ammonia or nitrate ions as initial nitrogen-containing reactants.

Figure 5.10 *Many U.S. farmers routinely apply liquid ammonia fertilizer to their fields. The label* anhydrous *on the tank means that the fertilizer contains no water.*

When organic matter decays, much of the released nitrogen recycles among plants and animals. Some returns to the atmosphere. Thus, some nitrogen gas removed from the atmosphere through nitrogen fixation eventually cycles back to the atmosphere.

The **nitrogen cycle** consists of the following steps:

1. Nitrogen-fixing bacteria that live in legume root nodules or in soil convert atmospheric nitrogen (N_2) to ammonia molecules or ammonium ions. Also, lightning converts N_2 to nitrogen oxides (NO_x).

2. Various soil bacteria oxidize ammonia and ammonium ions to nitrite ions and then to nitrate ions.

3. Plants take in nitrate ions from the soil, and then incorporate nitrogen atoms in synthesizing proteins and other nitrogen-containing compounds.

4. The nitrogen passes along the food chain to animals that feed on these plants and to animals that feed on other animals.

5. When those plants and animals die, bacteria and fungi take up and use some of the nitrogen from plant/animal protein and other nitrogen-containing molecules. The remaining nitrogen atoms are released from the decaying matter as ammonium ions and ammonia gas.

6. Denitrifying bacteria convert some ammonia, nitrite ions, and nitrate ions back to nitrogen gas, which returns to the atmosphere.

■ MAKING DECISIONS

A.3 PLANT NUTRIENTS

Consider and answer the following questions:

1. Why do some farmers alternate plantings of legumes and grain crops over several growing seasons (Figure 5.11)?
2. Why is it beneficial to return unused parts of harvested crops to the soil?
3. How might research on new ways to fix nitrogen help lower farmers' operating costs?
4. What consequences might result from over fertilizing?

Figure 5.11 *A contour strip farm in Wisconsin. Legumes (green) and grain (gold) are rotated over several growing seasons. How might this practice be beneficial?*

■ MODELING MATTER

A.4 THE NITROGEN CYCLE

In Unit 4, you learned how the hydrologic cycle can purify water (page 482). You also learned in Unit 3 that carbon-containing molecules change as carbon atoms cycle among living and nonliving components of Earth (page 345). For instance, plants transform carbon found as CO_2 into complex molecules, such as carbohydrates. Carbon is also chemically bound as dissolved CO_2 in the oceans and as carbonate rocks in Earth's crust (Figure 5.12).

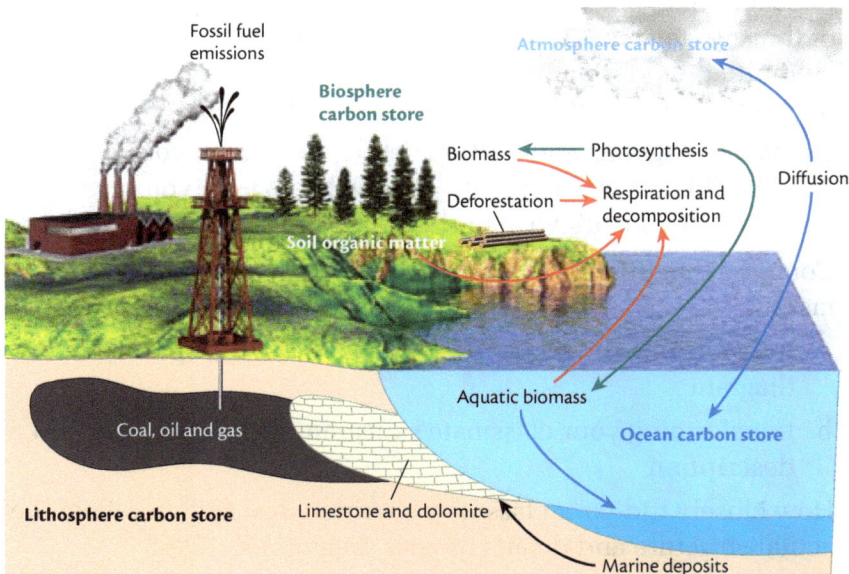

Fossil fuel emissions

Biosphere carbon store

Atmosphere carbon store

Biomass

Photosynthesis

Deforestation

Respiration and decomposition

Diffusion

Soil organic matter

Aquatic biomass

Coal, oil and gas

Ocean carbon store

Lithosphere carbon store

Limestone and dolomite

Marine deposits

Figure 5.12 *The carbon cycle.*

In addition to studying the hydrologic and carbon cycles, you examined figures illustrating those processes. Such visual models help you organize related information, allowing you to easily trace interactions and connections.

You have just learned that nitrogen atoms also cycle among the atmosphere, soil, and organisms. However, a visual model depicting this nitrogen-cycle information has not been presented. In this activity, you will create that missing diagram.

Look at Figure 5.12 as well as the figures on pages 346 and 483. Notice how they depict the carbon and hydrologic cycles. Then review Section A.2 Fertilizer and the Nitrogen Cycle (pages 512–514) in this unit (and earlier textbook material if needed) to guide your completion of the following steps.

1. Construct your own diagram of the nitrogen cycle.

   a. Use arrows to show the direction of flow as nitrogen atoms cycle among the atmosphere, soil, and living organisms.

   b. Include chemical formulas and names for key molecules and ions at each cycle stage.

   c. Use pictures and color as needed to clarify details in your model.

   d. Make your model easy to follow. A classmate should be able to summarize the steps correctly by studying your diagram.

2. Exchange your model with a classmate's model.

3. Select an appropriate starting point on your classmate's diagram and trace nitrogen through its cycle.

4. Repeat Step 3, but use your classmate's diagram to write a description of the key steps in the nitrogen cycle. Your written description should be limited to information in your classmate's diagram, even if some features are different from those in your diagram.

5. Exchange diagrams and written descriptions with your classmate. You should now have the nitrogen-cycle diagram you originally drew and your classmate's written description based on it.

6. Compare the difficulty you experienced in completing these two tasks:

   a. transforming the book's description of the nitrogen cycle into a diagram

   b. transforming your classmate's diagram of the cycle into a written description

7. How closely did your classmate's written description reflect the actual structure and details of your diagram? Explain.

8. Compare the description that your classmate wrote about your diagram with the description of cycle steps on page 514.

   a. Compared to the textbook description, did your classmate's description omit or add any details or steps? Explain.

   b. Which description is more detailed? Explain.

9. Based on your classmate's description,

   a. how easy and convenient was your diagram to interpret and follow? Explain.

   b. how could you modify your diagram to improve its accuracy or clarity? Explain.

10. Considering your answers to Question 9, make any needed changes to your diagram so it more clearly illustrates the nitrogen cycle.

So far, you have learned about the components of fertilizer and how they are utilized by plants. What are some natural sources of fertilizer, and how does chemistry apply to the production of synthetic fertilizer?

concept check 2

1. Is the nitrogen cycle more similar to the carbon cycle or to the water cycle? Why?
2. What does it mean for a substance to be oxidized?
3. In a nitrate ion, are the electrons that make up the covalent bonds between N and O shared equally by the atoms? Explain.

A.5 FIXING NITROGEN BY OXIDATION–REDUCTION

Fertilizer Sources

Before modern methods of manufacturing ammonia were developed, nitrogen-containing fertilizers came from either animal waste or nitrate compounds. Large quantities of such compounds came from Chilean guano beds (Figure 5.13). At the turn of the 20th century, speculation arose that guano beds would be depleted by about 1930, raising fears of an agricultural crisis and the specter of world famine. Actually, the development of commercial ammonia production largely eliminated dependence on natural nitrate sources.

Using commercial ammonia for fertilizer has had a huge impact on agriculture and world food supplies. World ammonia production has increased dramatically over the last 65 years, as farmers have increased their use of fertilizer to meet the food needs of growing populations. The U.S. chemical industry produces about 20 billion pounds of ammonia every year. Most is dedicated to making fertilizer.

Figure 5.13 *Guano (seabird dung) deposits were one of the first commercial sources of fertilizer.*

You will learn more about the Haber–Bosch process in Section B.

In seeking ways to fix nitrogen gas artificially, scientists in 1780 first combined atmospheric nitrogen and oxygen by exposing them to an electric spark. However, the cost of electricity made this too expensive for commercial use. A less-expensive method, the **Haber–Bosch process**, replaced it. Fritz Haber and Karl Bosch first demonstrated this technique for making ammonia from hydrogen gas and nitrogen gas in Germany in 1909, according to this equation:

$$N_2(g) + 3\,H_2(g) \rightleftharpoons 2\,NH_3(g)$$

Oxidation–Reduction

In the Haber–Bosch reaction of nitrogen gas and hydrogen gas, electrons are involved in an oxidation–reduction reaction. As you know, atoms that apparently lose their share of electrons are involved in the process called *oxidation* (see page 102). For example, the conversion of metallic sodium atoms (Na) into sodium ions (Na^+) is oxidation because electrically neutral sodium atoms are oxidized to +1 sodium ions by loss of one electron per atom. Recall that the opposite process—the apparent gaining of electrons—is called *reduction*. The formation of chloride ions (Cl^-) from electrically neutral chlorine atoms is an example of reduction. Electrons can be transferred to or from particular atoms, molecules, or ions. The products of such oxidation–reduction reactions also include atoms, molecules, or ions.

Regardless of how the electron-transfer occurs, the same oxidation–reduction principles apply.

You can judge the relative tendency of a covalently bonded atom to attract electrons in compounds from that element's *electronegativity* (see page 401). Nonmetallic elements typically have higher electronegativities than do metallic elements. Figure 5.14 shows the electronegativity values for some common elements.

Increasing Electronegativity →

Increasing Electronegativity ↑

H 2.1																	
Li 1.0	Be 1.5												B 2.0	C 2.5	N 3.0	O 3.5	F 4.0
Na 0.9	Mg 1.2												Al 1.5	Si 1.8	P 2.1	S 2.5	Cl 3.0
K 0.8	Ca 1.0	Sc 1.3	Ti 1.5	V 1.6	Cr 1.6	Mn 1.5	Fe 1.8	Co 1.8	Ni 1.8	Cu 1.9	Zn 1.6	Ga 1.6	Ge 1.8	As 2.0	Se 2.4	Br 2.8
Rb 0.8	Sr 1.0	Y 1.2	Zr 1.4	Nb 1.6	Mo 1.8	Tc 1.9	Ru 2.2	Rh 2.2	Pd 2.2	Ag 1.9	Cd 1.7	In 1.7	Sn 1.8	Sb 1.9	Te 2.1	I 2.5
Cs 0.7	Ba 0.9	Lu 1.2	Hf 1.3	Ta 1.5	W 1.7	Re 1.9	Os 2.2	Ir 2.2	Pt 2.2	Au 2.4	Hg 1.9	Ti 1.8	Pb 1.8	Bi 1.9	Po 2.0	At 2.2
Fr 0.7	Ra 0.9															

Figure 5.14 *Electronegativity values of selected elements.*

Oxidation States

A convenient, yet arbitrary, way to express the degree of oxidation or reduction of particular atoms is by assigning each atom an **oxidation state**. The higher (more positive) the oxidation state becomes, the more an atom has become oxidized. The lower (less positive) the atom's oxidation state becomes, the more the atom has become reduced.

In binary compounds (compounds composed of two elements), we assign atoms of the element with the lower electronegativity a **positive oxidation state**, corresponding to an apparent loss of electrons. Likewise, we assign atoms of the more electronegative element a **negative oxidation state**, corresponding to an apparent gain of electrons.

Consider the key chemical change in the Haber–Bosch process, as depicted with electron-dot formulas and space-filling models:

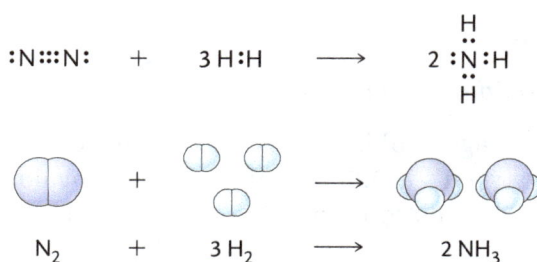

Note that each nitrogen atom in N_2 shares six electrons with another nitrogen atom, resulting in a triple covalent bond. Both nitrogen atoms exert equal attraction for the shared electrons. Each nitrogen atom in N_2 has a **zero oxidation state**. (This is true of any atom of an element that is not bonded to any other element.) The two nitrogen atoms share their bonding electrons equally. There is no separation of electrical charge. As the reaction progresses, each nitrogen atom becomes covalently bonded to three hydrogen atoms. Each bonded nitrogen and hydrogen atom shares an electron pair, but they do not share the pair equally. Nitrogen atoms (electronegativity value = 3.0) have a greater attraction for these shared electrons than do hydrogen atoms (electronegativity value = 2.1). Thus, the nitrogen atom in each NH_3 molecule acquires a greater portion of shared electrons than does each hydrogen atom.

Consequently, the nitrogen atom in NH_3 is assigned a negative oxidation state. Likewise, each hydrogen atom in NH_3 has lost some share of bonding electrons due to the reaction, so hydrogen is assigned a positive oxidation state. Thus, the oxidation state of an atom in a particular substance depends on the identity of neighboring atoms to which it is covalently bonded.

The Haber–Bosch process converts difficult-to-use nitrogen gas molecules into more useful ammonia molecules, a form of fixed nitrogen. Once nitrogen is chemically combined with another element, it can be readily converted to other nitrogen-containing compounds. For example, under proper conditions, ammonia will react with oxygen gas, forming nitrogen dioxide:

$$4\ NH_3(g)\ +\ 7\ O_2(g)\ \longrightarrow\ 4\ NO_2(g)\ +\ 6\ H_2O(g)$$

This is an oxidation–reduction reaction. In forming NO_2, the nitrogen atom in ammonia has apparently been oxidized; it has lost part of its share of electrons. Why does this occur? Because oxygen is more electronegative than nitrogen (O = 3.5; N = 3.0), oxygen attracts bonding electrons more strongly than nitrogen does. Each oxygen atom is considered to be reduced. Each oxygen atom has gained more control of its bonding electrons than it originally had in O_2.

DEVELOPING SKILLS

A.6 DETERMINING OXIDATION STATES

Sample Problem 1: *Which element in sulfur dioxide (SO_2) has a positive oxidation state?*

Figure 5.14 (page 518) indicates that sulfur's electronegativity value is 2.5 and oxygen's is 3.5. In sulfur–oxygen covalent bonding, sulfur has the weaker electron-attracting ability. Therefore, sulfur is assigned a positive oxidation state in SO_2.

Sample Problem 2: *The following equation represents the reaction of sulfur with oxygen gas to produce sulfur dioxide gas:*

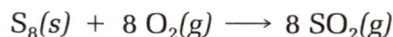

$$S_8(s) \; + \; 8\,O_2(g) \longrightarrow 8\,SO_2(g)$$

Why do chemists consider this reaction as an oxidation–reduction reaction in which sulfur is oxidized?

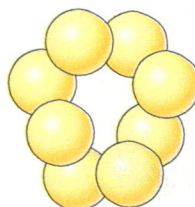

The element sulfur is found in eight-atom rings, S_8.

The oxidation state of sulfur changes from zero (in the pure element) to a positive value in the product, sulfur dioxide (see Sample Problem 1)—sulfur has been oxidized. By contrast, oxygen gas becomes reduced; it has become reduced from zero oxidation state to a negative oxidation state.

1. Consider each of the following covalent compounds. Using
    electronegativity values from Figure 5.14, decide which element in
    each compound has (i) a positive oxidation state and which has
    (ii) a negative oxidation state:

    a. ammonia, NH_3

    b. hydrogen chloride, HCl

    c. hydrogen fluoride, HF

    d. oxygen difluoride, OF_2

    e. iodine trifluoride, IF_3

    f. phosphorus trifluoride, PF_3

2. Each of these compounds includes a metallic element and a
    nonmetallic element. Select the element in each compound
    possessing (i) a positive oxidation state and (ii) a negative oxidation
    state.

    a. sodium iodide, NaI

    b. lead(II) fluoride, PbF_2

    c. lead(II) sulfide, PbS

    d. potassium oxide, K_2O

    e. iron(III) chloride, $FeCl_3$

    f. sodium phosphide, Na_3P

    g. sodium chloride, NaCl

3. Consider your answers to Questions 1 and 2. What conclusions can
    you draw about the oxidation states of metals and nonmetals in
    binary compounds?

4. Consider this chemical equation: $8\ Ni\ +\ S_8\ \longrightarrow\ 8\ NiS$.

    a. Does this equation represent an oxidation–reduction reaction?

    b. If so, identify the element oxidized and the element reduced. If
       not, explain why.

5. The element iron is part of an essential system of energy transfer
    within human cells. In that system, Fe^{2+} ions are converted to Fe^{3+}
    ions. Does that change represent an oxidation or a reduction?

6. Within the nitrogen cycle (see page 514), nitrogen gas (N_2) undergoes
    particular chemical reactions in which it is oxidized and other
    reactions in which it is reduced. Identify, by name and formula, a
    nitrogen-cycle product that forms by nitrogen gas being

    a. oxidized.

    b. reduced.

A.7 PRODUCTS OF THE CHEMICAL INDUSTRY

By now, you know there are very few things that we use regularly that have not been modified in some way. This is the chemical industry's focus—transforming natural resources into useful products to meet a wide variety of needs and purposes. The chemical industry also creates new substances and materials as replacements for natural ones. For example, plastics often replace wood or metals, and synthetic fibers often replace cotton or wool.

Even though the chemical industry is a worldwide, multibillion-dollar enterprise that affects everyone's life daily through its products and economic impact, most people are not aware of what happens when new materials are produced. This raises questions about how chemical industries operate, how they manufacture new products, and what those products contain.

The modern chemical industry employs well over a million people worldwide. Over the past 80 years, it has grown through mergers of smaller companies and creation of new companies. During that time, the industry's focus has expanded from a limited range of basic products to more than 70 000 products. Hundreds of chemical companies form the third-largest manufacturing industry in the United States. Only industries that produce machinery and electrical equipment are larger. In fact, if we include the food and petroleum industries in the chemical-industry category, this represents the world's *largest* industry.

Most chemical products reach the public indirectly because they are used to produce other consumer materials. For instance, the automobile and home-construction industries use enormous supplies of industrial chemicals. They use paints and plastics for automobile body parts such as bumpers, dashboard panels, upholstery and carpeting, and synthetic rubber in tires. Home construction involves large quantities of plastics for carpeting and flooring, insulation, siding, window frames, piping, and appliances. It also involves using paints, metals, and air-conditioning coolants. Figure 5.15 shows a range of products from various chemical industries.

Figure 5.15 *The chemical industry produces many materials that have useful properties.*

In Riverwood, the two industrial companies under consideration manufacture products involving nitrogen and lithium-ion batteries. Among the products manufactured by EKS Nitrogen Products Company are nitric acid and ammonia, which are often used in chemical reactions that produce other materials. By contrast, the batteries produced by WYE Battery Technology Corporation are used in the assembly of electric and hybrid-electric vehicles directly in their manufactured form.

EKS is committed to producing high-quality nitrogen-based fertilizer in Riverwood at reasonable cost, using the best available technologies. The production and sale of fertilizer represents a multimillion-dollar business that employs thousands of people worldwide and affects the lives of nearly everyone, from farmers and gardeners to food producers and consumers. This is why some Riverwood residents are eager to invite EKS into their community.

■ MAKING DECISIONS

A.8 CHEMICAL PROCESSING IN YOUR LIFE

To become more aware of how pervasive the products of chemical processing are in everyday life, list five items or materials around you that have *not* been manufactured, processed, or altered from their natural form. Start by considering everyday items, such as clothes, objects in your home, modes of transportation, books, foods, communication devices such as phones and computers, and sports and recreation equipment—all things you routinely encounter.

Now answer the following questions. Be prepared to share and discuss your answers in class.

Figure 5.16 *Are the materials used to package these products manufactured? Is this packaging necessary? What other packaging options exist?*

1. a. Which items on your list were wrapped, boxed, or shipped in materials that had been manufactured (Figure 5.16)? Explain.

    b. Is the packaging or shipping material necessary or simply a convenience? Why?

2. In what ways might each item or material on your list be better than or inferior to a manufactured, processed, or synthetic alternative? Consider factors such as cost, convenience, availability, and quality.

3. If a product is "100% natural," does that necessarily mean it was not involved in any processing or chemical or physical changes? Why? Support your answer with at least one example.

4. Many people avoid using synthetic fertilizers in their homes and gardens. Instead, they may use composted materials (food scraps and other plant and animal debris that have been decomposed).

    a. How does the practice of composting fit with the nitrogen cycle?

    b. Do you think that industrial composting could completely replace the use of synthetic fertilizers? Why or why not?

SECTION A SUMMARY

Reviewing the Concepts

> Fertilizers provide nutrients necessary for plant growth.

1. List the three main elements in fertilizer.

2. What would the expression *20–10–15* mean if found on a fertilizer label?

3. Why is potassium content expressed as percent K_2O in fertilizer?

4. List each test performed in Investigating Matter A.1 (page 507), along with the ions that the test identified.

5. Describe the role in plant growth of each of the three key ingredients in a typical fertilizer.

6. Why is it useful for fertilizers to be available with different compositions, such as 7–7–7 and 20–10–10?

> Nitrogen is transformed chemically as it cycles through living systems and the physical environment.

7. Why do plants need nitrogen?

8. Given the fact that nitrogen is abundant in the atmosphere, why is it included in fertilizers?

9. What does it mean to *fix* nitrogen gas?

10. List three ways in which atmospheric nitrogen can be fixed.

11. List two nitrogen-containing ions that are useful to plant growth.

12. How do plants and animals differ in the ways they obtain
    a. nitrogen?   b. amino acids?

13. What is one role of denitrifying bacteria in the environment?

14. Summarize the steps of the nitrogen cycle.

> The tendency of an atom to attract electrons within a covalent chemical bond can be expressed by the electronegativity of that element.

15. Referring to Figure 5.14 (page 518), identify the element that is most electronegative.
    a. List the element's symbol and name.
    b. Is this element a metal or nonmetal?

16. Referring to Figure 5.14 (page 518), identify the element that is least electronegative.
    a. List the element's symbol and name.
    b. Is this element a metal or nonmetal?

17. Describe how electronegativity values for elements change as one moves
    a. from left to right across any period of the periodic table.
    b. down a group of the periodic table.

18. Arrange each of the following sets of elements in order of their increasing attraction for electrons within a bond:
    a. silicon, sodium, and sulfur
    b. nitrogen, phosphorus, and potassium
    c. bromine, fluorine, lithium, and potassium

> Assigning oxidation states is a convenient way to identify oxidation or reduction of atoms in reactions.

19. What is the oxidation state of an atom in its elemental form (that is, not combined with an atom of another element)?

20. How does the oxidation state of an atom change when the atom is
    a. oxidized?   b. reduced?

21. How is it possible for the same element to be oxidized in one reaction and reduced in another?

22. What type of element (metal or nonmetal) is more often found in negative oxidation states when bonded to atoms of other elements?

23. Write the chemical formula for each of the following compounds and identify which element in each compound has a positive oxidation state and which has a negative oxidation state.
    a. water   c. carbon dioxide
    b. ammonia   d. magnesium chloride

> The chemical industry transforms elements and compounds into other useful materials.

24. List four different ways that the chemical industry is involved in the production of a box of breakfast cereal.

25. Look around wherever you are sitting right now and identify four things you see that are products of the chemical industry.

26. How is most of the ammonia produced in the United States used?

27. What was the most common source of ammonia before the 20th century?

28. Write the chemical equation that represents the main reaction in the Haber–Bosch process.

What role does nitrogen chemistry play in agriculture?

In this section, you investigated nutrients that plants need to grow and thrive, constructed a model of the nitrogen cycle, and applied your understanding of oxidation–reduction reactions to ways in which nitrogen compounds are converted, both naturally and industrially, to form useful products for agriculture. Think about what you have learned, then answer the question in your own words in organized paragraphs. Your answer should demonstrate your understanding of the key ideas in this section.

Be sure to consider the following in your response: role of nitrogen compounds in plant growth, forms of nitrogen that are most useful to plants, and natural and industrial methods for converting nitrogen compounds.

Connecting the Concepts

29. Describe one advantage and one disadvantage of using commercial fertilizer instead of guano or other animal waste to fertilize crops.

30. The quantity of phosphorus in fertilizer is reported as percent P_2O_5.

    a. What is the percent phosphorus in P_2O_5?

    b. Fertilizers are burned (oxidized) for analysis and the burning produces P_2O_5. In what form is the phosphorus actually found in the original fertilizer?

31. Describe a procedure for determining whether a soil sample contains any fixed nitrogen.

32. Compare the fertilizer tests conducted in this unit to the ion tests in Unit 4 (page 452). In what ways are these tests similar and in what ways are they different?

33. A magazine article claims that "oxygen is needed for all oxidation reactions." Do you agree or disagree with that statement? Support your answer with evidence.

34. What does the electronegativity of an electrically neutral atom indicate about its tendency to become oxidized?

35. In general, how do the electronegativities of metallic and nonmetallic elements compare?

36. Consider the Haber–Bosch process.

    a. Write the Lewis-dot structure for each reactant and product involved.

    b. Using the concept of electronegativity, determine which atoms in that reaction have a positive oxidation state and which have a negative oxidation state.

37. How does the concept of a limiting reactant apply to the use of fertilizers?

Extending the Concepts

38. You have considered three major natural cycles: water, carbon, and nitrogen. Compare these cycles with respect to

    a. conservation of mass,

    b. types of chemical change, and

    c. participating organisms.

39. Why do some vegetarians claim that their diets make more economical use of resources than the diets of non-vegetarians?

40. Some historians claim that development of the Haber–Bosch process prolonged World War I. Explain.

41. a. Magnesium is a key component of chlorophyll in green plants. Explain why magnesium is generally not included in commercial fertilizers.

    b. Identify some other substances required by growing plants that are not included in fertilizers.

42. Review the list of ingredients in a multipurpose vitamin capsule for humans and compare this to the ingredients in a typical commercial fertilizer. Suggest reasons for the similarities and differences you find.

43. How does lightning fix nitrogen?

44. Research and report on denitrifying bacteria. Include their typical habitats and any unusual characteristics.

SECTION B INDUSTRIAL PRODUCTION OF AMMONIA

What chemical principles can be used in the responsible manufacture of ammonia?

As you have learned, many industrial raw materials are extracted from Earth's crust (such as minerals, precious metals, sulfur, and petroleum), oceans (e.g., magnesium and bromine), and atmosphere. Nitrogen gas and oxygen gas, both obtained by low-temperature distillation from liquefied air, are valuable starting materials in the production of ammonia and nitric acid. As you will soon learn, producing ammonia also depends on understanding the implications of reversible reactions and chemical equilibrium.

GOALS

- Explain how temperature, reactant concentration(s), and the absence or presence of a catalyst affect reaction rates.
- Describe characteristics of a system in dynamic equilibrium.
- Use Le Châtelier's Principle to predict shifts in equilibria caused by perturbations to a system.
- Describe the Haber–Bosch process for industrial production of ammonia.
- Explain why many nitrogen-based compounds are effective chemical explosives.
- Describe how the production of ammonia can illustrate the goals and principles of green chemistry.

✓ concept check 3

1. In some chemical equations a double arrow, $\rightleftharpoons$, is used. What does this notation represent?
2. According to collision theory, how does an increase in temperature affect collisions between gas particles?
3. Is the following statement an accurate description of a chemical system at equilibrium? "At equilibrium, all reactions have stopped." Explain.

B.1 KINETICS AND EQUILIBRIUM

Producing ammonia from nitrogen gas and hydrogen gas is a chemical challenge. As you learned in Section A, molecular nitrogen (N_2) is very stable. This means that nitrogen fixation, the chemical combination of nitrogen gas with other elements, has a substantial activation-energy barrier. As you learned in Unit 2 (page 211), a reaction with a large energy barrier requires either that the reactant particles have substantial kinetic energy or that a catalyst reduces the energy required to initiate the reaction.

The reaction of nitrogen gas with hydrogen gas is also difficult because some ammonia molecules decompose back to nitrogen gas and hydrogen gas during the synthesis reaction. As you learned in Unit 2 (page 251), this kind of reaction—one in which products re-form reactants at the same time that reactants form products—is known as a *reversible reaction*. The double arrows used below indicate that both forward and reverse reactions occur simultaneously:

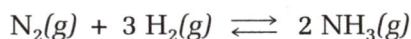

$$N_2(g) + 3\,H_2(g) \rightleftharpoons 2\,NH_3(g)$$

How do chemists and chemical engineers, whether at EKS Nitrogen Products Company or elsewhere, overcome these obstacles to produce ammonia?

Kinetics: Producing More Ammonia in Less Time

Chemical kinetics is the study of how fast chemical reactions occur. The **reaction rate** expresses how fast a particular chemical change occurs. To determine a reaction rate requires an experiment in which the amount (often measured as concentration) of one reacting substance is measured as a function of time. For instance, in synthesizing ammonia, if you know how fast nitrogen is being used and how long the reaction has been going on, you can calculate the amount of ammonia produced.

> Any rate is defined as the change in some quantity divided by the time required for the change. A common example of a rate is speed: miles per hour.

> You first learned about collision theory in Unit 2. According to this theory, the reaction rate also depends on the orientation of the colliding molecules.

For chemical reactions to occur, reactant molecules, atoms, or ions must collide with one another. According to **collision theory**, the reaction rate depends on the collision frequency and the energy involved in each collision (Figure 5.17). Increasing the concentrations of reactants speeds up reaction rates by increasing the number of collisions. High temperatures increase the reaction rate by providing more reacting molecules with sufficient energy to overcome the activation-energy barrier. Catalysts, on the other hand, increase the reaction rate by lowering the activation-energy barrier required for the reaction to occur, as shown in Figure 5.18.

Figure 5.17 *For a chemical reaction to occur, reactant particles must collide with proper orientation and sufficient energy. What are the benefits and limitations of using billiard balls to model simple molecular collisions?*

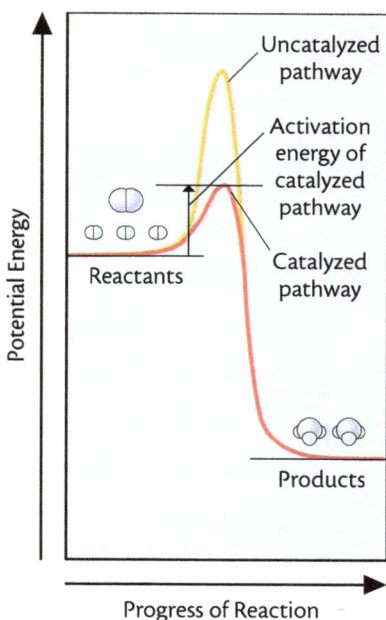

Figure 5.18 *A catalyst reduces the size of the activation energy barrier involved in synthesizing ammonia from its elements.*

Although a higher reaction temperature increases the average kinetic energies of the nitrogen and hydrogen molecules that react to form ammonia, ammonia itself becomes increasingly unstable at higher temperatures. The result is that ammonia decomposes back to nitrogen gas and hydrogen gas. If the reaction takes place at lower temperatures, fewer nitrogen and hydrogen molecules have enough energy to overcome the activation energy barrier, thus slowing the net rate of ammonia formation even though less ammonia decomposes at that lower temperature. So the formation of ammonia, the forward reaction, requires high temperatures, but high temperatures also increase the rate of decomposition of ammonia, the reverse reaction. What can we do to increase the rate of the forward reaction or decrease the rate of the reverse reaction, thus increasing the rate and yield of ammonia production?

The major breakthrough that led to profitable ammonia production was the discovery of a suitable catalyst. Catalysts made it possible to produce ammonia at lower temperatures (450–500 °C), thus slowing the rate of ammonia decomposition. Fritz Haber and his colleagues spent a great deal of time and energy in the early 1900s systematically searching for good catalysts. Today, the ammonia industry commonly employs iron as an ammonia synthesis catalyst.

Haber was awarded the 1918 Nobel Prize in Chemistry for his work on synthesizing ammonia.

Equilibrium: Favoring the Forward Reaction

Any reversible reaction appears to stop when the rate at which product forms equals the rate at which product reverts back to reactants—that is, when reactants and products attain **dynamic equilibrium**. At equilibrium, both the forward and the reverse reactions continue, but there is no further change in the amounts of reactants or products. At the point of dynamic equilibrium, the two opposing chemical changes are in exact balance, as modeled in Figure 5.19 and 5.20 (page 530).

Figure 5.19 *A system at dynamic equilibrium involves two ongoing processes acting in opposition to one another.*

Scientific American Conceptual Illustration

Figure 5.20 *Dynamic Equilibrium In the stoppered flask (left), we observe no overall change in the water level because its evaporation rate equals its rate of condensation. Water contained in the open flask (right) slowly escapes from the flask as water vapor. This happens because the rate of evaporation of water molecules is greater than the rate of condensation—this system is not in equilibrium. The closed system in the stoppered flask is, by contrast, an example of dynamic equilibrium.*

If the reverse reaction proceeds at an appreciable rate, then the amount of product that can be produced from the chemical reaction is decreased. Recall that in Unit 1 you learned how to predict the amount of product that would be produced by a known amount of reactant. When you made these calculations, you did not consider the possibility of a reverse reaction.

For many reactions, this is a valid simplification and it is not necessary to consider the reverse reaction. In these cases, the reverse reaction has a very slow rate, so we can assume that the forward reaction is the only important reaction.

However, the reverse reaction cannot be ignored in the synthesis of ammonia, so the net amount of ammonia formed from a given amount of nitrogen gas and hydrogen gas at a fixed temperature is limited by the competition between the forward and reverse reactions. One way to increase the amount of ammonia produced is to cool the ammonia as soon as it forms until it turns to a liquid and remove it from the reaction chamber. This prevents ammonia from decomposing back into nitrogen gas and hydrogen gas. If ammonia is continuously removed, the rate of the reverse reaction (decomposition of ammonia) is significantly decreased because there is less gaseous ammonia available to decompose. This causes the overall reaction (see equation on page 528) to favor the production of more ammonia because the rate of the forward reaction is greater than the rate of the reverse reaction.

> All reversible reactions in closed containers eventually reach equilibrium if conditions such as temperature remain constant.

Le Châtelier's Principle: Shifting the Equilibrium

The example above illustrates that a system at equilibrium can often be disturbed by changing the concentration of either reactants or products or by changing the temperature of the system. When one of these disturbances occurs, it causes either the forward or reverse reaction rate to become larger than the other and thus to be favored over the other. The reaction system will eventually reestablish dynamic equilibrium. However, the amounts of reactants and products present after equilibrium has been reestablished will be different from the amounts present before the disturbance.

For example, if more reactant was added to a system at equilibrium, the extra reactant will cause the forward reaction to increase, using up some of the added reactant, and produce more product. After the system returns to equilibrium, the concentration of the reactant that was added will be larger than the concentration of this reactant before the disturbance. The system has responded to the disturbance by undergoing a change that partially counteracts the initial effect of the disturbance. Thus, the initial equilibrium position is shifted. This effect, first described by the French chemist Henry Louis Le Châtelier in 1884, is commonly summarized in *Le Châtelier's principle.*

The external disturbance imposed on a system at equilibrium, sometimes called a *stress*, may be a change in the concentration of a particular reactant or product, a change in the temperature of the system, or (for a system including gases) a change in the total pressure. According to **Le Châtelier's principle**, the predicted shift in the equilibrium position is always in the direction that partially counteracts the imposed change in conditions. In the industrial production of ammonia, the removal of ammonia (a change in its concentration in the reaction vessel) results in the initial equilibrium position being shifted in favor of products. In the language of Le Châtelier's principle, the removal of ammonia is partially counteracted by the system, thus producing *more* ammonia.

> Shifting the position of an equilibrium system is often described by the direction of the favored reaction. If the disturbance causes the forward reaction to be favored, the equilibrium is shifted to the right side of the equation. If the reverse reaction is favored, the equilibrium is shifted to the left side of the equation.

Another external disturbance (stress) used to increase ammonia production is to add reactant molecules (nitrogen gas and hydrogen gas) continuously at high pressure. This higher pressure of reactant gases means the number of nitrogen and hydrogen gas molecules per unit volume is increased, thereby increasing the concentration of the reactants. The frequency of molecular collisions increases, which favors the forward reaction, and thus increases the amount of ammonia formed. This change can be viewed as partially counter-acting the initial increased pressure because the total number of gas molecules has been decreased. (Four molecules of gas—three molecules H_2 and one molecule N_2—are replaced by two molecules of NH_3 gas.)

In many cases, changing the system's temperature can also cause an equilibrium system to shift. The direction of that effect can be predicted based on whether the forward reaction is exothermic or endothermic. For example, the synthesis of ammonia is exothermic:

$$N_2(g) + 3\,H_2(g) \rightleftharpoons 2\,NH_3(g) + \text{Thermal energy}$$

Thermal energy can be regarded as a product of the forward (left to right) reaction. Raising the temperature would tend to favor the reverse (right to left) reaction, a chemical change that absorbs thermal energy. Consequently, less ammonia is formed at equilibrium at conditions of higher temperatures. Remember, though, that the temperature must be high enough to provide the nitrogen and hydrogen molecules with adequate kinetic energy to react. A delicate balance is needed. The temperature must be high enough to produce significant amounts of ammonia, but not so high that it promotes an excessive rate of ammonia decomposition.

INVESTIGATING MATTER
B.2 LE CHÂTELIER'S PRINCIPLE

Preparing to Investigate

In this investigation of a system at equilibrium, you will use what you have learned about Le Châtelier's principle to explore the effects of changing concentration and temperature on the position of an equilibrium system. The chemical system you will investigate is described by the following equilibrium equation:

$$\text{Thermal energy} + [Co(H_2O)_6]^{2+}(aq) + 4\,Cl^-(aq) \rightleftharpoons [CoCl_4]^{2-}(aq) + 6\,H_2O(l)$$

This system involves two complex ions. A **complex ion** is a chemical species composed of a single central atom or ion, usually a metal ion, to which other atoms, molecules, or ions are attached. One of your objectives in this investigation is to decide which complex ion above—$[CoCl_4]^{2-}$ or $[Co(H_2O)_6]^{2+}$—is blue and which is pink. See Figure 5.21.

Before you begin, read *Making Predictions* and *Gathering Evidence* to learn what you will need to do and note safety precautions. You will also need to create a data table that is appropriate for recording your data.

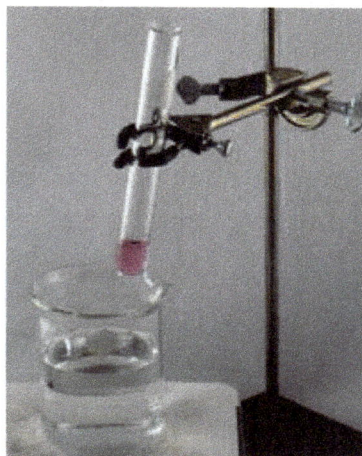

Figure 5.21 *Which complex ion is in this test tube?*

Making Predictions

Use the equation for this equilibrium system and what you know about solution chemistry to answer the following questions. Record the answers on your data sheet in a section labeled predictions.

1. How would the equilibrium system respond if chloride ion were removed from the system? Specifically, what would happen to the concentration of $(CoCl_4)^{2-}(aq)$?

2. How would the equilibrium system respond to an increase in temperature? Specifically, what would happen to the concentrations of $Cl^-(aq)$ and $(CoCl_4)^{2-}(aq)$?

3. In *Gathering Evidence*, you are told to add silver nitrate to the system. Which substance or substances present in the equilibrium will be affected by the addition of this reagent? Explain.

Gathering Evidence

1. Before you begin, put on your goggles, and wear them properly throughout the investigation.

2. Prepare a hot-water bath (60–70 °C) to use in Step 6.

3. Add 20 drops of 0.1 M cobalt(II) chloride $(CoCl_2)$ solution to a clean, dry test tube. Record the color.

4. Add 7 drops of 0.1 M silver nitrate $(AgNO_3)$ solution. *(**Caution:** AgNO₃ solution can stain skin and clothing. Handle with care.)*

5. Gently swirl the tube to ensure good mixing. Record the color.

6. Heat the tube in a hot-water bath for 30 seconds. Record the color.

7. Remove the tube from the hot-water bath. Add ~0.3 g sodium chloride (NaCl).

8. Gently agitate the tube. Heat the solution for 30 seconds in the hot-water bath. Record the color.

9. Place the test tube in a beaker containing ice water for 30 seconds. Record the color.

10. Reheat the test tube in the hot-water bath. Record the color.

11. Dispose of the mixture in the test tube as directed by your teacher.

12. Wash your hands thoroughly before leaving the laboratory.

Interpreting Evidence

To answer the following questions, refer to your observations and the equilibrium equation given in *Preparing to Investigate*.

1. What happened when you added $AgNO_3$ solution to the solution in the test tube?

2. What is the identity of the white precipitate that formed in Step 5? (*Hint:* Refer to Unit 4, page 448.)

3. Why did adding $AgNO_3$ solution affect the equilibrium, even though neither Ag^+ ions nor NO_3^- ions appear in the equilibrium equation? Write an equation to support your explanation.

4. Which way did the equilibrium shift after adding the $AgNO_3$ solution, toward reactants or products?

5. Which way did the equilibrium shift after adding NaCl in Step 7, toward reactants or products?

6. a. Which complex ion is pink: $[CoCl_4]^{2-}$ or $[Co(H_2O)_6]^{2+}$?

    b. Which complex ion is blue?

7. What was the effect of cooling the solution in Step 9?

8. Why did the solution's color change after heating in Step 8, but not in Step 6?

Making Claims

9. Review your results from Steps 5, 8, 9, and 10. Write a statement for each of these steps explaining how your results support or contradict Le Châtelier's principle.

Reflecting on the Investigation

10. Revisit the predictions you made before doing the experiment. Describe how well you were able to predict the behavior of the equilibrium system.

11. What would have been the effect if you had used hydrochloric acid in Step 7 instead of sodium chloride? Explain.

DEVELOPING SKILLS

B.3 CHEMICAL SYSTEMS AT EQUILIBRIUM

Sample Problem: *For the following equilibrium system, describe three changes you could make to increase the formation of nitric oxide gas, NO(g), at equilibrium.*

$$\text{Thermal energy} + 2\,NO_2(g) \rightleftharpoons 2\,NO(g) + O_2(g)$$

We need to identify factors that would cause the equilibrium to shift to the right, favoring NO production. Based on Le Châtelier's principle, any of the following changes would shift the equilibrium to the right: (i) increasing the temperature of the system, (ii) increasing the concentration of NO_2, or (iii) decreasing the concentration of either O_2 or NO.

1. For each equilibrium system, describe three changes you could make to favor the forward reaction.

    a. $2\,SO_2(g) + O_2(g) \rightleftharpoons 2\,SO_3(g) + \text{Thermal energy}$
    b. $H_2(g) + Cl_2(g) \rightleftharpoons 2\,HCl(g) + \text{Thermal energy}$

2. Examine the graph in Figure 5.22:

    a. What generalization can you make about the effect of temperature on the yield of ammonia?

    b. Do you think that generalization would remain valid for temperatures much lower than 400 °C? Explain your answer.

    c. What generalization can you make about the effect of total pressure on the yield of ammonia?

    d. What combination of temperature and pressure results in the highest ammonia yield?

3. For Questions 1a and 1b, describe three different changes you could make to favor the reverse reaction.

Figure 5.22 *Ammonia production in the system $N_2(g) + 3\,H_2(g) \rightleftharpoons 2\,NH_3(g) + \text{Thermal energy}$. The graph depicts the effect of pressure and temperature changes on the percent NH_3 present at equilibrium.*

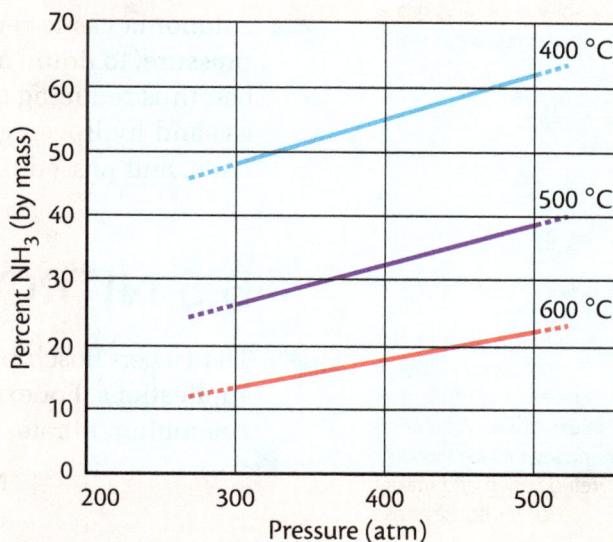

B.4 INDUSTRIAL SYNTHESIS OF AMMONIA

Large-scale ammonia production involves much more than allowing nitrogen gas and hydrogen gas to react in the presence of a catalyst. First, of course, the plant must obtain a continuous supply of the reactants. Nitrogen gas, which represents 78% of Earth's atmosphere, is liquefied from air through a series of steps involving cooling and compression. As you will soon learn, hydrogen gas can be obtained chemically from natural gas (mainly methane, CH_4).

To produce the hydrogen gas, chemical engineers first treat natural gas to remove sulfur compounds; then they allow methane to react with steam:

$$\text{Thermal energy} + CH_4(g) + H_2O(g) \rightleftharpoons 3\,H_2(g) + CO(g)$$

In modern ammonia plants, this endothermic reaction takes place at 200–600 °C and at pressures of 200–900 atm. Technicians, such as the one shown in Figure 5.23, must carefully control the ratio of methane to steam to prevent the formation of carbon compounds that would lower the yield of hydrogen gas.

Carbon monoxide, a product of the hydrogen-generating reaction shown above, is converted to carbon dioxide, which is accompanied by the production of additional hydrogen gas:

$$CO(g) + H_2O(g) \rightleftharpoons H_2(g) + CO_2(g)$$

All the hydrogen gas produced is then separated from carbon dioxide and from any unreacted methane.

In the Haber–Bosch process (see page 518), the reactants (hydrogen gas and nitrogen gas) are first compressed to high pressures (150–300 atm). Ammonia forms as the hot gases (at about 500 °C) flow over an iron catalyst. Ammonia gas is removed by converting it, by means of cooling and added pressure, to liquid ammonia, which is then removed from the reaction chamber, thus reducing the rate of ammonia decomposition. Nonreacted nitrogen gas and hydrogen gas are recycled, mixed with additional supplies of reactants, and passed through the reaction chamber again.

Figure 5.23 *This technician checks conditions within an industrial chemical storage system. What evidence suggests that safety is a priority?*

The CO_2 formed in this reaction can be removed in several ways, including allowing it to react with calcium oxide (CaO or lime), which forms solid calcium carbonate ($CaCO_3$), or by dissolving the CO_2 gas at high pressure in water.

About 80% of NH_3 produced is used in fertilizer and 5% is used in explosives. Ammonia's other uses include industrial refrigeration and making ice for hockey rinks.

B.5 NITROGEN'S OTHER FACE

The Haber–Bosch process has provided fairly cheap ammonia for a variety of applications. For example, ammonia reacts directly with nitric acid to produce ammonium nitrate, a substitute for natural nitrates once used as fertilizers.

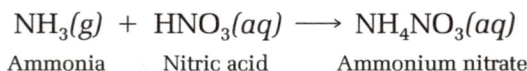

$$\underset{\text{Ammonia}}{NH_3(g)} + \underset{\text{Nitric acid}}{HNO_3(aq)} \longrightarrow \underset{\text{Ammonium nitrate}}{NH_4NO_3(aq)}$$

The widespread availability of ammonia and nitrates has changed the course of warfare as well as agriculture. Ammonia is a reactant in the production of explosives, most of which are nitrogen-containing compounds (see Figures 5.24 and 5.25). The development of the Haber–Bosch process provided a convenient source of ammonia for making both fertilizers and

Figure 5.24 *Explosive substances, such as those used here, are used in many mining and construction operations.*

military munitions. This allowed Germany to continue fighting in World War I even after its shipping connections to Chilean nitrate deposits were cut off by the British Navy.

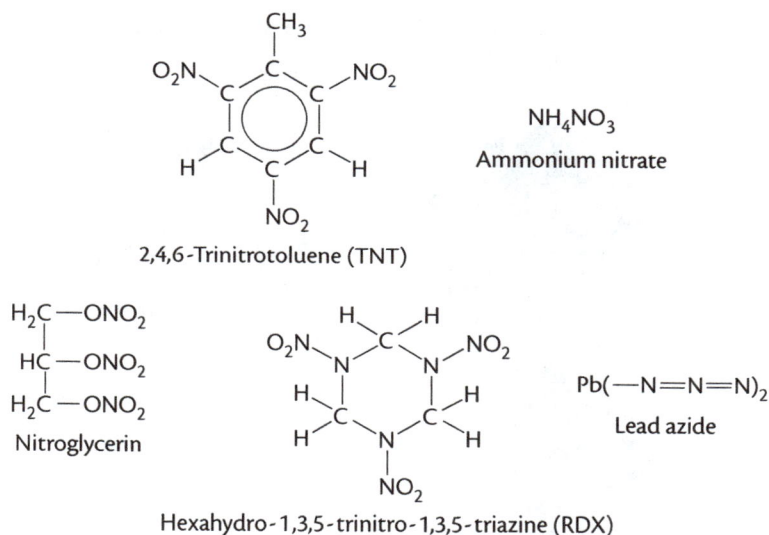

2,4,6-Trinitrotoluene (TNT)

NH_4NO_3
Ammonium nitrate

Nitroglycerin

$Pb(-N=N=N)_2$
Lead azide

Hexahydro-1,3,5-trinitro-1,3,5-triazine (RDX)

Figure 5.25 *Structural formulas of some common substances used as explosives. What similarities are shared by these chemical structures?*

Nitrogen-based explosives also have non-hostile, and even life-saving, uses. Air bags in automobiles are one such modern application. An air bag quickly inflates like a big pillow during a collision to reduce injuries to the driver and passengers. The uninflated air-bag assembly contains solid sodium azide, NaN_3. In a collision, sensors initiate a sequence of events that rapidly decompose the sodium azide to form a large volume of nitrogen gas:

$$3 \, NaN_3(s) \longrightarrow Na_3N(s) + 4 \, N_2(g)$$

Nitrogen gas quickly inflates the air bag (to about 50 L) within 50 ms (0.05 s) after the collision starts (see Figure 5.26).

   The forces released by nitrogen-based explosives also blast road-cuts through solid rock during highway construction. To cut through the stone faces of hills and mountains, road crews drill holes, drop in explosive canisters, and then detonate the explosives.

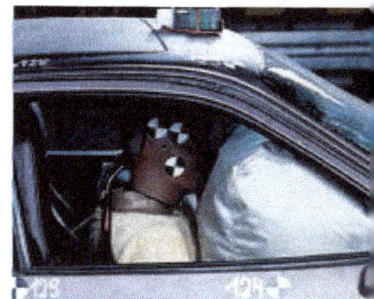

Figure 5.26 *Automobile air bags deploy due to nitrogen gas released from a nitrogen-based explosive compound.*

Explosions in general result from the rapid formation of gaseous products from liquid or solid reactants. Detonating an explosive such as sodium azide, dynamite, or nitroglycerin produces gases that occupy more than a thousand times the volume of the original solid or liquid.

Many compounds used as explosives involve nitrogen atoms in a positive oxidation state and carbon in a negative oxidation state within the same reactant molecule. This creates conditions for the release of vast quantities of energy. The energy released in this type of explosive reaction is due in part to the formation of N_2, a highly stable molecule.

The powerful explosive nitroglycerin (see Figure 5.25, page 537) was invented in 1846. However, it was too sensitive (unstable) to be useful. Workers never knew when it was going to explode. The Nobel family built a laboratory in Stockholm to explore ways to control this unstable substance (Figure 5.27). Although the father and four sons were interested in explosives, one son, Alfred, was the most persistent experimenter.

> Chemical explosions are often rapid, exothermic oxidation–reduction reactions that produce large volumes of gaseous product.

Figure 5.27 *The production of nitroglycerin, devised by Alfred Nobel, was a dangerous process that needed careful monitoring. Note the one-legged stool that ensured that the attendant didn't fall asleep on the job.*

Carelessness led to accidental explosions. One killed Alfred's brother Emil. The city of Stockholm finally insisted that Alfred move his experimenting elsewhere. Determined to continue research to make nitroglycerin less unpredictable and dangerous, Alfred rented a barge and performed experiments in the middle of a lake.

Alfred finally discovered that mixing oily nitroglycerin with a finely divided solid (diatomaceous earth) caused nitroglycerin to become stable enough for safe transportation and storage. However, the nitroglycerin would still explode if a blasting cap activated it. This new, more stable form of nitroglycerin carried a new name—*dynamite*.

A new era in explosives had begun. At first, dynamite served peaceful uses in mining and in road and tunnel construction (see Figure 5.24, page 537). By the late 1800s, dynamite also found destructive use in warfare.

Military use of his invention caused Alfred Nobel considerable anguish and motivated him to use his fortune to benefit humanity. His will specified that his money be dedicated to annual international prizes for advances in physics, chemistry, physiology and medicine, literature, and peace. (The Swedish parliament later added economics as an award category.) Nobel Prizes, first awarded in 1901, are still regarded as the highest honors individuals can receive in these fields. Table 5.1 lists recent Nobel Prize recipients in chemistry, together with their contributions to chemical science.

Nobel Laureates in Chemistry 2005–2010		
Year	**Awardees**	**Contributions**
2010	Richard F. Heck, United States, University of Delaware; Ei-ichi Negishi, Japan, Purdue University; Akira Suzuki, Japan, Hokkaido University	Development of palladium-catalyzed cross couplings in organic synthesis
2009	Venkatraman Ramakrishnan, India, MRC Laboratory of Molecular Biology; Thomas A. Steitz, United States, Yale University, Howard Hughes Medical Institute; Ada E. Yonath, Israel, Weizmann Institute of Science	Studies of the structure and function of the ribosome
2008	Osamu Shimomura, Japan, Marine Biological Laboratory, Woods Hole, and Boston University Medical School; Martin Chalfie, United States, Columbia University; Roger Y. Tsien, United States, University of California, San Diego, Howard Hughes Medical Institute	Discovery and development of the green fluorescent protein, GFP
2007	Gerhard Ertl, Germany, Fritz-Haber-Institut der Max-Planck-Gesellschaft	Work on chemical processes on solid surfaces
2006	Roger D. Kornberg, United States, Stanford University	Studies of the molecular basis of eukaryotic transcription
2005	Yves Chauvin, France, Institut Français du Pétrole; Robert H. Grubbs, United States, California Institute of Technology; Richard R. Schrock, United States, Massachusetts Institute of Technology	Development of the metathesis method in organic chemistry, reducing potentially hazardous waste through green-chemistry synthesis methods

Table 5.1 *Alfred Nobel's will established an annual award for those who "shall have made the most important chemical discovery or improvement."*

DEVELOPING SKILLS

B.6 EXPLOSIVE NITROGEN CHEMISTRY

Sample Problem: *Another nitrogen-based explosive is trinitrotoluene (TNT). The following reaction describes the explosion of TNT.*

$$2\ C_7H_5N_3O_6(s) \longrightarrow 6\ N_2(g)\ +\ 5\ H_2O(g)\ +\ 7\ CO(g)\ +\ 7\ C(s)$$

How many moles of CO are produced by the explosion of 1.5 moles of TNT?

$$1.5\ \text{mol TNT}\ \left(\frac{7\ \text{mol CO}}{2\ \text{mol TNT}}\right) = 5.3\ \text{mol CO}$$

The following equation describes the explosion of nitroglycerin:

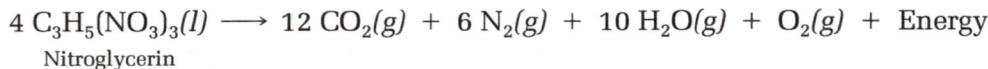

$$4\ C_3H_5(NO_3)_3(l) \longrightarrow 12\ CO_2(g)\ +\ 6\ N_2(g)\ +\ 10\ H_2O(g)\ +\ O_2(g)\ +\ \text{Energy}$$
Nitroglycerin

1. How many total moles of gaseous products form in the explosion of one mole of nitroglycerin?

2. One mole of gas at standard temperature and pressure occupies a volume of 22.4 L. One mole of liquid nitroglycerin occupies approximately 0.1 L. By what factor does the volume increase when one mole of nitroglycerin explodes? (Assume temperature remains constant.)

3. In fact, when nitroglycerin explodes, the temperature increase causes the gas volume to increase eight times more than the factor you just calculated in Question 2. By what combined factor does the total volume suddenly increase during an actual nitroglycerin explosion?

4. How do your answers to Questions 2 and 3 help explain the destructive power of a nitroglycerin explosion?

concept check 4

1. What would be the effect of removing a product from a system at equilibrium? Explain your answer.
2. When water is left in a glass it evaporates completely. Is this a violation of Le Châtelier's principle? Why or why not?
3. One advantage of the Riverwood location for EKS is the availability of relatively inexpensive energy. Why is access to energy an important consideration for EKS when picking a site for ammonia production? What parts of the production of ammonia require energy?

B.7 FROM RAW MATERIALS TO PRODUCTS

Some chemical reactions you have observed in this course are essentially the same as reactions used in industry to synthesize chemical products. However, chemical reactions in industry must be scaled up to produce very large quantities of high quality products at low cost.

Four considerations become crucial in attempting to scale up chemical reactions: *engineering, profitability, waste,* and *safety*. Chemical engineers face many challenges in designing manufacturing systems for industry. One challenge is the management of energy, usually thermal energy. In a classroom laboratory, the energy released or taken in by a reaction is relatively small. When a reaction is scaled up to industrial quantities, the management of this energy transfer becomes crucial. If thermal energy is not removed appropriately, potentially dangerous, costly, and destructive situations can occur. In the case of ammonia production, many of the reactions are endothermic, so thermal energy must be added to the reaction system. If this energy input is not managed correctly, the efficiency of the reaction process will decrease, increasing costs and decreasing profitability. You will learn more about the energy requirements of ammonia production in the next section.

The chemical industry also faces challenges in dealing with unwanted materials that result from chemical processes. When reactions occur on an industrial scale, large quantities of waste can accumulate. The definition of waste is anything that is not the desired product of the reaction. So waste can refer to excess reactants, other substances (byproducts) produced by the reaction, or excess thermal energy.

> A major responsibility of the U.S. Environmental Protection Agency (EPA) is managing the cleanup of hundreds of chemical-waste sites in the United States.

Once laws were passed that regulated the handling of waste, many chemical industries discovered that, with some additional processing, they could turn some previously unwanted materials or products into valuable commodities. Some "waste" compounds can become intermediates in producing useful substances. Such wastes-turned-resources offer new sources of income. In the production of ammonia, carbon dioxide is produced as a waste product during the conversion of methane into hydrogen gas. However, this carbon dioxide can be used to produce urea, another key agricultural chemical. As a result, ammonia production facilities often manufacture urea as well.

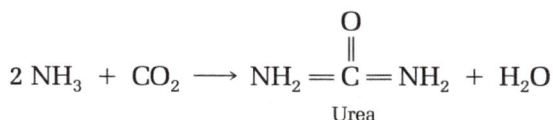

> The caffeine that is removed in the process of making decaffeinated coffee and teas is used in the manufacturing of caffeinated soft drinks and energy drinks.

$$2\,NH_3 + CO_2 \longrightarrow \overset{\displaystyle O}{\underset{\text{Urea}}{NH_2 = \overset{\|}{C} = NH_2}} + H_2O$$

Additionally, catalysts and modified processes have allowed manufacturers to increase their efficiency and decrease the amounts of starting materials (reactants) they need. For example, refinements to the Haber–Bosch process have increased the efficiency of the process. This efficiency has come from improvements in reactor design, improved use of thermal energy transfer, and the development of new catalysts for each step in the process. In the next section, you will look at the role of energy in the production of ammonia. This information will help you decide whether EKS should be invited to open an ammonia production facility in Riverwood.

B.8 AMMONIA AND ENERGY USE

By now, you know that the Haber–Bosch process, developed in the early 20th century, is still the dominant method of ammonia production. Even with improvements in efficiency, the production of ammonia and nitrogen-based fertilizers remains among the most energy intensive of common industrial processes. The fertilizer industry uses approximately 1.2% of all the human-generated power in the world. Production of ammonia (see Figure 5.28) represents 87% of this total.

As you learned in Section B.4, high temperatures and pressures are required for the successful reaction of hydrogen and nitrogen to form ammonia. In addition to this obvious use of energy, there are several less visible uses of energy in the overall ammonia production process. For instance, the preparation of each reactant requires energy before the actual reaction can take place. Nitrogen is widely available and would seem to be fairly economical to use, since it makes up about 80 percent of the air around us. However, nitrogen must be separated from air and purified before use in the Haber–Bosch process. This process requires a series of cooling and compression steps. Each of these steps requires energy.

Energy is also required to produce hydrogen. Of the several potential sources for hydrogen gas, methane or natural gas is most common. The first step in producing hydrogen from natural gas is the reaction of methane and steam.

Figure 5.28 *In ammonia production plants like the one shown here, large quantities of energy are required to maintain the high temperatures and pressures needed to synthesize ammonia.*

$$\text{Thermal energy} + CH_4(g) + H_2O(g) \rightleftharpoons 3\,H_2(g) + CO(g)$$

Note that this is an endothermic reaction, which means that energy must be added for this process to take place. Besides this energy input, this reaction consumes natural gas (methane), which is a fuel source. By using methane for this process, the methane is made unavailable for use as a fuel.

Reliance on methane as a source of hydrogen in ammonia production has another drawback. The cost associated with ammonia production is directly related to the price of natural gas. As natural gas prices have increased in Europe and the United States, the production of ammonia has decreased. Production has moved to areas of the world with more abundant supplies of, and lower prices for, natural gas.

As the price of natural gas continues to increase, interest in alternative sources of hydrogen for ammonia production grows. Other sources include coal gasification, biomass conversion, and electrolysis of water. However, none of these processes are as efficient as using methane.

One way to compensate for the increase in materials cost is to make the process more efficient. Recall that ammonia production facilities often are located with facilities that produce urea. This combination of production facilities is an example of how chemical industry can optimize the production process to minimize costs and waste generation. In the next section, you will learn more about the underlying principles of this type of optimization.

B.9 GREEN CHEMISTRY

Now that you know about the chemistry involved in the production of nitrogen-based fertilizers, you have part of the knowledge necessary to evaluate the EKS proposal to the town council. Another aspect to be considered is the relationship of the ammonia plant to the surrounding community.

An important global initiative, green chemistry, addresses the chemistry involved in an industrial process and its effects on humans and the environment. The goal is to make chemical products and production less hazardous to human health and to the environment. This initiative is sometimes termed *Benign by Design*. The word *benign* might be new to you. In medicine, something that is not harmful is called *benign*. In the context of industrial chemistry, benign by design means safety and environmental concerns are considered at the beginning of the development and design of an industrial process. Doing so might seem like common sense, but in years past, the harmful effects of chemical wastes and industrial processes were not as well understood as they are today. In meeting the objectives of green chemistry, companies also try to make their processes more efficient and profitable.

The American Chemical Society's Green Chemistry Institute promotes education and research that encourages green-chemistry practices.

As you read the 12 Principles of Green Chemistry below, keep in mind what you have learned about how ammonia is produced and how these principles should be addressed in the EKS proposal to the Riverwood town council.

Principles guiding the green chemistry movement include these general points:

1. It is better to prevent waste than to treat it or clean it up after it forms.

2. Synthetic methods should be designed so that the desired product contains as much of the material used in the process as possible.

3. Whenever possible, reactants used and waste generated should be benign.

4. The products produced should be useful for their intended purpose but also as benign as possible.

5. Solvents and other potentially hazardous substances should be made unnecessary wherever possible and benign where not possible.

6. Energy requirements should be recognized for their environmental and economic impact and should be minimized wherever possible.

7. Raw materials (reactants) should be obtained from renewable resources wherever possible.

8. The industrial process used to make a product should contain as few steps as possible. The more steps involved, the more waste that can be generated.

9. Catalysts should be used whenever appropriate.

10. Chemical products should be designed so that if they decompose, the resulting products are not harmful.

11. Pollution detection should be incorporated into as many steps of a process as possible to detect and control the production of hazardous substances.

12. Chemical processes should be designed to reduce the chance for chemical accidents, including accidental releases, explosions, and fires.

You have learned how the production of ammonia has changed over the years. Can you see how the changes in the production process of ammonia reflect the chemical industry's move toward green chemistry? In the next Making Decisions activity, you will use what you know about the production of ammonia and these ideas to identify which green chemistry principles will be most important when considering the EKS proposal.

■ MAKING DECISIONS

B.10 WHAT DOES RIVERWOOD WANT?

The first of several town meetings to discuss the possibility of a chemical plant in Riverwood is coming up soon. Representatives from EKS and WYE and town council members will attend. All interested local citizens are also encouraged to attend.

Some possible questions for the EKS representatives are listed below.

A. What are some wastes that EKS might generate in the manufacture of ammonia?

B. What is the efficiency of the process used by EKS to produce ammonia?

C. What hazards are associated with the substances that will be used by EKS?

D. Are there other options for starting materials that would more benign?

E. What are the contingency plans for accidental releases of materials?

F. What role will the price of energy have in the sustainability of the EKS facility in Riverwood?

G. What special safety training is needed by workers at the facility?

1. Identify the Principles of Green Chemistry that are addressed by each question.

2. Propose at least two other questions that could be used to evaluate EKS's commitment to the Principles of Green Chemistry.

3. It is not likely that one proposal will address all of these questions. Which questions (either in the list above or drafted by you or your classmates) must be addressed by the EKS proposal and which would be useful but not mandatory?

SECTION B SUMMARY

Reviewing the Concepts

> The rate of a particular reaction depends on temperature, reactant concentration(s), and the influence of a catalyst.

1. What is meant by the *rate* of a reaction?
2. What does *chemical kinetics* mean?
3. Explain why reactions tend to speed up
   a. with increased temperature.
   b. with increased concentration.
   c. when a suitable catalyst is added.

> When a system is in dynamic equilibrium, the rate of the forward reaction equals, and is thus balanced by, the rate of the reverse reaction.

4. What is equal about equilibrium?
5. What is dynamic about dynamic equilibrium?
6. How can you tell if chemical equilibrium is represented in a chemical equation?
7. What do *forward* and *reverse* mean when speaking about an equilibrium system?

> Le Châtelier's principle can be used to predict a shift in the equilibrium position of a reversible reaction.

8. Summarize Le Châtelier's principle.
9. List three types of stress that can be applied to an equilibrium system.
10. Consider the following system at equilibrium:

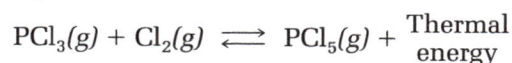

$$PCl_3(g) + Cl_2(g) \rightleftharpoons PCl_5(g) + \text{Thermal energy}$$

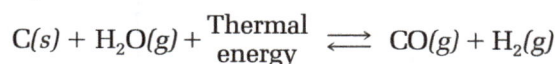

What effect (if any) will each of the following changes have on the position of that equilibrium system?
   a. adding more Cl_2
   b. lowering the temperature
   c. removing some PCl_5 as it forms
   d. decreasing the total pressure

11. Consider the following reaction at equilibrium:

$$C(s) + H_2O(g) + \text{Thermal energy} \rightleftharpoons CO(g) + H_2(g)$$

What effect (if any) will each of the following changes have on the position of that equilibrium system?
   a. lowering the temperature
   b. adding steam at constant volume to the equilibrium system
   c. adding a catalyst
   d. increasing the total pressure

12. What is the effect of removing some thermal energy from an exothermic reaction that is at equilibrium?

Ammonia is commonly produced industrially by the Haber–Bosch process.

13. What are the sources of the reactants for the Haber–Bosch process?

14. The Haber–Bosch process works most effectively under particular conditions. What are the optimal conditions for

   a. pressure?

   b. temperature?

15. What is the advantage of removing ammonia from the reaction mixture as it forms in the Haber–Bosch process?

16. Why does the yield of ammonia decrease if the Haber–Bosch process is conducted at too high a temperature?

Nitrogen compounds are commonly used in explosives.

17. What are the characteristics of a chemical explosion?

18. What two characteristics of nitrogen are responsible for the particular effectiveness of nitrogen-based explosives? Explain.

19. What were Alfred Nobel's contributions to chemistry?

20. Consider the equation for the principal reaction that causes an air bag to quickly inflate

$$3\ NaN_3(s) \longrightarrow Na_3N(s) + 4\ N_2(g)$$

What volume of nitrogen gas (at standard temperature and pressure, see Unit 2, page 188) would an air bag reaction involving 1.0 g NaN_3 produce?

Modern management practices in the chemical industry stress conservation, safety, and pollution prevention in decisions about manufacturing, storing, transporting, and disposing of chemical materials.

21. "An ounce of prevention is worth a pound of cure." Explain how this relates to the green-chemistry initiative.

22. Provide an example of how following green chemistry principles can help a chemical-manufacturing company become more profitable.

What chemical principles can be used in the responsible manufacture of ammonia?

In this section, you learned about equilibrium and explored how it is affected by changes in a system, considered some nonagricultural uses of nitrogen, and studied the industrial production of ammonia. Think about what you have learned, then answer the question in your own words in organized paragraphs. Your answer should demonstrate your understanding of the key ideas in this section.

Be sure to consider the following in your response: kinetics, equilibrium, Le Châtelier's principle, energy use and management, scaling up, and green chemistry.

Connecting the Concepts

23. Consider the following equilibrium system:

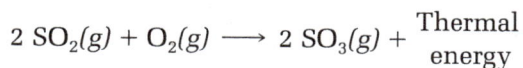

$$2\ SO_2(g) + O_2(g) \longrightarrow 2\ SO_3(g) + \text{Thermal energy}$$

Explain how this equilibrium system would be affected by:

a. adding more molecules of oxygen gas at constant volume.

b. increasing the temperature.

c. increasing the volume of the reaction vessel.

d. increasing the total pressure on the system.

24. Scaling up a reaction to production levels involves many challenges not necessarily apparent at the lab scale. Discuss three such challenges faced by chemical engineers at EKS Nitrogen Products Company.

25. Refrigerated food lasts longer than food stored at room temperature. Explain.

26. Many explosive compounds contain nitrogen atoms in a positive oxidation state and carbon atoms in a negative oxidation state. What happens to the oxidation states of each type of atom if these compounds explode?

Extending the Concepts

27. Compare the nitrogen-supply crisis of the early 20th century to current concerns about petroleum supplies.

28. Before chemists can convert methane to nitrogen gas, they must remove sulfur-containing compounds. How is sulfur removed from natural gas? Why is this necessary? Include and explain relevant chemical equations as part of your answer.

29. In several situations, an industrial by-product has become a valuable commodity. Discuss, in detail, one example.

30. Describe some examples or analogies, other than chemical reactions, of dynamic equilibria.

31. The Toxics Release Inventory (TRI) is available from the EPA. Use the TRI to judge how much your state or metropolitan area has accomplished in reducing emissions from manufacturing facilities.

32. A pressure cooker reduces the total time needed to cook foods. Investigate the design of a pressure cooker and explain why it speeds up cooking times.

33. Select three green-chemistry principles and, for each, describe how you could apply and follow a comparable principle in your daily activities.

SECTION C GENERATING ELECTRICAL ENERGY FROM CHEMICAL REACTIONS

How is chemical energy transformed into electrical energy?

You have learned how EKS Nitrogen Products Company produces ammonia using the Haber–Bosch process. Such a chemical plant could affect Riverwood both positively and negatively. You will now learn more about WYE Battery Technology Corporation. Then you will be able to determine if EKS or WYE (or neither of the companies) should be invited to locate a chemical plant within the old Riverwood Corporation building.

In producing batteries for electric vehicles, WYE specializes in **electrochemistry**, chemical changes that produce or are caused by electrical energy. The following discussions and laboratory investigation provide background on electrochemistry principles. This information will help you understand how WYE's proposed new plant would operate.

GOALS

- Describe the design of a voltaic cell.
- Use the activity series of metals to predict the direction of electron flow within a particular voltaic cell.
- Use illustrations and half-reactions describing oxidation and reduction to explain chemical processes by which voltaic cells convert chemical energy to electrical energy.
- Apply the concepts of equilibrium and spontaneity to voltaic cells.

✓ concept check 5

1. What happens to an atom that is oxidized? To an atom that is reduced?
2. How can a metal activity series be used to predict chemical reactions?
3. How can energy produced by a voltaic cell be measured?

C.1 CONVERTING CHEMICAL ENERGY INTO ELECTRICITY

WYE Battery Technology Corporation proposes to manufacture batteries to be used for hybrid and electric vehicles (see Figure 5.29). To create the batteries, the company will need to harness the energy from oxidation–reduction (redox) reactions. You studied redox reactions at the beginning of the course in Unit 1. These are reactions that involve the gain and loss of electrons. You explored (in Unit 1, page 71) the reaction between a metal and the ionic solution of another metal. Through this investigation, you learned that some metals lose electrons (become oxidized) more readily than others; that is, some metals are more chemically active than others. The relative tendencies of metals to release electrons can be summarized in the activity series of metals (see Table 5.2). A metal that is higher in the activity series will give up electrons more readily than a metal that is lower. For example, according to Table 5.2, aluminum atoms are oxidized (lose electrons) more easily than iron atoms.

Figure 5.29 *The batteries produced by WYE will power electric vehicles much like this one.*

Table 5.2

Activity Series of Common Metals		
Metal	**Products of Metal Reactivity**	
Li(s) $\longrightarrow$	Li$^+$(aq) +	e$^-$
Na(s) $\longrightarrow$	Na$^+$(aq) +	e$^-$
Mg(s) $\longrightarrow$	Mg^{2+}(aq) +	2 e$^-$
Al(s) $\longrightarrow$	Al^{3+}(aq) +	3 e$^-$
Mn(s) $\longrightarrow$	Mn^{2+}(aq) +	2 e$^-$
Zn(s) $\longrightarrow$	Zn^{2+}(aq) +	2 e$^-$
Cr(s) $\longrightarrow$	Cr^{3+}(aq) +	3 e$^-$
Fe(s) $\longrightarrow$	Fe^{2+}(aq) +	2 e$^-$
Ni(s) $\longrightarrow$	Ni^{2+}(aq) +	2 e$^-$
Sn(s) $\longrightarrow$	Sn^{2+}(aq) +	2 e$^-$
Pb(s) $\longrightarrow$	Pb^{2+}(aq) +	2 e$^-$
Cu(s) $\longrightarrow$	Cu^{2+}(aq) +	2 e$^-$
Ag(s) $\longrightarrow$	Ag$^+$(aq) +	e$^-$
Au(s) $\longrightarrow$	Au^{3+}(aq) +	3 e$^-$

Table 5.2 *The higher a metal is positioned in an activity series such as this, the more readily it gives up electrons.*

Connecting two metals from different positions on the activity series creates an electrical potential between the metals. Electrical potential (volts) is somewhat like water pressure in a pipe. Just as pressure causes water to flow in the pipe, **electrical potential** "pushes" electrons through the wire connecting the two metals. The greater the difference between the chemical activities of the two metals, the greater the electrical potential that the cell generates.

The differing tendency of metals to lose electrons allows an oxidation–reduction reaction to generate electrical energy. You can make a simple device called a **voltaic cell** from two half-cells connected in a circuit (see Figure 5.30). Each **half-cell** contains a metal partially immersed in a solution of ions of that metal. For example, one half-cell could contain copper metal immersed in a solution of Cu^{2+} ions. Another could contain a piece of zinc metal immersed in a solution of Zn^{2+} ions. You know from Unit 1, page 71, that the oxidation–reduction reaction between Cu^{2+} and Zn readily occurs. By separating these reactants into half-cells, the electrons are forced to flow through a wire and provide electrical energy to an external circuit.

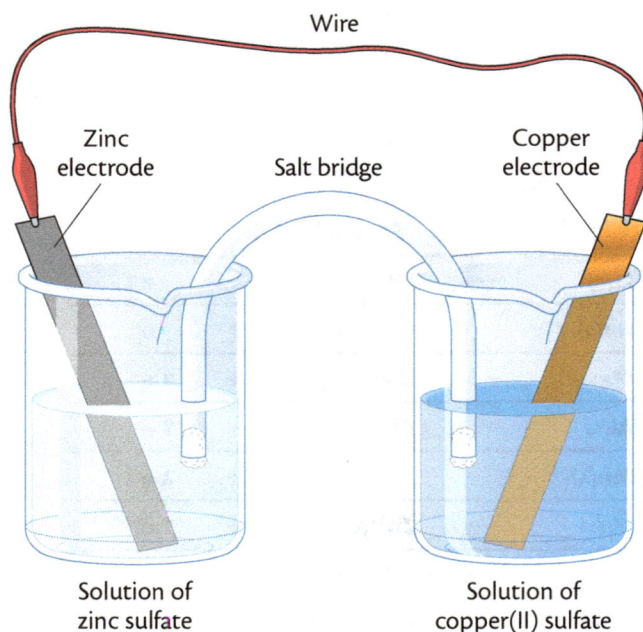

Figure 5.30 *A voltaic cell. Which metal is oxidized? Which metal is reduced? In which direction do electrons flow?*

A barrier prevents the solutions in the two half-cells from mixing. A wire connects the two metals, which act as **electrodes**, and allows electrons to flow between them, as shown in Figure 5.30. Such an electron flow constitutes an **electric current**. To complete the circuit and maintain a balance of electrical charges within the system, the two half-cells must be connected by a **salt bridge**. Dissolved ions in the salt bridge complete the internal electrical circuit by allowing ions to move freely between the half-cells, preventing build-up of electrical charge near the electrodes. Without the flow of ions, a positive electrical charge would build up in one half-cell and a negative charge would build up in the other half-cell. That situation would prevent any further flow of electrons in the wire.

INVESTIGATING MATTER
C.2 VOLTAIC CELLS

In this investigation, you will construct several voltaic cells and measure and compare the electrical potentials that they generate. You will also explore factors that may help determine the actual electrical potential a particular voltaic cell generates. You will set up the apparatus, formulate one or more testable questions, then design and perform an investigation to answer your question or questions.

Preparing to Investigate

1. Before you begin, put on your goggles, and wear them properly throughout the investigation.

2. Using scissors, make five notches in the filter paper. See Figure 5.31. Place the filter paper in the bottom of a Petri dish or other shallow container.

3. Label each "arm" with the symbol for the metal sample that will be placed on that arm.

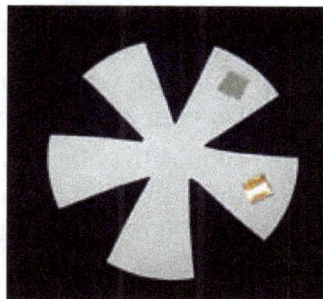

Figure 5.31 *The apparatus you will build is similar to the one depicted here. Voltaic cells result in spontaneous flow of electrons from the metal of higher activity to ions of the less-active metal.*

4. Saturate the center portion of the filter paper with KNO_3 solution from the dropper bottle.

5. Place 2–3 drops of $Cu(NO_3)_2$ on the end of one of the arms. The wet portions of the filter paper should meet.

6. Clean a copper strip with steel wool and place it on the area where the $Cu(NO_3)_2$ was dropped.

7. Repeat Steps 5 and 6 with the remaining metals and metal nitrates. You now have a completed apparatus.

8. Measure the electrical potential of each pair of metals by touching one electrode to each of the metal samples and reading the display. Record your results.

Asking Questions

Examine the apparatus and the other materials that your teacher provided. Think about what you learned about cell potential and electrodes in Section C.1 and develop one or more scientific questions that you can answer in this investigation. Share these questions with your laboratory group or classmates as directed by your teacher.

Gathering Evidence

Develop a procedure to answer your question or questions. Recall the discussion in Unit 2 (page 166) about experimental design and variables. Make sure that each investigation changes only one variable at a time. What variables will be controlled in all situations? Once the procedure has been approved by your teacher, carry out your investigations.

Interpreting Evidence

1. What changes in the apparatus provided the largest voltage? The smallest voltage?
2. List the cells (metal pairs) in order of lowest to highest potential.
3. How does the size of the electrode change the electrical potential of the cell?

Making Claims

4. Compare your results to the activity series in Table 5.2 (page 549). Do your results agree?
5. Would the electrical potential generated by cells composed of each of the following pairs of metals be larger or smaller than that of the Zn–Cu cell? Refer to Table 5.2 (page 549).
   a. Zn and Cr
   b. Zn and Ag
   c. Sn and Cu

6. How did changing the size of the zinc and copper electrodes affect the measured electrical potential? Explain and provide evidence supporting your answer.
7. Would an Ag–Au cell be a commercially feasible voltaic cell? Explain and provide evidence for your answer.

C.3 VOLTAIC CELLS AND HALF-REACTIONS

In the voltaic cells you constructed in the previous investigation, each metal sitting on filter paper soaked in a solution of its ions represented a half-cell. The activity series predicts that zinc is more likely to be oxidized (lose electrons) than copper. Thus, in the zinc–copper cell you investigated, oxidation (electron loss) occurred in the half-cell containing zinc metal immersed in zinc nitrate solution. Reduction (electron gain) took place in the half-cell consisting of copper metal in copper(II) nitrate solution. The half-reactions (individual electron-transfer steps) for that cell are:

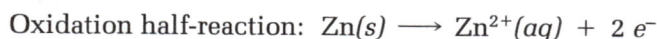

$$\text{Oxidation half-reaction: } Zn(s) \longrightarrow Zn^{2+}(aq) + 2\ e^-$$

$$\text{Reduction half-reaction: } Cu^{2+}(aq) + 2\ e^- \longrightarrow Cu(s)$$

The electrode at which oxidation takes place is the **anode**. Reduction occurs at the **cathode**. (See Figure 5.32.)

The overall reaction in the zinc–copper voltaic cell is the sum of the two half-reactions, added so that the electrical charges balance—the total electrons lost and gained are the same—resulting in a net electrical charge of zero.

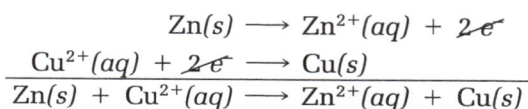

$$Zn(s) \longrightarrow Zn^{2+}(aq) + 2e^-$$
$$\underline{Cu^{2+}(aq) + 2e^- \longrightarrow Cu(s)}$$
$$Zn(s) + Cu^{2+}(aq) \longrightarrow Zn^{2+}(aq) + Cu(s)$$

Because a barrier separates the two reactants (Zn and Cu^{2+}) in the cell, the electrons released by zinc must travel through the wire to reach (and reduce) the copper ions.

The greater the difference in reactivity of the two metals in a voltaic cell, the greater the tendency for electron transfer to occur, and the greater the electrical potential (volts) of the cell. A zinc–gold voltaic cell, therefore, would generate a larger electrical potential than a zinc–copper cell. (See Table 5.2, page 549, to compare the placement of these metals in the activity series.)

One way to remember these electrode processes: Note that anode and its associated process (oxidation) both begin with vowels, whereas cathode and its process (reduction) both start with consonants.

Figure 5.32 **RED**uction always occurs at the **CAT**hode.

CHEM**QUANDARY**

ENERGY CELLS: ARE THEY ALL CREATED EQUAL?

You studied fuel cells in Unit 3 Section D (page 376). They are used to power cars, buses, and businesses. How do fuel cells differ from voltaic cells?

DEVELOPING SKILLS

C.4 GETTING CHARGED BY ELECTROCHEMISTRY

Each of the following questions deals with voltaic cells, their properties, and equations to describe them.

Sample Problem: Consider a voltaic cell containing lead metal (Pb) immersed in lead(II) nitrate solution, $Pb(NO_3)_2$, and a half-cell containing silver metal (Ag) in silver nitrate solution, $AgNO_3$.

   a. Predict the direction of electron flow in the wire connecting the two metals.

   b. Write equations for the two half-reactions and the overall reaction.

   c. Which metal is the anode and which is the cathode?

The answers are as follows:

   a. Table 5.2 (page 549) shows that lead is a more active metal than silver is. Therefore, lead will be oxidized, and electrons will flow from lead to silver.

   b. One half-reaction involves forming Pb^{2+} from Pb, as shown in Table 5.2. The other half-reaction produces Ag from Ag^+, which can be written by reversing the equation in Table 5.2 and doubling it so electrons lost and gained are the same:

$$Pb(s) \longrightarrow Pb^{2+}(aq) + 2e^-$$
$$\underline{2\ Ag^+(aq) + 2e^- \longrightarrow 2\ Ag(s)}$$
$$Pb(s) + 2\ Ag^+(aq) \longrightarrow Pb^{2+}(aq) + 2\ Ag(s)$$

> The total number of electrons consumed in reduction equals the total number of electrons liberated in oxidation. Overall, the voltaic cell based on these two half-reactions involves twice as many silver ions reduced as lead atoms oxidized.

   c. In the cell reaction, each Pb atom loses two electrons. Pb is thus oxidized, making it the anode. Each Ag^+ ion gains one electron, which is a reduction reaction. Because reduction takes place at the cathode, Ag must be the cathode.

Now answer the following questions.

1. Predict the direction of electron flow in a voltaic cell made from each specified pair of metals partially immersed in solutions of their ions.

   a. Al and Sn

   b. Pb and Mg

   c. Cu and Fe

2. A voltaic cell uses tin (Sn) and cadmium (Cd) as the electrodes. The overall equation for the cell reaction is

$$Sn^{2+}(aq) \ + \ Cd(s) \longrightarrow Cd^{2+}(aq) \ + \ Sn(s)$$

   a. Write the two half-reaction equations for this cell.

   b. Which metal, Sn or Cd, loses electrons more readily?

3. Sketch a voltaic cell composed of a Ni–Ni$(NO_3)_2$ half-cell linked to a Cu–Cu$(NO_3)_2$ half-cell.

4. For each voltaic cell designated below, identify the anode and the cathode. Assume that each voltaic cell uses appropriate ionic solutions.

   a. Cu–Zn cell

   b. Al–Zn cell

   c. Mg–Mn cell

   d. Au–Ni cell

concept check 6

1. Why won't a voltaic cell operate if the salt bridge is missing?
2. What factors affect a cell's electrical potential?
3. Consider the discussion of equilibrium in Section B. Write a statement about the relationship between equilibrium and a voltaic cell.

C.5 EQUILIBRIUM IN ELECTROCHEMICAL SYSTEMS

Voltaic cells create a flow of electrons by separating a pair of oxidation–reduction reactions and allowing both ions and electrons to move between the half-cells. When you built voltaic cells as part of Investigating Matter C.2, you compared the measured potential with the metal activity series. These reactions occurred without any input of additional information or stimuli. In chemical terms, the reactions were **spontaneous**. The direction that electrons flow is generally shown with an arrow. If the cell operates spontaneously, the arrow will be moving from anode toward the cathode. That means that the atoms of the anode will give up electrons, forming ions that enter the electrolyte solution. The concentration of ions will thus increase in this half-cell.

What would happen simultaneously to the electrolyte solution in the cathode half-cell? Consider what you learned about equilibrium in Section B.

A system is at equilibrium when the forward and reverse reactions occur at the same rate. For a voltaic cell to generate electricity, electrons must flow in only one direction. The voltaic cell will run only if the system is not at equilibrium. In fact, as the system gets closer to equilibrium, the cell's potential decreases.

The system can be reversed, but only if the electrons are forced to flow in a direction that is not spontaneous. When you conducted electroplating in Unit 1 (page 107), you forced electrons to flow in a direction that was not spontaneous. The energy to force the electrons to flow in reverse came from an outside source, in that case a 9-V battery. The electrons combined with metal ions to form elemental metal atoms, separating an ionic compound into the elemental components. This type of reaction transforms electrical energy into chemical energy and is called **electrolysis**.

Electrolysis can be used to extract alkali metals such as lithium from their ores by passing an electrical current through a molten mineral solution (an electrolyte), causing a chemical reaction. In this case, the flow of electrons reduces Li^+ ions in the electrolyte to lithium, $Li(s)$. This method of reducing metals is also called electrometallurgy. Electrolysis requires much electrical energy, so the cost of electricity is a factor in its use.

In the next activity, you will begin to visualize what is happening at a particulate level to each component in a voltaic cell.

> Electrolysis uses electrical energy to cause a non-spontaneous reaction. A voltaic cell generates electrical energy through a spontaneous reaction.

MODELING MATTER

C.6 VISUALIZING CHANGES WITHIN VOLTAIC CELLS

In Investigating Matter C.2 and sections following, you were asked to create and then draw voltaic cells. You identified the anode, cathode, electrolytic solutions, salt bridge, and direction of electron flow. In this activity, you will focus on the changes in each half-cell, including the flow of electrons and changes from atoms to ions and back.

1. Sketch a voltaic cell composed of a $Zn–Zn(NO_3)_2$ half-cell linked to a $Cu–Cu(NO_3)_2$ half-cell with a $NaNO_3$ salt bridge.

2. Consider the composition of each half-cell.

   a. What ions would be present in each of the electrolyte solutions?

   b. What ions are present in the salt bridge?

   c. Which way do electrons flow?

3. Consider the reaction at the cathode.

   a. Write the reaction that takes place at the cathode.

   b. What would happen to the cathode as the voltaic cell operates? Sketch three drawings of the cathode to represent

      i. Time 0: before the reaction begins

      ii. Time 1: after the cell has been operating for a short period of time

      iii. Time 2: at the point where the cell stops running

4. Now consider the reaction at the anode.

    a. Write the reaction that takes place at the anode.

    b. What would happen to the anode while the voltaic cell operates? Sketch three drawings of the anode to represent

        i. Time 0: before the reaction begins

        ii. Time 1: after the cell has been operating for a short period of time

        iii. Time 2: at the point where the cell stops running

5. What other term would describe the voltaic cell at Time 2, when the cell stops running?

6. You know that reduction occurs at the cathode. If the metal ions form metal atoms, how is charge conserved? In other words, what substances move into the cell? From where do these substances come?

7. What ions would be present in the cathode half-cell electrolyte solution at

    a. Time 0: before the reaction begins?

    b. Time 1: after the cell has been operating for a short period of time?

    c. Time 2: at the point where the cell stops running?

8. Answer question 7 for the anode half-cell electrolyte solution.

9. What would limit the length of time that the voltaic cell would operate? In other words, what is the limiting reactant or reactants?

10. Propose a way that the voltaic cell could operate again after it exhausts the limiting reactant.

11. Now that you have visualized what is happening at the atomic level in a voltaic cell, what do you think would happen to the electrodes if electricity were put into the system rather than taken out?

MAKING DECISIONS

C.7 WHAT DO WE NEED TO KNOW ABOUT WYE?

Green chemistry principles were listed and described in Section B.9 (page 543). Consider what you have just learned about voltaic cells, then answer the following questions.

1. Which of the 12 Green Chemistry Principles would be of most concern when considering manufacturing and use of voltaic cells?

2. Citizens of Riverwood must decide whether a new chemical plant will be invited to locate in the old Riverwood Corporation building. After reviewing the 12 principles, which 3 do you think are most important for making this decision?

3. What precautions must the Town Council implement to keep citizens of Riverwood safe if WYE is allowed to manufacture batteries in Riverwood?

SECTION C SUMMARY

Reviewing the Concepts

> Electrochemistry involves chemical changes that produce or are caused by electrical energy.

1. What is electrolysis?
2. What is a voltaic cell?
3. What is a half-cell?
4. Voltaic cells require a salt bridge.
   a. What is a salt bridge?
   b. Why is a salt bridge necessary for the operation of a voltaic cell?
   c. Describe one way to make a salt bridge.
5. Diagram a simple Ag–Cu voltaic cell. Label electrodes, solutions, and salt bridge.
6. Considering your results in Investigating Matter C.2 (page 551), does the electrical potential produced by a voltaic cell depend on the
   a. size of the electrodes? Explain.
   b. specific metals used? Explain.

> The activity series of metals can be used to predict the direction of electron flow within a particular voltaic cell.

7. In a voltaic cell, what process takes place at the
   a. anode?
   b. cathode?

8. Sketch a voltaic cell made from Ni and Zn in solutions of their ions. Label anode and cathode (identifying each metal) and show the direction of electron flow. See Table 5.2 (page 549).
9. Predict the direction of electron flow in a voltaic cell made from each of these metal pairs in solutions of their ions. See Table 5.2 (page 549).
   a. Ag and Sn
   b. Cr and Ag
   c. Cu and Pb

> Redox reactions can be used to convert chemical energy to electrical energy by harnessing the flow of electrons.

10. Using Table 5.2 (page 549), write the overall reaction and the two half-reactions for a voltaic cell made from each of these metal pairs in solutions of their ions.
    a. Ag and Sn
    b. Cr and Ag
    c. Cu and Pb
11. Sketch a voltaic cell made from Sn and Zn in solutions of their ions. Label anode and cathode (identifying each metal) and write a half-reaction for the process that occurs at each electrode. Explain, referring to the sketch, how electricity is generated by this system. See Table 5.2 (page 549).

Voltaic cells operate spontaneously, leading to constant changes within the system.

12. Consider the anode and cathode in a voltaic cell.

    a. How does the anode change during operation?

    b. How does the cathode change during operation?

13. What changes occur in the solution on the anode side of a voltaic cell?

14. How does the concentration of the cathode cell solution change?

15. Assuming that the voltaic cell is constructed properly, when will it stop working?

How is chemical energy transformed into electrical energy?

In this section, you have constructed, diagrammed, and studied voltaic cells and their particulate level changes, written half-reactions, learned about equilibrium and voltaic cells, and related metal reactivity to voltaic cell operation. Think about what you have learned, then answer the question in your own words in organized paragraphs. Your answer should demonstrate your understanding of the key ideas in this section.

Be sure to consider the following in your response: anode, cathode, salt bridge, voltaic cell, electrolysis, half-reactions, equilibrium, and spontaneous reactions.

Connecting the Concepts

16. Compare Table 1.3 (page 75) and Table 5.2 (page 549). What is included in Table 5.2 that was not part of Table 1.3? Why is this information more important in the current unit?

17. Consider the discussion of Le Châtelier's principle in Section B.1 (page 531). Applying these ideas to a specified voltaic cell, how could you change the quantity of electrical energy produced without changing the identity of the electrodes?

18. In Unit 3, you learned that all energy transformations lead to some loss of energy. Where might some energy be lost in the transformation of chemical energy to electrical energy within a voltaic cell?

19. In everyday life, decisions made on an impulse may be referred to as *spontaneous*. How is this use of the word *spontaneous* similar to and different from how it is used to describe chemical reactions?

Extending the Concepts

20. Investigate the concept of *entropy* and discuss some of the challenges entropy presents in voltaic cells.

21. Recently, fuel cells have become available as an alternative to electrical energy generated in power plants and transmitted through wires. Investigate these cells and consider whether you would want them to be located in your neighborhood. Consider, for example, whether they are a "greener" alternative for electrical energy and what types of fuels they require and wastes they produce.

CHEMISTRY *AT WORK*
Q&A

Khalil Amine, Materials Scientist and Group Leader at Argonne National Laboratory in Argonne, Illinois

The small energy storehouses known as batteries have made it possible to power up electronic machines anyplace, not just near an electrical outlet. But as useful as batteries are, they're far from perfect. Take the battery in a car. Even though a car battery can charge up a vehicle's lights and start the engine, it doesn't provide enough energy to move the car very far—that's why cars still need gasoline. As Americans' interest in electric and hybrid gas–electric cars has grown, the need for better batteries has increased. Read on to see how one chemist's work is fueling the next generation of high-tech batteries for these cars.

An electric car developed recently in Norway.

Q. What is materials science, and why do you need it to develop new types of batteries?

A. Materials science is the study of matter and new ways to use matter to meet our needs. It's a field made up of many other fields, including physics and engineering, but it's mostly chemistry. To create new kinds of batteries, we have to invent new types of materials. We use materials science to try to understand everything about these materials: their structures, physical properties, electrochemical properties, and the impact that all these factors have on what we would like these materials to do, such as store energy.

Q. Why did you choose to go into materials science, and why did you focus on batteries?

A. When I was growing up in Morocco, I was always good at math and chemistry. I had to choose between these two fields before I left to go to college in France. Chemistry was much more interesting to me because there are so many different fields of chemistry to study. I earned a master's degree in chemistry and a doctoral degree in materials science, since it has such a heavy chemistry focus. I used my degree in materials science to get a job in Japan in 1990. In Japan, energy use is an important issue. The country has no oil, gas, or coal reserves, so it has to import almost all of its energy sources from overseas. Eventually, I started working on developing new types of batteries to fill this energy need. Now, I focus my own energy on the same kind of work for Argonne.

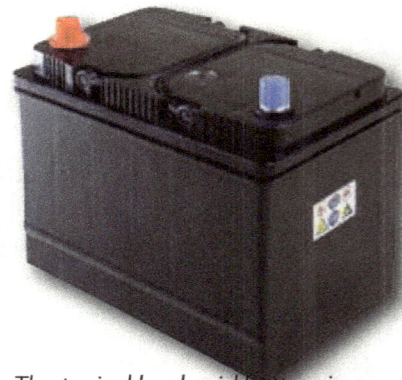

The typical lead-acid battery in a gasoline-powered car weighs more than thirty pounds.

Q. What's wrong with the batteries we already have in cars?

A. The batteries in gasoline-powered cars are lead-acid batteries. This type of technology is almost 90 years old, and these batteries are trustworthy, durable, and cheap. There's nothing wrong with using them the way we do now, to start our cars and power our headlights and radios. But cars that run only on electricity will need a battery that powers everything, including motion. Lead-acid batteries can't store enough energy for this purpose. A car would need 50 lead-acid batteries to run on electricity!

A lithium-ion cell-phone battery would not provide nearly enough energy to power a car.

Q. What types of batteries will replace lead-acid batteries in electric cars?

A. The best current candidate to replace lead-acid batteries is the lithium-ion battery. These batteries are the kind that we use in cell phones and computers now. The most common lithium-ion batteries have a cathode made of a lithium compound and an anode made of carbon. A solution of lithium ions in between allows current to pass from one end to the other. These batteries are small and light, so they are a good choice for cars.

Q. So, will we use cell-phone batteries for our cars in the future?

A. We won't use the same batteries because they still have a few kinks we'd like to work out. The electrolyte solution currently in use is flammable, so my colleagues and I are working on making a safer version. We're also experimenting with new materials for the anode. These new materials change the types of chemical reactions that happen in batteries, which can vastly increase the quantity of energy these batteries can store. Different types of metal alloys, such as nickel, manganese, and cobalt compounds, are all possibilities. We're especially interested in another kind of battery in development called a lithium-air battery. These batteries have a cathode made of porous carbon, which allows oxygen from the air to flow into the battery. The oxygen participates in chemical reactions that allow lithium-air batteries to store 12 times the quantity of energy that regular lithium-ion batteries can! These batteries might be powering our cars in a few decades.

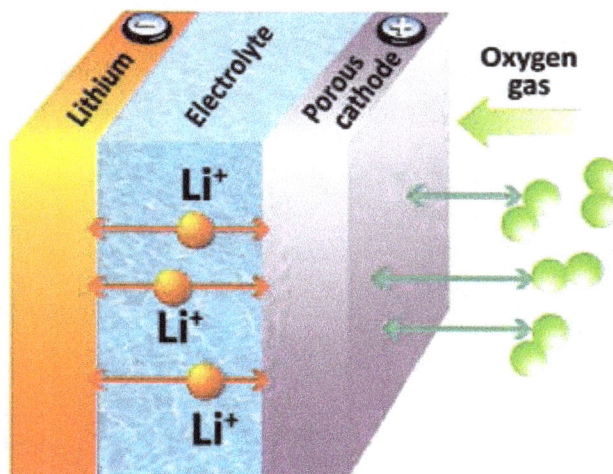

The lithium-air battery has a cathode of porous carbon (at right). The anode (at left) is usually nonporous lithium foil.

Q. What advice do you have for students interested in battery technology?

A. Being innovative and having a broad knowledge base is really important for this field. You should take classes in chemistry, electrochemistry, and materials science in college, along with other fields of science too—you never know what might spur an idea that will lead to a new or better battery. Right now, battery technology has electric potential for students with the right spark.

SECTION D INDUSTRIAL PRODUCTION OF BATTERIES

What challenges must be met to optimize production and use of batteries?

You have learned how voltaic cells produce electrical energy from chemical reactions. In this section, you will learn more about the batteries that WYE Battery Technology Corporation proposes to manufacture in Riverwood. Could these advanced vehicle batteries contribute to the economy of Riverwood and the United States? What risks are posed by manufacturing these batteries? Would EKS or WYE best suit the needs and capabilities of the Riverwood community? Read on to gather more evidence to support your decision.

GOALS

- Describe the function and composition of primary and secondary batteries.
- Calculate the electrical potential produced by a voltaic pile or battery.
- Analyze the life cycle of a battery.
- Use burden–benefit analysis in decisions about implementing chemical technologies.

concept check 7

1. How does a voltaic cell produce electrical energy?
2. In Investigating Matter C.2, you considered the effect of electrode size on a voltaic cell.
   a. Does the size of the electrode affect the electrical potential of a voltaic cell? Explain.
   b. What characteristics of the cell might be affected by the size of the electrode?
3. What type of substance would you expect to find in an alkaline battery?

■ INVESTIGATING MATTER
D.1 BUILDING A VOLTAIC PILE

Preparing to Investigate

In 1800, Alessandro Volta constructed a *voltaic pile* by stacking pairs of zinc and copper plates, each pair separated by cloth or cardboard that had been soaked in a solution of salt or sulfuric acid. Since concentrated sulfuric acid is corrosive, you will use a more benign electrolyte, copper(II) sulfate, in this investigation.

Gathering Evidence

Part I: Volta's Pile

1. Before you begin, put on your goggles, and wear them properly throughout the investigation.

2. Place each piece of zinc on top of a copper piece. Assemble four pairs of Cu–Zn pieces.

3. Prepare pieces of filter paper soaked in copper(II) sulfate as directed by your teacher.

4. Build a stack or pile of these pairs of discs as shown in Figure 5.33, with a piece of electrolyte soaked filter paper separating each pair from its neighbors. The sequence of the components is as follows: Cu, electrolyte, Zn, Cu, electrolyte, Zn, etc. (*Note:* Be careful that the electrolyte solution does not drip down the side of the stack as this can cause a short circuit between the elements of the pile.)

Figure 5.33 *Each pair of zinc and copper pieces is separated from the adjacent pair by a piece of filter paper soaked in electrolyte.*

5. When the device is arranged as specified, measure the voltage between the bottom-most copper piece and the top-most zinc piece. Record your data.

Part II: Constructing to Specifications

6. Based upon your results in Part I, build a pile that will produce an electrical potential of ~6.6 volts.

7. Describe the construction of your pile.

8. Try to power a small LCD electronic device such as a clock, thermometer, or calculator. Describe your results.

Interpreting Evidence

1. What was the measured potential of the pile in Part I?

2. What was the measured potential of your pile in Part II?

3. Were you able to power an electronic device in Part II? Explain.

Making Claims

4. What factors in the design of the pile would affect the potential generated? Explain.

Reflecting on the Investigation

5. How did you decide how to construct the pile in Part II?

6. How is the pile similar to commercial voltaic cells, i.e., batteries?

D.2 CELL POTENTIAL

Cell potential is sometimes informally referred to as voltage, since it is measured in volts.

As you observed, the cell potential of a voltaic pile varies with the number of "units" in the pile. You learned earlier that connecting two metals in a voltaic cell creates an electrical potential between the metals. In the voltaic pile, several cells are placed in series, so the total electrical potential is the sum of the individual cell potentials.

Volta used silver discs in place of copper in some of the voltaic piles that he constructed. Unfortunately for Volta, the only way he could compare the potential produced by different piles was by evaluating the shock he received when he completed the circuit. Fortunately for us, we can now accurately measure cell potential and know that silver–zinc cells produce a different potential than copper–zinc cells.

The potential of a cell reflects its tendency to move electrons through the cell. A cell with a small potential does not have much power to move electrons. A cell at equilibrium has no electron movement, and thus no potential.

Cell potentials for many electrochemical cells have been measured at standard conditions. Standard conditions refer to pure gases at 1 atm pressure, pure metals, and electrolyte solutions with concentrations of 1.0 mol/L. To make comparisons easier, these standard cell potentials are often split into contributions from each of their electrodes. These electrode potentials are known as standard potentials and denoted $E°$. Since it is not possible to measure the potential of an electrode in isolation, electrodes are compared against the standard hydrogen electrode ($2\ H^+(aq)\ +\ 2\ e^- \longrightarrow H_2(g)$), which has a potential defined as 0 V.

Electrode potentials are often organized within a table of standard reduction potentials. Given such a table, the standard potential of any cell can be calculated by first determining which metal will be reduced. The reduction half-reaction of the other electrode reaction (the oxidized species) is then reversed, and the sign of its potential is switched. The half-reaction equations and their potentials are added together to determine the overall standard cell potential. One or both of the half-reaction equations may need to be multiplied to ensure that the number of electrons lost and gained is equivalent, but this multiplication does *not* affect the standard potentials.

In the next activity, you will apply these ideas to some voltaic cells that you have investigated in Investigating Matter C.2 and D.1. As you look at and use Table 5.3, think about the metals listed and their everyday applications. What metals might you choose to use if you wanted to maximize electrical potential? Why?

> Recall that the less active metal is more likely to be reduced, so Table 5.3 lists metals in the opposite order from a metal activity series.

Table 5.3

Standard Reduction Potentials

Species	Half-reaction	Standard Reduction Potential, $E°$ (volts)
Au^+/Au	$Au^+(aq) + e^- \longrightarrow Au(s)$	+1.61
Ag^+/Ag	$Ag^+(aq) + e^- \longrightarrow Ag(s)$	+0.80
Cu^{2+}/Cu	$Cu^{2+}(aq) + 2\ e^- \longrightarrow Cu(s)$	+0.34
H^+/H_2	$2\ H^+(aq) + 2\ e^- \longrightarrow H_2(g)$	0 (by definition)
Pb^{2+}/Pb	$Pb^{2+}(aq) + 2\ e^- \longrightarrow Pb(s)$	−0.13
Sn^{2+}/Sn	$Sn^{2+}(aq) + 2\ e^- \longrightarrow Sn(s)$	−0.14
Ni^{2+}/Ni	$Ni^{2+}(aq) + 2\ e^- \longrightarrow Ni(s)$	−0.23
Cd^{2+}/Cd	$Cd^{2+}(aq) + 2\ e^- \longrightarrow Cd(s)$	−0.40
Fe^{2+}/Fe	$Fe^{2+}(aq) + 2\ e^- \longrightarrow Fe(s)$	−0.44
Zn^{2+}/Zn	$Zn^{2+}(aq) + 2\ e^- \longrightarrow Zn(s)$	−0.76
Al^{3+}/Al	$Al^{3+}(aq) + 3\ e^- \longrightarrow Al(s)$	−1.66
Mg^{2+}/Mg	$Mg^{2+}(aq) + 2\ e^- \longrightarrow Mg(s)$	−2.36
Li^+/Li	$Li^+(aq) + e^- \longrightarrow Li(s)$	−3.05

DEVELOPING SKILLS

D.3 DETERMINING POTENTIAL IN ELECTROCHEMICAL CELLS

Sample Problem: *Consider a voltaic cell containing lead metal (Pb) immersed in lead(II) nitrate solution, Pb(NO₃)₂, and a half-cell containing silver metal (Ag) in silver nitrate solution, AgNO₃.*

   a. Determine which metal will be oxidized and which metal will be reduced.

   b. Write equations for the two half-reactions, including standard potentials.

   c. Write the equation for the overall reaction, including the standard potential.

The answers are as follows:

   a. Table 5.3 (page 565) shows that silver has a positive reduction potential (lies higher in the table). Therefore, silver will be reduced and lead will be oxidized.

   b. One half-reaction involves forming Pb^{2+} from Pb, which can be written by reversing the equation in Table 5.3. The other half-reaction produces Ag from Ag^+, as shown in Table 5.3, which must be doubled so that electrons lost and gained are the same. Standard potentials are not affected by this multiplication but do change signs when the equation is reversed.

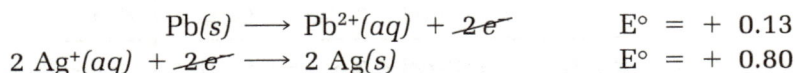

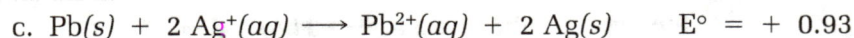

$$Pb(s) \longrightarrow Pb^{2+}(aq) + 2e^- \qquad E° = + 0.13$$
$$2\,Ag^+(aq) + 2e^- \longrightarrow 2\,Ag(s) \qquad E° = + 0.80$$

   c. $Pb(s) + 2\,Ag^+(aq) \longrightarrow Pb^{2+}(aq) + 2\,Ag(s) \qquad E° = + 0.93$

1. How can the electrical potentials you measured in Investigating Matter D.1 be explained in terms of the values in Table 5.3 (page 565)?

2. A voltaic cell uses magnesium (Mg) and zinc (Zn) as the electrodes.

   a. Which metal will be the cathode?

   b. Write the cathode half-reaction, including the standard potential.

   c. Write the anode half-reaction, including the standard potential.

   d. Write the overall reaction equations for this cell, including the standard potential.

3. In Developing Skills C.4, you considered the flow of electrons and identified the anode and cathode in several voltaic cells. Now use Table 5.3 to calculate the expected electrical potential and identify the anode and cathode for each of the following voltaic cells. (Assume that each voltaic cell uses appropriate ionic solutions.)

   a. Cu–Fe cell   c. Pb–H_2 cell   e. Al–Sn cell

   b. Ni–Cd cell   d. Li–Zn cell

4. Suppose you wanted to make a voltaic cell with a potential of at least 1.5 V. Use Table 5.3 (page 565) to propose one possible cell.

D.4 PRIMARY BATTERIES

Voltaic cells are a convenient way to convert chemical energy to electrical energy. Such cells can fit in small, portable containers. Commercial voltaic cells, called **batteries**, can be constructed from various combinations of metals and ions. In an ordinary zinc–carbon battery, zinc is the anode, and a graphite rod surrounded by a water-based paste of manganese(IV) oxide (MnO_2) and graphite serves as the cathode. A mixture of ammonium chloride and zinc chloride in an aqueous paste serves as the electrolyte. The fact that the electrolyte is in paste form, rather than liquid, accounts for other name of this type of battery—the **dry cell**.

Alkaline batteries, common in portable music players, have similar zinc and graphite–MnO_2 electrodes, but the electrolyte is an alkaline aqueous potassium hydroxide (KOH) paste. See Figure 5.34. Alkaline batteries are also considered dry cells, as are most modern consumer batteries. Zinc–carbon and alkaline batteries are **primary batteries**, that is, they are designed for a single use and cannot be recharged.

> The term *battery* once referred to two or more electrochemical cells connected together. In common use, though, a battery is any portable source of electrical energy from chemical energy.

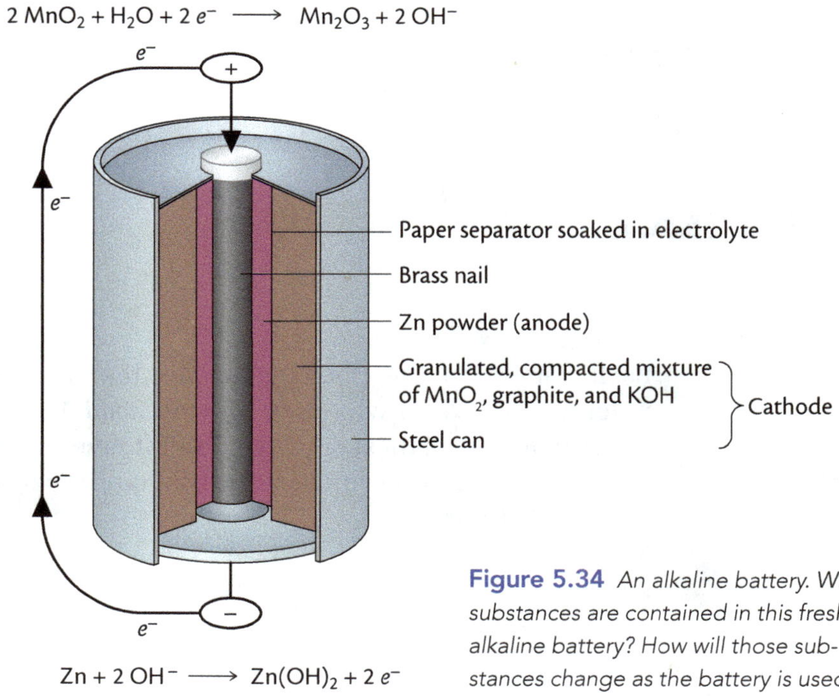

$$2\, MnO_2 + H_2O + 2\, e^- \longrightarrow Mn_2O_3 + 2\, OH^-$$

- Paper separator soaked in electrolyte
- Brass nail
- Zn powder (anode)
- Granulated, compacted mixture of MnO_2, graphite, and KOH ⎫
- Steel can ⎭ Cathode

$$Zn + 2\, OH^- \longrightarrow Zn(OH)_2 + 2\, e^-$$

Figure 5.34 *An alkaline battery. What substances are contained in this fresh alkaline battery? How will those substances change as the battery is used?*

Both zinc–carbon and alkaline batteries generate an electrical potential of 1.54 V. The following oxidation–reduction equations describe the chemical changes involved.

Zinc–Carbon Battery

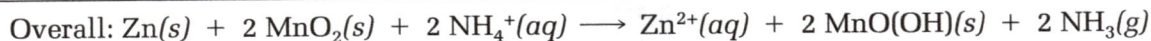

Oxidation: $Zn(s) \longrightarrow Zn^{2+}(aq) + 2e^-$

Reduction: $2\,MnO_2(s) + 2\,NH_4^+(aq) + 2e^- \longrightarrow 2\,MnO(OH)(s) + 2\,NH_3(g)$

Overall: $Zn(s) + 2\,MnO_2(s) + 2\,NH_4^+(aq) \longrightarrow Zn^{2+}(aq) + 2\,MnO(OH)(s) + 2\,NH_3(g)$

Alkaline Battery

Oxidation: $Zn(s) + 2\,OH^-(aq) \longrightarrow Zn(OH)_2(s) + 2e^-$

Reduction: $2\,MnO_2(s) + H_2O(l) + 2e^- \longrightarrow Mn_2O_3(s) + 2\,OH^-(aq)$

Overall: $n(s) + 2\,MnO_2(s) + H_2O(l) \longrightarrow Zn(OH)_2(s) + Mn_2O_3(s)$

While zinc–manganese dioxide systems are the most common dry cells, there are several other types of primary dry cells in use. Lithium cells have an anode made of lithium foil and a cathode constructed from a metal oxide, commonly manganese dioxide. A lithium salt in an organic solvent serves as the electrolyte.

Zinc–air and zinc–silver oxide cells have the same anode (zinc powder) and electrolyte (KOH) as alkaline batteries, but use either air and carbon or silver oxide and carbon, as their cathodes. Zinc–air batteries are commonly used in hearing aids but suffer from a high self-discharge rate. Self-discharge is the natural loss of energy within stored batteries from chemical reactions within the cell. Zinc–silver oxide cells have fewer problems with self-discharge, but are more expensive.

CHEMQUANDARY

BATTERY SIZES

Each battery in the photo generates the same electrical potential, 1.5 V. So why does anyone need 1.5-V batteries larger than the smallest one? Wouldn't it save space, weight, and perhaps even resources to restrict consumer use to the smallest batteries?

Primary batteries generate electrical potential only as long as all starting materials (reactants) remain. When the reactants reach equilibrium or the limiting reactant (see Unit 3, page 343) is depleted, primary batteries are at the end of their useful life and cannot be recharged.

D.5 SECONDARY BATTERIES

Nickel-Based Batteries

Unlike primary batteries, secondary batteries are rechargeable. We can return their systems to their original states and reuse the batteries. Nickel–cadmium (NiCd) cells and lead-acid automobile batteries are common examples of such rechargeable batteries. The NiCd rechargeable battery is based on a nickel–iron battery that Thomas Edison developed in the early 1900s. The NiCd battery anode is cadmium, and the cathode is nickel oxide. These electrodes have a rolled design (Figure 5.35) that increases surface area and, thus, increases the generated current (electron flow).

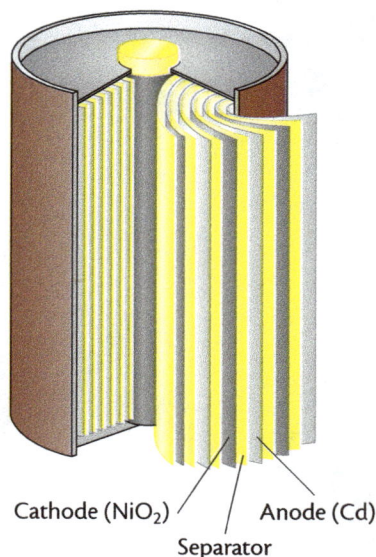

Cathode (NiO_2)   Anode (Cd)
Separator

Figure 5.35 *Cross-section of a NiCd battery.*

When this battery operates, Cd and NiO_2 are converted to $Cd(OH)_2(s)$ and $Ni(OH)_2(s)$. These solid products cling to the electrodes, allowing them to convert back to reactants when the used battery is connected in a recharging circuit. In recharging, an external electrical potential causes electrons to flow in the opposite direction, thus reversing the reaction. Although you can recharge NiCd batteries many times, some of the same processes that prevent recharging of primary batteries can take their toll and eventually reduce the efficiency of recharging.

NiMH or nickel–metal hydride batteries (see Figure 5.36, page 570) use a metal–metal compound that absorbs hydrogen instead of the cadmium in NiCd batteries, improving both their performance and their environmental impact. Most gasoline–electric hybrid vehicles built before 2010 use NiMH cells for energy storage.

NiCd batteries should be recycled to prevent the cadmium, which is toxic, from leaching into ground water.

Scientific American Working Knowledge Illustration

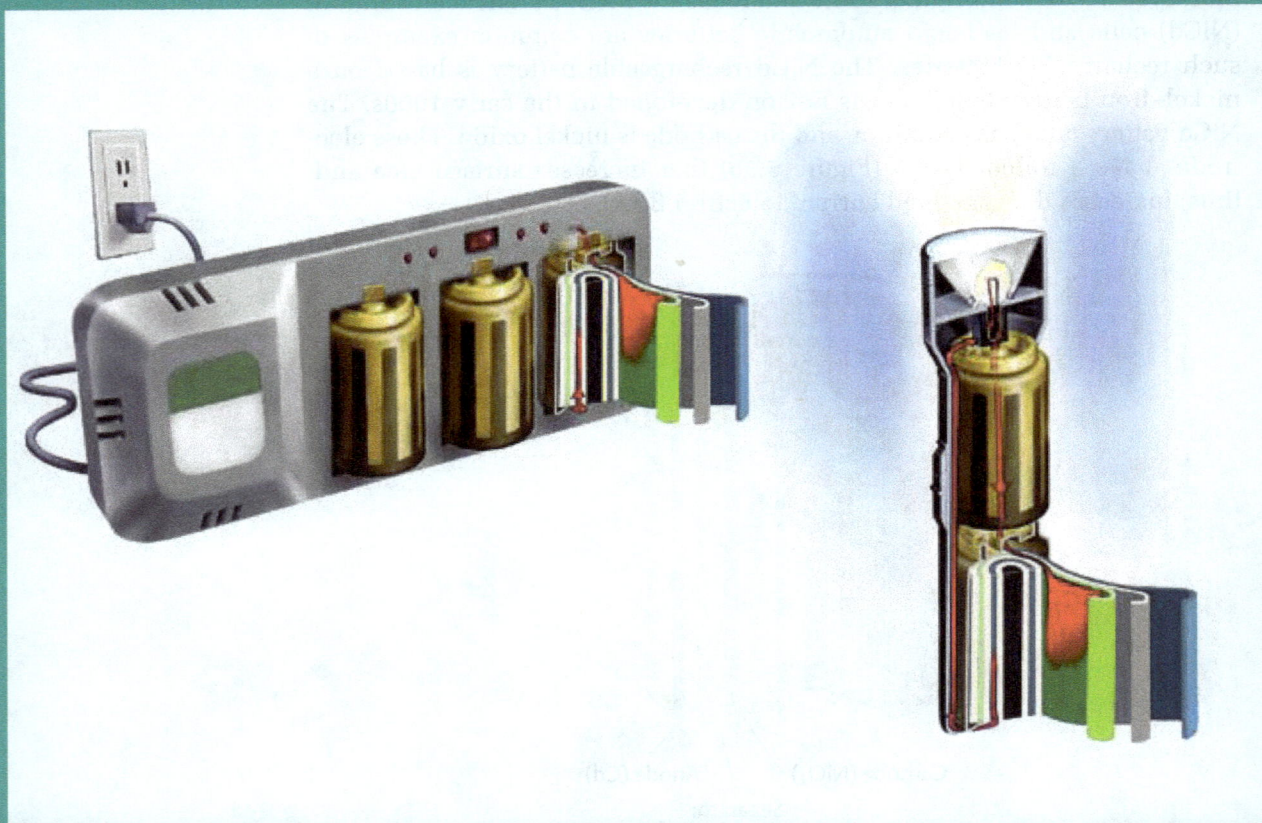

Figure 5.36 *Rechargeable batteries When a nickel–metal hydride (NiMH) battery is recharged, the charging circuit removes electrons (red) from the positive electrode (green) composed of nickel(II) hydroxide. This process oxidizes the Ni^{2+} to Ni^{3+}. Electrons are supplied to the negative electrode (blue). This causes a hydrogen-absorbing metal alloy to become reduced. When this battery is discharged during use, electron flow is reversed. The nickel(II) hydroxide gains electrons, while the metal hydride loses electrons.*

Lead-Acid Batteries

One everyday example of a rechargeable wet cell (a battery with a liquid electrolyte) is the 12-V automobile battery, known as a *lead-acid battery*. This battery, often used for starting, lighting, and ignition in a motor vehicle, is composed of a series of six electrochemical cells. As illustrated in Figure 5.37, each cell consists of an anode of uncoated lead (Pb) plates and a cathode of lead plates coated with lead(IV) oxide (PbO_2). The electrodes are immersed in a dilute solution of sulfuric acid, H_2SO_4, the electrolyte in this system.

When the vehicle's ignition is turned on, the electrical circuit is completed. As the electrons travel from one electrode to the other, they provide energy for the car's electrical systems. Metallic lead at the anode is oxidized to Pb^{2+}.

Lead-acid batteries are found in trucks, aircraft, motorcycles, self-starting lawnmowers, and motorized golf carts.

DISCHARGING

PbO$_2$ (cathode):
PbO$_2(s) + $ SO$_4{}^{2-}(aq) + 4$ H$^+(aq) + 2\ e^- \longrightarrow$
  PbSO$_4(s) + 2$ H$_2$O(l)

CHARGING

PbO$_2$ (anode):
PbSO$_4(s) + 2$ H$_2$O$(l) \longrightarrow$
  PbO$_2(s) + $ SO$_4{}^{2-}(aq) + 4$ H$^+(aq) + 2\ e^-$

Pb (anode):
Pb$(s) + $ SO$_4{}^{2-}(aq) \longrightarrow$ PbSO$_4(s) + 2\ e^-$

Pb (cathode):
PbSO$_4(s) + 2\ e^- \longrightarrow$ Pb$(s) + $ SO$_4{}^{2-}(aq)$

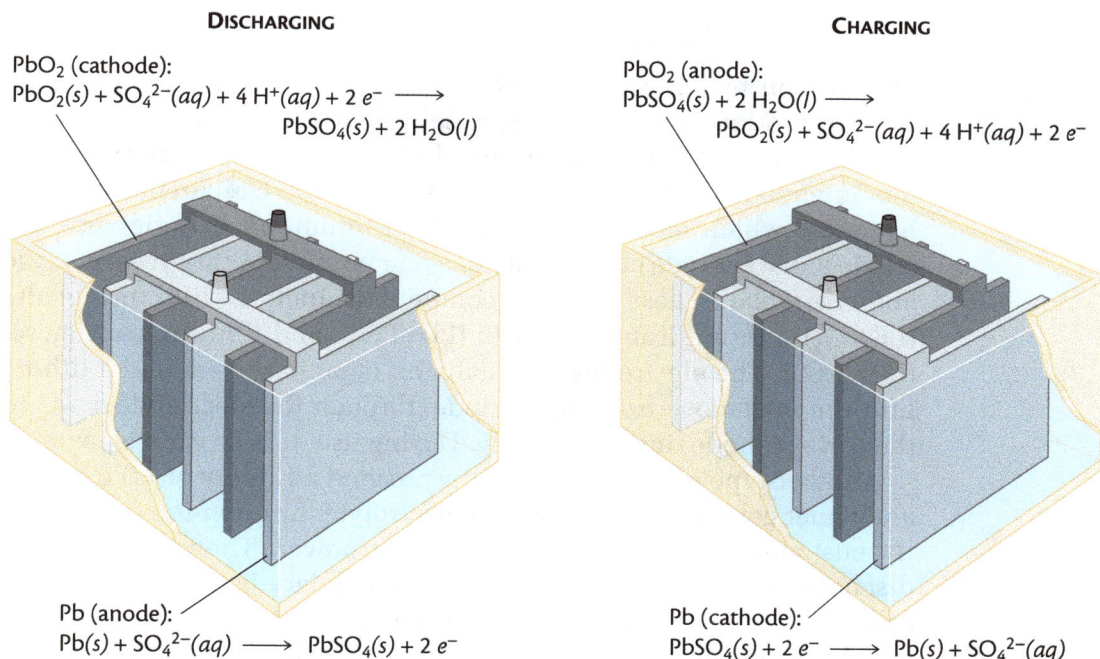

Figure 5.37 *Discharging and recharging an automobile battery. What furnishes the energy to recharge this battery?*

The freed electrons travel through the wire to the lead(IV) oxide cathode, reducing Pb^{4+} in PbO$_2$ to Pb^{2+}. Lead ions (Pb^{2+}) produced at both electrodes then form PbSO$_4$ by reacting with the electrolyte, as shown in the following equations.

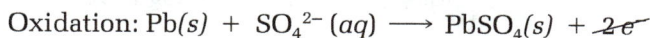

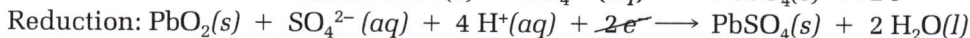

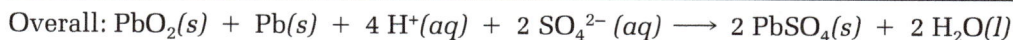

$$\text{Oxidation: Pb}(s) + \text{SO}_4{}^{2-}(aq) \longrightarrow \text{PbSO}_4(s) + \cancel{2e^-}$$
$$\underline{\text{Reduction: PbO}_2(s) + \text{SO}_4{}^{2-}(aq) + 4\ \text{H}^+(aq) + \cancel{2e^-} \longrightarrow \text{PbSO}_4(s) + 2\ \text{H}_2\text{O}(l)}$$
$$\text{Overall: PbO}_2(s) + \text{Pb}(s) + 4\ \text{H}^+(aq) + 2\ \text{SO}_4{}^{2-}(aq) \longrightarrow 2\ \text{PbSO}_4(s) + 2\ \text{H}_2\text{O}(l)$$

If an automobile battery is used too long without being recharged, it runs down; that is, the redox reaction stops. Lead(II) sulfate eventually coats the electrodes, which reduces their ability to react and produce a current and electrical potential. In an automobile, recharging is accomplished by an alternator or generator, which converts some mechanical energy from the vehicle's engine into electrical energy that forces electrons to move in the opposite direction through the battery. This reverses the direction of the battery's chemical reactions (see Figure 5.37).

Lead-acid batteries can pose dangers if rapidly charged by an outside source of electrical energy. Hydrogen gas, formed by the reduction of H$^+$ in the acidic electrolyte, is released at the lead electrode. A spark or flame can ignite the hydrogen gas, causing an explosion. This is one reason you must be careful to avoid sparks when using jumper cables to start a vehicle with a dead battery.

You can sometimes recognize a run-down or dead car battery by observing layers of solid white PbSO$_4$ coating the electrodes.

Lithium-Ion Batteries

Lithium-ion batteries are a type of rechargeable batteries now commonly used in products such as laptops, cell phones, and MP3 players. They are also increasingly used in gasoline-electric vehicles and electric vehicles—the intended application for WYE's batteries. Unlike lithium (primary) batteries, which are not rechargeable, lithium-ion batteries use a carbon anode and a cathode constructed of a compound of lithium with oxygen and another metal. These lithium compounds include lithium cobalt oxide ($LiCoO_2$), lithium iron phosphate ($LiFePO_4$), and lithium manganese oxide ($LiMn_2O_4$). The electrolyte is usually an organic solvent and a lithium salt. Lithium ions move from the cathode, through the electrolyte, and attach to the carbon anode during charging. During use, this process is reversed.

Another type of lithium-ion battery, called a Li-poly or LiPo battery, uses a polymer gel instead of the liquid electrolyte. Lithium-ion batteries are easily customized to the devices that they power. They are slower to self-discharge and lighter than most other secondary batteries. However, these cells lose up to 20% of their capacity each year due to deposits that form in the electrolyte during use and charging. They are also subject to overheating and must contain safety features that increase production costs.

concept check 8

1. Distinguish between primary and secondary batteries.
2. How do lithium and lithium-ion cells differ?
3. What types of batteries are commonly recycled? Why do you think that these batteries (and not other types) are recycled?

D.6 MANUFACTURING AND RECYCLING BATTERIES

Battery Manufacturing

The manufacture of commercial batteries varies depending upon the type of battery, but all have several common features, including anode, cathode, electrolyte, and container. The composition of several types of batteries is summarized in Table 5.4.

Primary batteries are often constructed within a metallic (usually nickel-plated steel) can. For an alkaline battery, this can functions as part of the battery's cathode. Alkaline batteries are constructed by grinding the cathode materials, including manganese oxide and graphite, into small particles, then compacting this material into a hollow cylinder and inserting it into the steel can. A paper separator soaked in potassium hydroxide electrolyte is inserted inside the can/tablet assembly (see Figure 5.34, page 567), and a gel containing zinc powder is used to fill the inner portion of the battery. This zinc-containing gel acts as the anode. The battery is completed with a brass nail for current collection and a seal to contain the ingredients and extend battery life.

Lead-acid car batteries, by contrast, have containers made from polymers, often polypropene (polypropylene). These containers are divided into six sections, each of which will contain one cell. The electrodes are prepared on grids made from lead or a lead alloy, which are coated with pastes. The cathode paste contains lead oxide, sulfuric acid, and water, whereas the anode paste adds powdered sulfates to this mixture. A porous insulating material is used to separate the positive and negative plates. One element or cell (anode, cathode, and separator) is placed into each section of the container. The cells are connected with conductors, the container is filled with a mixture of sulfuric acid and water (battery acid), and the battery is sealed.

Nearly 90% of U.S. used car batteries are recycled, leading to the recovery of lead metal, sulfuric acid, and battery-casing polymer.

Table 5.4

Composition of Common Primary and Secondary Batteries

	Type	Anode	Electrolyte	Cathode
			Principal Components	
PRIMARY	Zinc–carbon	Zn (sheet)	NH_4Cl or $ZnCl_2$	MnO_2/C
	Alkaline	Zn (powder)	KOH	MnO_2/C
	Lithium	Li (foil)	Organic solvent and Li salt	Metal oxide
	Zinc–air	Zn (powder)	KOH	Air/C
	Zinc–silver oxide	Zn (powder)	KOH	Ag_2O/C
SECONDARY	NiCd	Cd	KOH	NiO_2
	NiMH	Metal-metal compound	KOH	NiO_2
	Lead-acid	Pb	H_2SO_4	PbO_2
	Lithium ion	C	Organic solvent and Li salt	Li/metal/oxygen compound
	Li–poly or LiPo	C	Polymer gel and Li salt	Li/metal/oxygen compound

Adapted from Smith, M. J.; Gray, F. M. Batteries, from Cradle to Grave. *J. Chem. Ed.* 2010, 87, 162–167.

Lithium-ion batteries, like those WYE Battery Technology Corporation produces, are also constructed in individual cells. Each cell contains a graphite (carbon) anode and a manganese-based cathode, as well as a separator designed to allow for ion flow. The electrodes are made by creating a *slurry* (a suspension containing fine particles of an insoluble material) from a solvent and particles of active electrode materials and binders. The slurry is then applied to a surface to form electrodes, much like paste is applied to plates in the lead-acid battery construction process. The electrodes and separator are then assembled into a cell, an electrolyte consisting of an organic solvent and a lithium salt is added, and the cell is encased in a polymer-coated aluminum cover. Several cells are needed to construct an electric-vehicle battery.

Battery Recycling

Although recycling of lead-acid batteries is very common, recycling of other battery types is far less widespread in the United States. Recycling of lead-acid batteries makes sense both economically and ecologically. The lead can be reprocessed and reused, but it cannot be safely discarded. The situation is less clear cut for other batteries, particularly common primary alkaline and zinc–carbon cells. Since these batteries no longer contain mercury or other heavy metals, they are fairly safe for disposal. In addition, their recycling requires energy and time and results only in the recovery of common metals including zinc, steel, and manganese compounds.

The outlook is better for recycling of secondary batteries. Many of these batteries contain nickel, cadmium, or cobalt, which are scarcer than zinc and iron and may have sources located in areas of political unrest. While lithium supplies are also limited, the economic need to recover lithium does not yet play a large role in secondary battery recycling. The process of recycling any battery is energy-intensive. The combustible parts of the battery are burned off, then the remainder is cut up and thermal energy is added to liquefy the metals, which can then be separated.

Recycling is likely to be more practical for larger rechargeable batteries, such as those used in electric or hybrid-electric vehicles. Also, new batteries can be designed with the intention of recycling.

▌MODELING MATTER

D.7 LIFE CYCLE OF A BATTERY

Earlier, in Modeling Matter A.4, you represented the nitrogen cycle using a diagram. Throughout this course, you have explored natural cycles, including the carbon and nitrogen cycles, as well as life cycles of human-made products. In many ways, because of the nature of atoms and conservation, these cycles are similar. In this activity, you will consider the life cycle of the product manufactured by WYE Battery Technology Corporation—the lithium-ion battery.

1. List the main components of a lithium-ion battery. Consider the electrodes, electrolyte, and casing or container materials.

2. What are the most common sources of the materials listed in Question 1?

3. Using Figure 1.49 on page 110 as a model, create a diagram representing the life cycle of a lithium-ion battery for an electric or hybrid-electric vehicle. Be sure to include at least one image for each of the following: materials acquisition, manufacturing, use and maintenance, and recycling or waste management.

4. Which steps in the life cycle of lithium-ion batteries consume energy? For each, explain the particular energy needs.

5. Which steps generate wastes or emissions? Explain.

6. What will happen to the materials when the batteries are depleted or the vehicle is removed from service?

CHEM**QUANDARY**

COMMODITY OR SPECIALTY

A commodity is a good or product with an identity that does not depend on its source or producer. Its price is a function of supply and demand, often set by global factors. On the other hand, a specialty product may have different characteristics depending upon its producer, or it may be available only from one or a few producers or sources. The price of a specialty product depends upon the desirability and availability of its features. The ammonia produced by EKS Nitrogen Products is an example of a commodity. Would the vehicle batteries produced by WYE Battery Technology Corporation be considered a specialty product or a commodity? Why? How, if at all, does this classification of products affect the decision facing Riverwood's citizens?

D.8 MOVING TOWARD GREENER METHODS AND PRODUCTS

As you learned earlier in this unit, industries, governments, and citizens in the United States and elsewhere have recognized their joint responsibility to ensure that chemical products are manufactured with maximum benefit and minimum risk to society. The EPA's Design for the Environment program, which includes green chemistry (pages 543–544), has already influenced the way chemical-processing, printing, and dry cleaning industries conduct business.

The chemical industry has developed new synthesis methods, based on using safe starting materials to replace toxic or environmentally unsafe substances. Green chemistry focuses on preventing environmental pollution directly at the point of manufacturing. In this approach, the chemical industry works as a social partner to sustain development and international trade without damaging the environment.

New green chemistry-based processes use more environmentally benign reactants than do traditional processes and create waste products that are less damaging to the air and water. For example, chemical processes can now use D-glucose, ordinary table sugar, to replace benzene, a known carcinogen. The D-glucose serves as a raw material for synthesizing other reactants that can be used to produce nylon and medicinal drugs. Also, in some processes, chemists can substitute nontoxic food dyes for catalysts composed of toxic metals such as lead, chromium, and cadmium.

Isocyanates are substances used to make polyurethanes, which in turn are used to produce seat cushions, insulation, and contact lenses. Carbon dioxide can replace phosgene, a toxic gas, in manufacturing isocyanates. Green chemistry has led to methods for synthesizing industrial chemicals using water as a solvent instead of toxic alternatives, whenever possible, and for employing materials that can be recycled and reused, thus significantly reducing waste-disposal problems.

The chemical industry has the responsibility to make products in ways that are as hazard-free as possible. The chemical industry must deal honestly with the public to ensure that people clearly understand risks and benefits of chemical operations. Chemistry companies also must assure consumers that their products are safe when used as intended.

Chemical industries must comply with relevant laws and regulations, as well as with voluntary standards set by many manufacturers themselves. Independent, outside organizations, such as the International Organization for Standardization in Geneva, Switzerland, encourage worldwide compliance.

No initiatives of the chemical industry or the government can eliminate all risks involved in manufacturing chemical substances, any more than we can completely eliminate the risks of automobile travel. However, knowing the risks, continuing to explore the sources of and alternatives to those risks, and making prudent decisions all remain essential.

A chemical feedstock is a starting material from which we can make many other materials. For example, crude oil is a chemical feedstock for fuels and plastics.

As users of the industry's chemical products, all consumers share these responsibilities. Having studied some basic concepts about the manufacturing of chemical products, you can now turn to using such knowledge to weigh the risks and benefits of particular decisions—to you, the community, and the environment.

■ MAKING DECISIONS
D.9 ASSET OR LIABILITY?

If you were a Riverwood citizen, what would be your view? Should the town invite a chemical plant to locate in Riverwood? Which one? On what basis should such a decision be made?

To clarify this challenge, it is helpful to consider and evaluate both benefits and burdens—positive and negative factors—associated with any choice. Then you will be ready to address key questions that confront Riverwood citizens as they decide on their choices.

Positive Factors

- The Riverwood economy would improve. Either plant would employ about 200 residents. This would add about $8 million to Riverwood's economy each year. Also, each plant employee would indirectly provide jobs for another four people in local businesses. This is very desirable because 15% of Riverwood's labor force of 21 000 is currently unemployed.

- Farming costs might go down. Each year farmers in the Riverwood area spread 700 tons of fertilizer. The fertilizer comes from a fertilizer plant 200 miles away. Transport costs increase farmers' expenses by $15 for each ton of ammonia-based fertilizer. Local farmers thus stand to save about $10 000 each year in transportation costs.

- New local industries could result. With local access to lithium-ion batteries, small companies interested in developing and producing products that use these batteries may decide to locate in Riverwood. Already, a few local entrepreneurs have discussed the possibilities of producing electric delivery trucks and cargo vans that would be able to drive 100 miles on one charge (see Figure 5.38). The lithium iron phosphate batteries that they want to use could be produced within a special division at WYE. Such an expansion in battery production and creation of a new company that builds electric vehicles would create another 25–30 jobs, which would add another $600 000 per year to the local economy. Yet another possibility is the development of a lithium battery recycling plant in Riverwood.

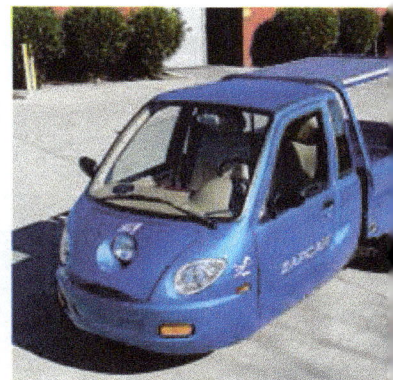

Figure 5.38 *A local source of lithium-ion batteries might entice other manufacturers, such as makers of solar-electric trucks like this one, to build facilities in Riverwood.*

- As a commercial refrigerant, ammonia is used to produce large quantities of ice. Ready access to ammonia supplies could support a commercial ice-making plant in Riverwood. Such a company might employ 25 to 35 individuals, mainly drivers and some plant workers, which could add $500 000 each year to Riverwood's economy.

- The tax base would improve, as either company would contribute to the town's tax base. This will provide a large increase in revenues for the community.

Negative Factors

- Rates of worker-related injuries and accidents could increase. Ammonia is manufactured at high pressure and high temperature. At ordinary temperature and pressure, ammonia gas is extremely toxic at high concentrations. Large amounts of ammonia released due to an accident on the road or at the plant could injure or kill workers and other community members within the vicinity. Several cases of work-related injury or illness or even death in ammonia-based fertilizer plants are reported each year. Although few worker injuries have been reported in battery production plants, a few large fires have occurred. Such fires are difficult to fight, and they may add contaminants to the local air and water.

- Demand for lithium-ion batteries may decrease. Although recent government incentives and new technologies have increased public awareness and demand for electric and hybrid-electric vehicles, this may be short-lived. Electric/hybrid-electric vehicle demand may decrease, subsidies may end, or new technology may replace lithium-ion batteries. Hybrid-electric vehicle battery technology is still new—replacement or recycling issues may render these vehicles not competitive.

- Water quality might suffer. If ammonia, ammonia-bearing wastewater, or wastes from battery production leaked into the Snake River, the resulting water contaminants could threaten aquatic life.

- Specialty battery or ammonia markets might decline.

- Sources for raw materials may be uncertain. The United States does not have large natural deposits of lithium minerals and currently relies on Chile and Argentina for 97% of lithium supplies. These countries may decide it is more advantageous to use their lithium supplies to produce lithium-based batteries for export rather than export the raw materials to other countries.

- The fertilizer industry is among the largest consumers of ammonia. Current fertilizer-intensive agricultural methods have created controversy. In some cases, crop yields have declined despite use of increased quantities of synthetic fertilizer. Some farmers have elected to use less synthetic fertilizer (or none at all if they are attempting to be certified as "organic," see Figure 5.39), so ammonia demand may decline in coming years.

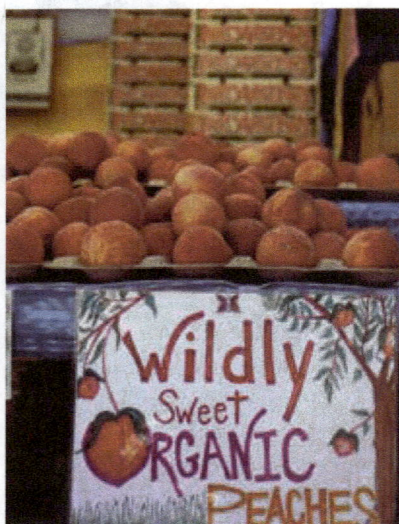

Figure 5.39 *An increase in organic farming could lead to decreased demand for synthetic fertilizers.*

Considering Burdens and Benefits

The poet William Wordsworth used the phrase, "Weighing the mischief with the promised gain . . ." in judging technological advancements (railroads, new in his time). Since the dawn of civilization, people have often accepted the burdens of new technologies to gain technology's benefits. Fire, one of civilization's earliest useful devices, gave people the ability to cook, warm themselves, and forge tools from metals. Yet fire out of control can destroy property and life. Every technology offers its benefits at a price.

One way to identify an acceptable new technology or venture is to evaluate it, finding an option that has a relatively low probability of producing harm—that is, delivers benefits that far outweigh the burdens. Unfortunately, benefit–burden analysis—weighing what Wordsworth called "mischief" against "promised gain"—is not an exact science.

For instance, some technologies may present high burdens immediately, while others may be associated with chronic, low-level risks for years or even decades. Many burdens are impossible to predict or assess with certainty. Individuals can control some potential burdens, but others must be addressed and controlled at regional or national levels. In short, it is quite difficult to conduct a thorough burden–benefit analysis. However, complete these activities:

> In many cases, electing not to make a decision is, in fact, also a decision—one accompanied by its own burdens and benefits.

1. Based on what you have learned in this unit, work in groups to create two lists for each chemical plant, one summarizing benefits and one summarizing burdens/risks.

2. Review your summaries of the burdens and risks. Are any negative factors completely unacceptable? If so, the plant associated with that risk or burden is probably not a viable option for Riverwood.

3. Within each list, mark the most valuable benefits with a plus symbol and the most serious burdens with a minus symbol.

4. Also consider the likelihood of occurrence of each benefit and burden. A burden that is fairly minor but almost certain to occur might merit more consideration than a burden that is more serious but extremely unlikely. Using a scale from 1 (highly unlikely) to 5 (extremely likely), rate each item on your lists in terms of its likelihood of occurrence.

5. Reread the opening commentaries.
    a. Identify the most convincing argument for a particular benefit or burden. Why do you find this argument convincing?
    b. Identify the least convincing argument for a particular benefit or burden. Why do you find this argument less convincing?

6. Based on your responses to Questions 3, 4, and 5, decide your own position on the question of locating a chemical plant near Riverwood. Discuss your view with your group.

7. As a class, discuss whether Riverwood should invite a chemical plant to the community. Be sure that the concerns of students who do not share the opinion of the majority are also heard and considered.

SECTION D SUMMARY

Reviewing the Concepts

> Batteries, which consist of one or more voltaic cells, provide convenient, portable ways to energize many common electrical devices.

1. What is the source of electrical energy in a battery?

2. Two types of voltaic cells are dry cells and alkaline batteries.

   a. List three similarities of these two cell types.

   b. How do these cells differ?

3. Write half-reaction equations for each of these oxidation-reduction processes. For each, identify (i) what is oxidized and (ii) what is reduced.

   a. $Pb(s) + Cu^{2+}(aq) \longrightarrow Pb^{2+}(aq) + Cu(s)$

   b. $Cr(s) + 3\ Ag^+(aq) \longrightarrow Cr^{3+}(aq) + 3\ Ag(s)$

4. Consider the following equation,

   $$PbO_2(s) + Pb(s) + 4\ H^+(aq) + 2\ SO_4{}^{2-}\ (aq) \longrightarrow 2\ PbSO_4(s) + 2\ H_2O(l)$$

   a. In what type of battery does this reaction occur?

   b. What is (i) oxidized and (ii) reduced?

   c. Does this equation represent the charging or discharging of the battery?

   d. Identify the substance that you might observe as a white coating on battery electrodes.

   e. Under what conditions could this battery produce hydrogen gas?

> The electrical potential of a voltaic pile or battery depends upon the identity of its electrodes and can be calculated from standard reduction potentials.

5. Use Table 5.3 (page 565) to identify the anode and cathode and calculate the expected electrical potential for each of the following voltaic cells. (Assume that each voltaic cell uses appropriate ionic solutions.)

   a. Cu–Cd cell   c. Mn–H_2 cell   e. Au–Li cell

   b. Sn–Zn cell   d. Pb–Cd cell

6. Use Table 5.3 (page 565) to propose a voltaic cell that would generate each of the following potentials:

   a. 0.47 V   b. 1.60 V   c. 3.97 V   d. 0.32 V

> Life cycle analysis of products such as batteries allows designers and users to make informed choices about starting materials, manufacturing, use, and disposal.

7. Think about alkaline batteries.

   a. What are some common starting materials in the manufacture of these batteries?

   b. Are these readily available in the form that is required for manufacturing? Explain.

8. Consider primary and secondary batteries of the same size and electrical potential.

   a. How do the life cycles of these batteries differ?

   b. How might this affect which battery you choose? Explain.

9. Why shouldn't batteries be disposed of in landfills?

10. What are some challenges in recycling household batteries?

Burden–benefit analysis is useful in weighing both positive and negative consequences when making decisions.

11. List two positive and two negative aspects of the technologies involved in producing
    a. ammonia.   b. batteries.
12. Describe a specific change that the chemical industry has implemented in response to the principles of green chemistry.

What challenges must be met to optimize production and use of batteries?

In this section, you have constructed a voltaic pile, used half-reactions to determine the electrical potential of batteries, examined the composition of batteries, compared primary and secondary batteries, and used life-cycle analysis to study environmental challenges in the manufacturing and disposal of batteries. Think about what you learned, then answer the question in your own words in organized paragraphs. Your answer should demonstrate your understanding of the key ideas in this section.

Be sure to consider the following in your response: electrode potentials, primary and secondary batteries, life-cycle, and green chemistry principles.

Connecting the Concepts

13. Why do batteries eventually stop operating? Explain in terms of limiting reactants.
14. Why are some batteries rechargeable and some not?

15. Choose one of the three natural cycles that you have explored (carbon, nitrogen, and water) and compare it to the life cycle of a battery.
16. Identify one benefit and one burden or risk associated with each of the following:
    a. playing high-school basketball
    b. driving a car
    c. jogging
    d. receiving a dental X-ray
    e. applying fertilizers to garden plants
    f. drinking bottled water

Extending the Concepts

17. Consider voltaic cells and electrolytic cells (cells where electrolysis occurs).
    a. How are they similar?
    b. How do they differ?
    c. Sketch a diagram of each type of cell and highlight the differences.
    d. Are these processes opposites? Explain.
18. Explain how you can test the condition of a lead storage battery with a *hydrometer*, a device that measures liquid density. (*Hint:* Sulfuric acid solutions are more dense than liquid water.)
19. Identify several technologies in your community. Use the risk-assessment table below to assign an appropriate letter to each activity. Discuss your decisions and compare your rankings with those made by others.

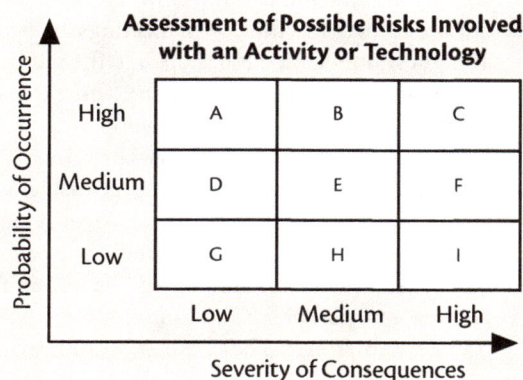

Assessment of Possible Risks Involved with an Activity or Technology

Probability of Occurrence	Low	Medium	High
High	A	B	C
Medium	D	E	F
Low	G	H	I

Severity of Consequences

PUTTING IT ALL TOGETHER

RIVERWOOD NEWS

TODAY'S
WEATHER:
mostly sunny

Breaking news at RiverwoodNewsLive.com

MORNING EDITION

A Chemical Plant for Riverwood

Decision Nears: Is a Chemical Plant in Riverwood's Future?

By Gary Franzen
RIVERWOOD NEWS STAFF REPORTER

After months of study and discussion, the Riverwood Town Council is prepared to act on separate proposals from EKS Nitrogen Products Company and WYE Battery Technology Corporation to locate a plant near Riverwood in the old Riverwood Corporation building. At tonight's special meeting, the council will decide which, if either, plant to approve. The meeting starts at 7:30 p.m. in town hall.

Mayor Cisko remarked, "I'm very pleased with the turnout we've had for the town meetings held on this issue. Tonight's council meeting is open to the public. I encourage all community members to attend and express their views about a chemical plant in Riverwood."

At the request of the Riverwood Town Council, both companies have prepared comprehensive summaries to inform citizens of their plans for a Riverwood plant. These summaries were circulated at previous town meetings and have been widely distributed in the community through newspapers, pamphlets, and on the town's Web site. Additional summaries can be obtained from the mayor's office.

Riverwood Corporation building

TOWN COUNCIL MEETING

Your class, with your teacher's guidance, will decide how to organize and conduct the town council meeting. Make sure that your format provides enough information so that Riverwood citizens will be able to consider benefits and burdens of each company's proposal and express their support for or opposition to each proposal. The meeting should conclude with a decision about whether a chemical plant will be allowed to operate in Riverwood, and, if so, which company will be invited to locate in Riverwood.

LOOKING BACK

Whether or not you decided to allow a chemical plant to locate in Riverwood, you have learned some valuable chemistry in the process of making that decision. You learned how key substances, such as ammonia, are produced and used, and how nitrogen cycles among the air, soil, and living organisms.

You also learned how chemists use electrochemical principles to harness chemical energy from spontaneous chemical reactions and also to provide energy enabling other reactions to occur. In addition, your acquired chemical knowledge and skills in analyzing burdens and benefits can help you to deal more effectively with future decision-making challenges.

ATOMS: NUCLEAR INTERACTIONS

?

A citizens group wants to ban all nuclear radiation from Riverwood and its surrounding area. *Turn the page to learn what you might contribute to the community's discussions of this proposal.*

CANE

CITIZENS AGAINST NUCLEAR EXPOSURE

invite you to attend an informative seminar on banning the use of all nuclear energy and materials in the Riverwood area.

SEE YOU ON FRIDAY, 7:00 P.M., RIVERWOOD TOWN HALL

DO YOU KNOW
whether each statement below is true or false?
If not, you could be in danger of serious nuclear exposure!

1. Home smoke detectors contain radioactive materials.

2. Radioactive materials and radiation are unnatural—they did not exist on Earth until created by scientists.

3. All radiation causes cancer.

4. Human senses can detect radioactivity.

5. Individuals vary widely concerning how they are affected by exposure to radiation.

6. Small amounts of matter change to immense quantities of energy released by nuclear weapons.

7. Physicians can distinguish cancer caused by radiation exposure from cancer resulting from other causes.

8. Medical X-rays are dangerous.

9. Nuclear power plants create serious hazards to public health and to the environment.

10. An improperly operated nuclear power plant can explode like a nuclear weapon.

11. Some nuclear wastes must be stored for centuries to prevent dangerous radioactivity from escaping.

12. New, dangerous elements are being invented every day.

13. Nuclear power plants produce material that could be converted into nuclear weapons.

14. All nuclear medical techniques are highly dangerous.

A local Riverwood organization, Citizens Against Nuclear Exposure (CANE), has organized to prevent all uses of nuclear power, the disposal of nuclear waste, food irradiation, and nuclear medicine in the Riverwood area. The flyer reproduced on the opposite page is a sample of the organization's effort to communicate with Riverwood residents.

Some Riverwood citizens are wary of the restrictions CANE proposes because they fear that such restrictions would hinder access to some options in medical diagnosis and treatment. The grandmother of Ms. Lynn Paulson, a Riverwood High School chemistry teacher, is among these citizens. She asked Ms. Paulson whether any high-school chemistry students could provide background about nuclear science and technology to the senior citizens in the community. Several senior citizens plan to attend the announced CANE meeting, but they would like to acquire some background information to be ready to question CANE representatives about their proposal. Ms. Paulson has agreed to help. In this unit, you and your classmates will assume the role of Riverwood High School chemistry students preparing a community presentation.

The chemistry you have learned thus far involves chemical changes due to sharing or transferring outer-shell electrons among atoms. You will encounter much different changes in this unit—changes associated with nuclei (rather than electrons) of atoms. This unit examines nuclear radiation, radioactivity, and nuclear energy, plus implications of their use and development.

As you progress through this unit, record the ideas and applications you decide to share in your presentation. In particular, focus on the assertions contained in the CANE flyer. You may also want to investigate issues that will be of interest to your audience.

SECTION A THE NATURE OF ATOMS

What evidence led to a modern understanding of the composition of atoms?

In this section, you will discover how both well-planned experiments and chance observations led to the current model of the atom's structure and to the discovery of radioactivity. Determining and describing the structure of atoms rank among the greatest scientific accomplishments of the past centuries. The stories of these great discoveries will help you understand atomic and nuclear chemistry and how researchers have used methods of science in their investigations.

GOALS

- Define and describe radioactivity.
- Distinguish between ionizing and non-ionizing radiation.
- Describe Rutherford's gold-foil experiment, including how its results led to a new model of the atom.
- Define isotope and radioisotope.
- Interpret a given isotope in terms of its atoms' protons and neutrons.
- Calculate the average molar mass of an element, given isotopic molar masses and abundance data.

concept check 1

1. Draw a representation of an atom of sodium. Label the components.
2. Are all atoms of sodium identical? Explain.
3. What are some sources of ionizing radiation?

◼ INVESTIGATING MATTER
A.1 EXPLORING IONIZING RADIATION

You are being constantly bombarded by radiation. Whether you are at school, in an airplane, at a park, or inside your house, you are always exposed to multiple forms of radiation. In this investigation, you will examine a variety of everyday (and less common) objects using an instrument designed to detect ionizing forms of radiation.

Preparing to Investigate

Before you begin, read *Gathering Evidence* to learn what you will need to do and note safety precautions. Create a plan for data collection that will allow you to make some claims about materials that emit ionizing radiation.

Gathering Evidence
Part I: Background Radiation

1. Before you begin, put on your goggles and gloves, and wear them properly throughout the investigation.

2. Turn on your detector. Let the detector warm up for ~2 minutes. You will notice that your detector will indicate the presence of ionizing radiation even though it is not near a sample or pointed toward a source. This radiation is called **background radiation**. It is always present, and results from a variety of sources, some of which are natural and others that result from human activity.

3. Move to several places within the classroom or laboratory and take readings.

4. If your teacher approves and your detector is portable, take readings outside of the classroom.

5. Record the readings collected in your data table.

Part II: Measuring Ionizing Radiation from Samples
You will be provided with several objects and materials. Handle all samples with forceps. Use your detector to measure ionizing radiation emitted by each sample (see Figure 6.1). Record all readings in your data table.

Part III: Pinpointing Sources of Ionizing Radiation
Choose one or two samples that emit ionizing radiation. Handle all samples with forceps. Use your detector to identify the source of the radiation (an area or areas where the radiation seems to be concentrated). Record your findings in your data table.

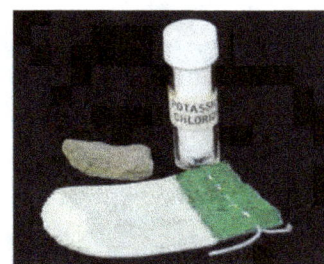

Figure 6.1 *Some common objects emit ionizing radiation.*

Interpreting Evidence

1. Did background radiation vary from place to place? Explain.
2. Identify any locations inside or outside the classroom that had background radiation readings that were
   a. unusually high.
   b. unusually low.
3. Which samples emitted ionizing radiation?
4. Were the radiation levels emitted by the various samples similar or different? Support your answer with data.
5. Compare your results with those of your classmates.
   a. Did areas within and outside the classroom give consistent readings for background radiation from group to group?
   b. Did the same samples emit radiation?
   c. How do you explain any differences between your results and those of your classmates?

Making Claims

6. What statement can you make about the level of background radiation in and around your classroom?
7. Consider Part III of the investigation.
   a. Did all parts of the samples you examined emit ionizing radiation?
   b. If not, could you identify a purpose for the source of the ionizing radiation? Explain.

Reflecting on the Investigation

8. What safety precautions did you take? Would someone using this object every day use precautions?
9. Do you think you could predict, just by looking at it, whether a sample will emit ionizing radiation? Explain.

A.2 THE GREAT DISCOVERY

You learned about electromagnetic radiation in Unit 2 (see page 221).

The history of modern scientific investigation of the atom began with the study of radiation. Scientists have long been interested in light—because life would not exist without it—and other types of radiation. By the end of the 19th century, scientists had already studied many types of radiation from a variety of sources.

In 1895, a series of observations significantly broadened scientific understanding of radiation. The German physicist W. K. Roentgen was studying **fluorescence**, a phenomenon in which certain materials emit light when struck by radiant energy, such as ultraviolet rays (Figures 6.2 and 6.3).

Figure 6.2 *French physicist Henri Becquerel's investigations of fluorescence led to the discovery of radioactivity.*

Figure 6.3 *Ultraviolet light shining on particular objects, such as certain fabrics, produces fluorescence—visible light emitted by material exposed to such electromagnetic radiation.*

Roentgen found that certain materials fluoresced when exposed to beams of **cathode rays** emitted from the cathode when electricity passed through an evacuated glass tube (see Figure 6.4). A few years after Roentgen's work, cathode rays were identified as beams of electrons.

Roentgen was working with a cathode-ray tube covered by black cardboard. He observed an unexpected glow of light on a piece of paper across the room. The paper was coated with a fluorescent material, and Roentgen expected it to glow when exposed to radiation. However, visible radiation could not pass through the black cardboard covering the cathode-ray tube, and the fluorescent paper was not in the path of electrons from the tube. Roentgen hypothesized that some other radiation passing through the black cardboard had been emitted by the cathode-ray tube. He named the mysterious radiation *X-rays*, where *X* represented the unknown radiation. Scientists now know that **X-rays** are a form of high-energy electromagnetic radiation.

Older computer monitors and TV screens were based on a version of the cathode-ray tube.

In modern X-ray devices, the X-rays are generated when an electron beam strikes a metal target, which is often made of tungsten.

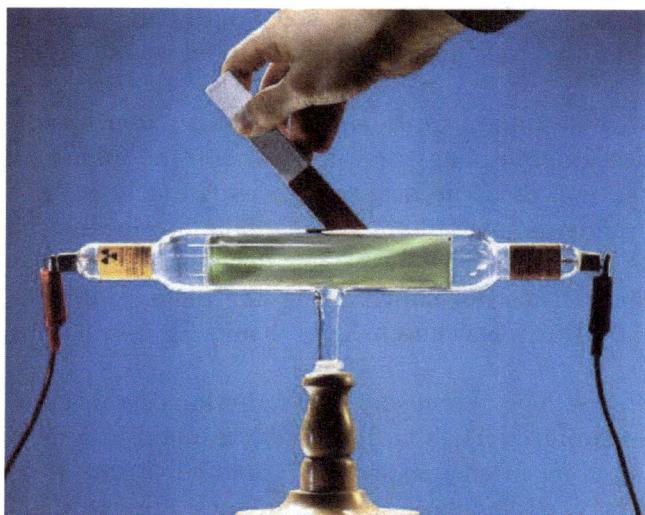

Figure 6.4 *A beam of electrons moves from the cathode (left) to the anode (right). The visible light emitted results from collisions of electrons with the fluorescent screen inside the tube. The deflection of the beam by a magnet indicates that the beam particles have a negative electrical charge. Collision of the electron beam with the glass or anode produces X-rays.*

Further experiments revealed that these X-rays could penetrate many materials, but could not easily pass through dense materials such as lead and bone. Scientists soon realized how useful X-rays could be in medicine. In fact, one early X-ray image Roentgen obtained was of his wife's hand. Figures 6.5 and 6.6 show some modern X-ray images.

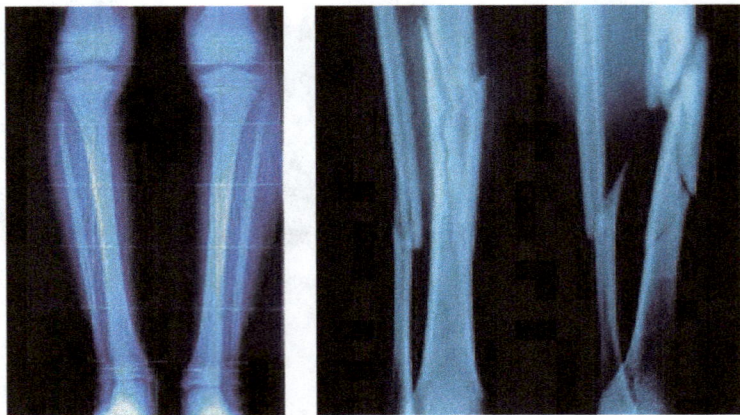

Figure 6.5 *These X-ray images reveal a pair of normal human legs (left) and a pair of broken legs (right). Development of this useful medical diagnostic tool emerged from the study of fluorescence.*

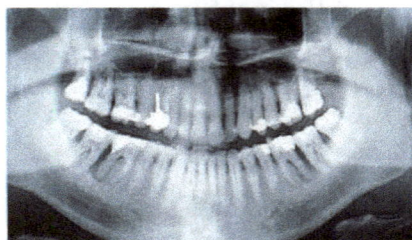

Figure 6.6 *An X-ray image of a human jaw. Such images help dentists detect cavities and other dental problems.*

Roentgen's discovery intrigued other scientists, including the French physicist Henri Becquerel. Because X-rays could produce fluorescence, Becquerel wondered if fluorescent minerals might give off X-rays as they fluoresce. In 1896, Becquerel placed in sunlight some crystals of a fluorescent mineral that contained uranium. He then wrapped an unexposed photographic plate in black paper and placed the mineral crystals on top of the wrapped plate. If the mineral did emit X-rays, they would penetrate the black paper and the exposed film would darken, even though it was shielded from light.

Cloudy weather prevented Becquerel from completing his experiments. He stored the wrapped photographic plates in a drawer with the uranium-containing mineral. After several days, he decided to develop some of the stored plates, thinking that perhaps some fluorescence might have persisted, causing some fogging of the photographic plates. When Becquerel developed the plates, he was astounded. Instead of faint fogging, the plates had been strongly exposed. Figure 6.7 illustrates the chain of events in Becquerel's investigation.

Fluorescence stops as soon as the external source of radiation (in this case, the Sun) is removed from the object. Thus, a fluorescent mineral in a dark drawer should not cause such an intense exposure. Scientists at that time could not offer a satisfactory explanation for Becquerel's observations. Becquerel suspected that the rays that exposed the photographic plates in the

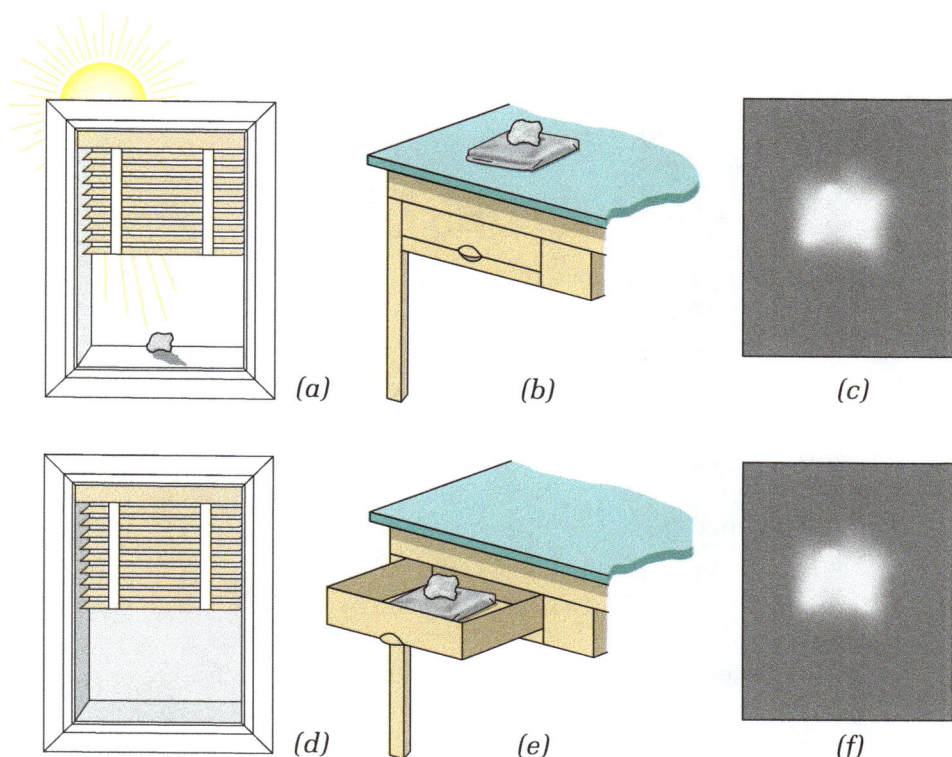

Figure 6.7 *Becquerel's investigation. Becquerel placed a fluorescent mineral in direct sunlight (a), then put it on an unexposed, wrapped photographic plate (b). Radiation exposed the plate (c). On a cloudy day (d), the wrapped plate was placed in a drawer with the mineral and kept from light (e); however, the mineral sample still exposed the photographic plate (f).*

drawer were more energetic and had much greater penetrating ability than X-rays. Thus, he interrupted his study of X-rays to investigate the mysterious radiation apparently given off by the uranium-containing mineral. Although he could not explain it, Becquerel had discovered **radioactivity**, which is now known to involve the spontaneous emission of particles and energy from atomic nuclei. This phenomenon is distinctly different from X-ray production or fluorescence (Figure 6.8).

Becquerel suggested that Marie Curie (Figure 6.9), a graduate student working with him, attempt to isolate the radioactive component of pitchblende, a uranium ore, for her research for her doctoral dissertation. Her preliminary work was successful. Her physicist husband, Pierre Curie, changed his research focus to join her on the pitchblende project. Working together, Marie and Pierre Curie discovered that the level of radioactivity in pitchblende was four to five times greater than expected from its known uranium content. The Curies suspected the presence of another radioactive element. After processing more than a thousand kilograms of pitchblende, they isolated tiny quantities (measured in milligrams) of two previously unknown radioactive elements. These elements later became known as polonium (Po) and radium (Ra).

Figure 6.8 *Many minerals exhibit fluorescence when exposed to ultraviolet light.*

Figure 6.9 *Marie Curie discovered two highly radioactive elements—radium (named for the radiation it emitted) and polonium (named for her native Poland). Her work earned her two Nobel Prizes.*

CHEM**QUANDARY**

SCIENTIFIC DISCOVERIES

What do the following events have in common with Becquerel's discovery of radioactivity?

1. As Charles Goodyear experimented with natural rubber (a sticky material that melts when heated and cracks when cold), a mixture of rubber and sulfur came in contact with a hot stovetop. He noted that the rubber-and-sulfur mixture did not melt. *Vulcanization*, a process that makes rubber more durable, resulted from this observation.

2. Roy Plunkett, a research chemist, used gaseous tetrafluoroethene ($F_2C{=}CF_2$) from a storage cylinder, but the gas flow stopped long before the cylinder should have completely emptied. He cut open the cylinder and discovered a new, white solid that is now known as polytetrafluoroethene, *Teflon*.

3. James Schlatter, a research chemist trying to produce an antiulcer drug, accidentally got some of the substance on his fingers. When he later licked his fingers to pick up a piece of paper, his fingers tasted very sweet, and he correctly linked the sweetness to the antiulcer drug. Instead of finding an antiulcer drug, he discovered *aspartame*, an artificial sweetener.

A.3 NUCLEAR RADIATION

The nuclei of unstable atoms are sources of nuclear radiation.

Many people respond with alarm and even panic when they hear the word *nuclear*. In addition, the general term *radiation*, which is sometimes used to refer particularly to nuclear radiation, can also cause anxiety. In reality, radiation falls into two general types: *ionizing radiation* and *non-ionizing radiation*.

Electromagnetic radiation in the visible and lower-energy regions of the spectrum (see Figure 6.10) is **non-ionizing** (long-wavelength) **radiation**. Non-ionizing radiation transfers its energy to matter, causing atoms or molecules to vibrate (infrared radiation), move their electrons to higher energy levels (visible radiation), or heat up (such as in microwave ovens). Although this radiation is generally considered safe, excessive exposure can be harmful. Sunburn, for example, results from an overexposure to non-ionizing radiation from the Sun (Figure 6.11). In fact, intense microwave and infrared radiation can cause lethal burns.

Frequency, ν (Hz)

Visible light

| 10^4 | 10^6 | 10^8 | 10^{10} | 10^{12} | 10^{14} | 10^{16} | 10^{18} | 10^{20} | 10^{22} | 10^{24} |

Microwaves

Ultraviolet

Gamma rays

X-rays

Infrared

Radio waves

| 10^4 | 10^2 | 1 | 10^{-2} | 10^{-4} | 10^{-6} | 10^{-8} | 10^{-10} | 10^{-12} | 10^{-14} | 10^{-16} |

Wavelength, λ (m)

Increasing energy

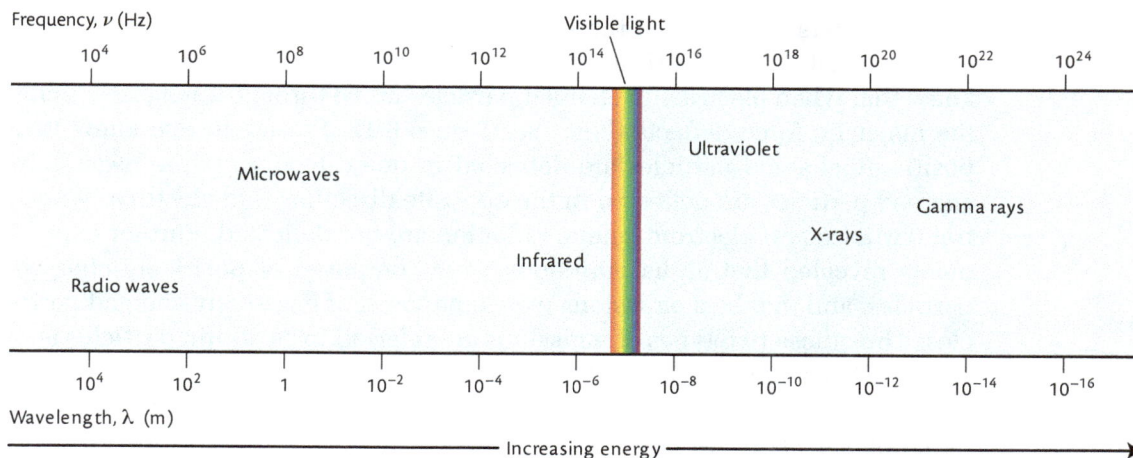

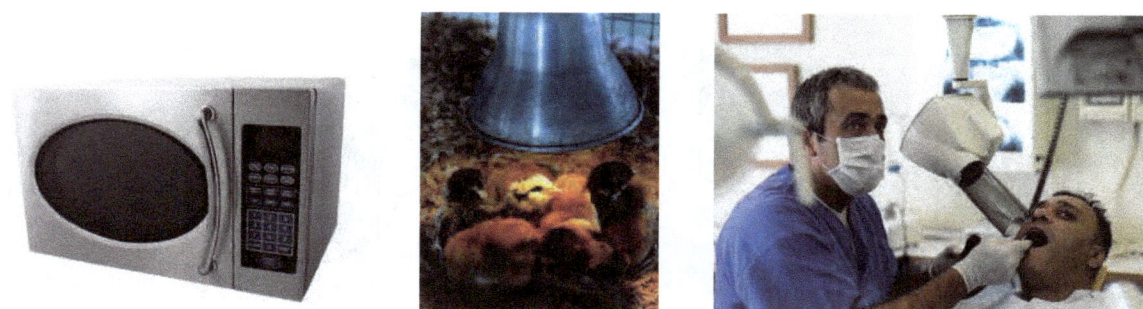

Figure 6.10 *The electromagnetic spectrum. Microwaves, infrared radiation (from heat lamps), and X-rays are all examples of electromagnetic radiation.*

Ionizing radiation, which includes high-energy electromagnetic radiation (short-wavelength ultraviolet radiation, X-rays, and gamma rays) and all nuclear radiation, carries more energy and potential for harm than non-ionizing radiation. Energy from ionizing radiation can eject electrons from atoms and molecules, forming molecular fragments and ions. These fragments and ions can be highly reactive. If formed within a living system, they can disrupt normal cellular chemistry, causing serious cell damage.

Nuclear radiation is a form of ionizing radiation that is caused by changes in the nuclei of atoms. In chemical reactions, the atomic number (number of protons) does not change. An atom of aluminum (13 protons) always remains an aluminum atom, and an iron atom (26 protons) always remains an iron atom. However, atoms with unstable nuclei—radioactive atoms—can spontaneously change their identities. A radioactive atom changes spontaneously through disintegration of its nucleus, which results in emission of high-speed particles and energy. When this happens, the identity of the radioactive atom often changes. An atom of a different element forms. This process is **radioactive decay**. The emitted particles and energy make up **nuclear radiation**.

Ernest Rutherford showed in 1899 that nuclear radiation included at least two different types of emissions, which he named *alpha rays* and *beta rays*. Shortly afterward, scientists discovered a third kind of nuclear radiation: *gamma rays*.

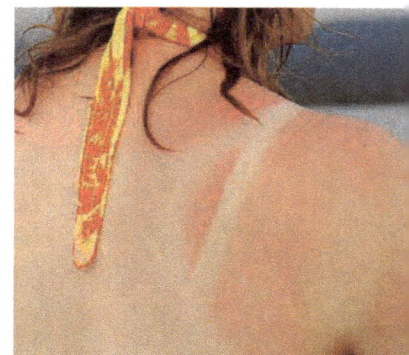

Figure 6.11 *Sunburn often results from overexposure of skin to ultraviolet radiation.*

Alpha (α), beta (β), and gamma (γ) are the first three letaters of the Greek alphabet.

Researchers allowed the three types of nuclear radiation to pass through magnetic fields to investigate their electrical properties. Scientists already knew that when electrically charged particles move through a magnetic field, the magnetic force deflects them. See Figure 6.12. Scientists also knew that positively charged particles are deflected in one direction, while negatively charged particles are deflected in the opposite direction, and electrically neutral particles and electromagnetic radiation are not deflected. Further experiments revealed that alpha emissions were composed of positively charged particles, and that beta emissions were composed of negatively charged particles. Thus these two types of emissions are referred to as **alpha particles** and **beta particles** (not rays, as originally named). **Gamma rays**, not deflected by a magnetic field, have no electric charge; they are high-energy electromagnetic radiation similar to X-rays.

> Alpha rays and beta rays are more commonly referred to as particles because they possess measurable mass.

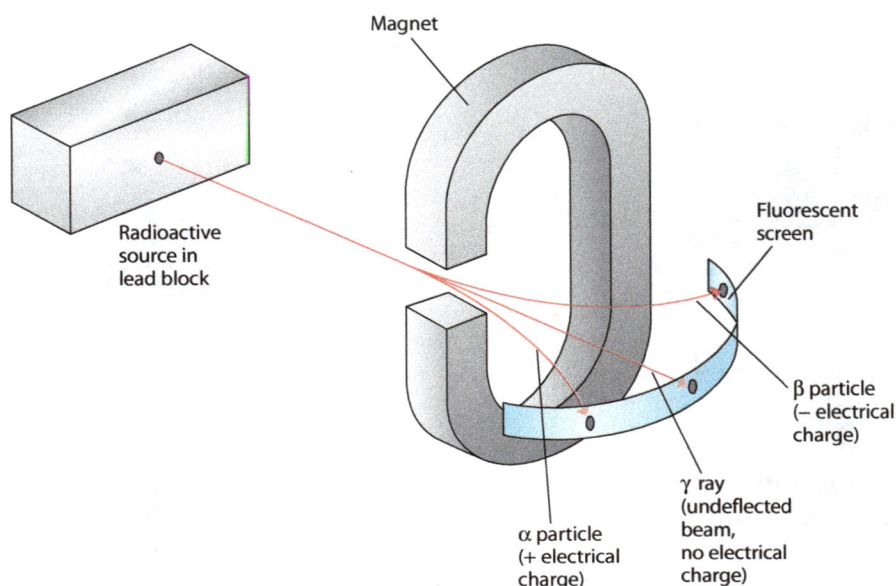

Figure 6.12 *Behavior of alpha (α) particles, beta (β) particles, and gamma (γ) rays passing through a magnetic field.*

By describing the nature of radioactivity, scientists toppled an old theory (a common event in scientific progress—new knowledge replaces an old body of knowledge). Once scientists knew about alpha particles, beta particles, and gamma rays, they became convinced that atoms, which were originally thought to be the smallest, most fundamental units of matter, must be composed of even smaller particles.

From the results of another experiment, the *gold-foil investigation*, Rutherford proposed a fundamental model of the atom that is still useful today. To do so, he developed an ingenious, indirect way to "look" at the structure of atoms.

A.4 THE GOLD-FOIL EXPERIMENT

Before Rutherford's research, scientists had tried to explain the arrangement of electrons and positively charged particles within atoms in several ways. In the most widely accepted model, an atom was viewed as a volume of positive electrical charge, with the negatively charged electrons embedded within, like peanuts in a candy bar. In the late 1800s this was known as the "plum pudding" model, because it resembled the distribution of raisins within that traditional English dessert.

About 1910, Rutherford decided to test the plum pudding model. Working in Rutherford's laboratory in Manchester, England, Hans Geiger and Ernest Marsden focused a beam of alpha particles—the most massive of the three types of nuclear radiation—at a thin sheet of gold foil only 0.000 04 cm (about 2000 atoms) thick (see Figure 6.13). Geiger and Marsden used a zinc sulfide-coated screen to detect the alpha particles after they passed through the gold foil (Figure 6.14). The screen emitted a flash of light where each alpha particle struck it. By observing the tiny light flashes at different positions with respect to the gold foil, Geiger and Marsden deduced the paths of the alpha particles as they interacted with the gold foil.

Figure 6.13 *A sheet of gold leaf is prepared for trimming. This very thin material was produced with machines in Germany, but gold foil or leaf was traditionally prepared by beating gold into sheets.*

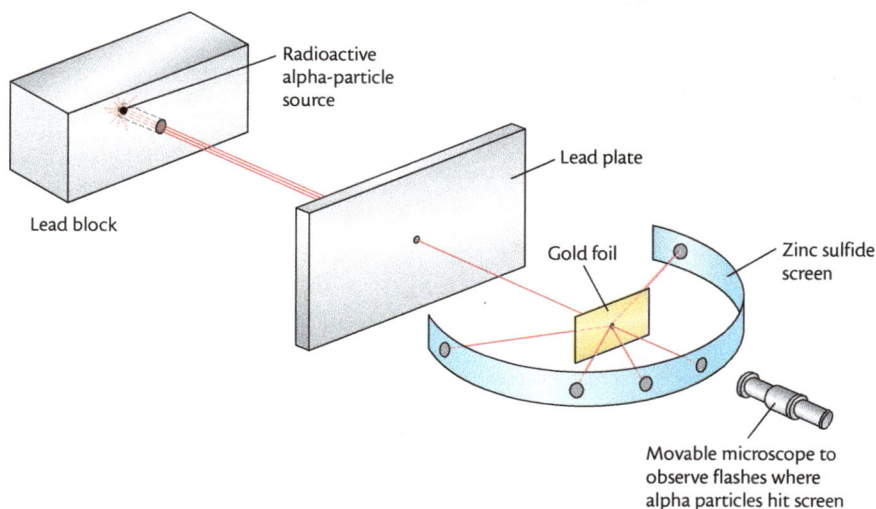

Figure 6.14 *The alpha-particle scattering experiment led Rutherford to conclude that the atom is largely empty space with an incredibly dense, positively charged nucleus at its center.*

Rutherford expected alpha particles to be scattered slightly as they were deflected by the gold atoms in the foil, producing a pattern similar to water being sprayed from a nozzle. However, he was in for quite a surprise. First, most of the alpha particles passed straight through the gold foil as if nothing were there (see Figure 6.15). This implied that most of the volume occupied by the gold atoms was empty space. But Rutherford was even more surprised that a few alpha particles, about 1 in every 20 000, bounced *back* toward the source. He described his astonishment this way: "It was almost as incredible as if you fired a 15-inch [artillery] shell at a piece of tissue paper and it came back and hit you."

Whatever repelled these deflected alpha particles must have been extremely small because most of the alpha particles went straight through the foil. Rutherford concluded, "On consideration, I realized that this scattering backward must be the result of a single collision, and when I made calculations I saw that it was impossible to get anything of that order of magnitude unless you took a system in which the greater part of the mass of the atom was concentrated in a minute nucleus. It was then that I had the idea of an atom with a minute massive centre, carrying a charge."

From these results, Rutherford developed the modern model of the nuclear atom. He named the tiny, dense, positively charged region at the center of the atom the *nucleus*. He envisioned that electrons orbited the nucleus, somewhat as planets orbit the Sun. Figure 6.15 illustrates how the nuclear model explains the results of the gold-foil experiment.

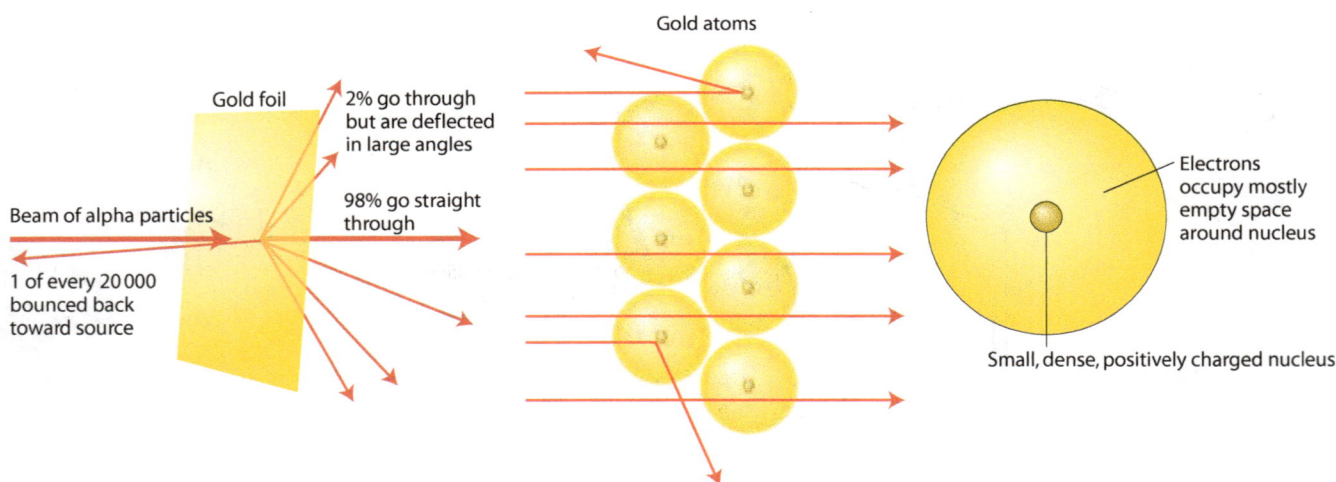

Figure 6.15 *Results of the alpha-particle scattering experiment. Most alpha particles passed through the foil but a few were deflected, some at very large angles (left). Proposed model (middle) to account for the results. Nuclear model of the atom (right).*

A.5 ARCHITECTURE OF ATOMS

Since Rutherford's time, scientific understanding of atomic structure has expanded and changed somewhat. Rutherford's model of a central, dense nucleus surrounded mostly by empty space is still valid. Later research has shown that the idea of orbiting electrons is incorrect. The general regions in which electrons are most likely to be found can be described, but particular movements or locations of electrons cannot be described.

Further research revealed that the nucleus is composed of two types of particles: neutrons, which are electrically neutral, and protons, which possess a positive charge. These particles, as well as electrons, are called **subatomic particles**. Protons and neutrons have about the same mass, 1.7×10^{-24} g. Although this mass is incredibly small, it is much greater than the electron's mass, which is 9.1×10^{-28} g. As shown in Table 6.1, one mole of electrons (6.02×10^{23} electrons) has a mass of only 0.0005 g. The same number of protons or neutrons would have a mass of about 1 g. In other words, a proton or a neutron is about 2000 times more massive than an electron. Thus, protons and neutrons account for nearly all the mass of every atom and also for nearly all of the total mass of every object that you encounter.

> Neutrons are actually slightly more massive than protons.

> *Protons:*
>   positive electrical charge
> *Neutrons:*
>   no electrical charge
> *Electrons:*
>   negative electrical charge

Table 6.1

Three Components of Atoms			
Particle	**Location**	**Charge**	**Molar Mass (g/mol)**
Proton	**Nucleus**	+1	1
Neutron	**Nucleus**	0	1
Electron	**Outside of nucleus**	−1	0.0005

The diameter of a typical atom is about 10^{-10} m, but an average nuclear diameter is 10^{-14} m, which is only one ten-thousandth ($10^{-14}/10^{-10}$) of the entire atom's diameter. Looking at this another way, the nucleus occupies only about one trillionth (10^{-12}) of an atom's total volume. Imagine that a billiard ball represents the diameter of an atom's nucleus. On that scale, electrons surrounding this billiard-ball nucleus would extend out in space more than one-half kilometer (about a third of a mile) away in all directions. This is consistent with the observation that most alpha particles, each the size of a helium nucleus, passed right through Rutherford's sheet of gold foil.

As you learned in Unit 1 (page 54), each atom of an element has the same number of protons in its nucleus, and each element has a unique number of protons. This number, called the *atomic number*, identifies the element. For example, each carbon atom nucleus contains six protons, Therefore, the atomic number of carbon is 6.

> In most periodic tables, the atomic number is written above the symbol of each element.

Yet all atoms of a given element do not necessarily have the same number of neutrons in their individual nuclei. Recall that atoms of the same element with different numbers of neutrons are called *isotopes* of that element. Naturally occurring carbon atoms, each containing six protons, may have six, seven, or even eight neutrons. The composition of these three carbon isotopes is summarized in Table 6.2. Figure 6.16 shows their nuclei.

Table 6.2

	Three Carbon Isotopes			
Name	Total Protons (Atomic Number)	Total Neutrons	Mass Number	Total Electrons
Carbon-12	6	6	12	6
Carbon-13	6	7	13	6
Carbon-14	6	8	14	6

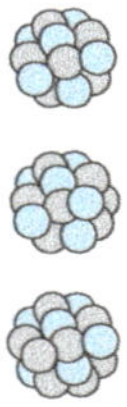

Carbon-12

Carbon-13

Carbon-14

Figure 6.16 *Nuclei of three carbon isotopes, containing, respectively, 6, 7, and 8 neutrons (gray) and 6 protons (blue).*

Isotopes are distinguished by their different mass numbers. The mass number, as you learned in Unit 1 (page 54), represents the total number of protons and neutrons in an atom. The three carbon isotopes in Table 6.2 have mass numbers of 12, 13, and 14, respectively. To specify a particular isotope, the atomic number and the mass number are written in front of the element's symbol in a particular way. For example, an isotope of carbon (C) with an atomic number 6 and a mass number 12 is written this way:

$$^{12}_{6}C$$

An atom of strontium with an atomic number 38 and mass number 90 would be symbolized as follows:

$$^{90}_{38}Sr$$

Another way to identify a particular isotope is to write the name or symbol of the element followed by a hyphen and its mass number. For example, an isotope for carbon may be called carbon-12, C-12, or ^{12}C. The symbols, names, and nuclear composition of some isotopes are summarized in Table 6.3.

Table 6.3

		Some Common Isotopes			
Symbol	Name	Total Protons (Atomic Number)	Total Neutrons	Mass Number	Total Electrons
Li	Lithium-7	3	4	7	3
O	Oxygen-16	8	8	16	8
Ne	Neon-20	10	10	20	10
Ga	Gallium-67	31	36	67	31
Ba	Barium-138	56	82	138	56
Pt	Platinum-195	78	117	195	78
Tl	Thallium-201	81	120	201	81
Pb	Lead-208	82	126	208	82

▍DEVELOPING SKILLS

A.6 INTERPRETING
ISOTOPIC NOTATION

Sample Problem: *Suppose you know that one product of a nuclear reaction is an isotope containing 85 protons and 120 neutrons. The atom therefore has a mass number of 205 (85 p + 120 n). What is the symbol of this element?*

$$^{205}_{85}?$$

Consulting the periodic table, you find atomic number 85 represents the element astatine (At).

$$^{205}_{85}\text{At}$$

1. Prepare a summary chart like Table 6.3 (previous page) for the following six isotopes. (Consult the periodic table for any needed information.)

   a. $^{12}_{?}\text{C}$ d. $^{24}_{12}?$

   b. $^{14}_{7}?$ e. $^{202}_{?}\text{Hg}$

   c. $^{16}_{?}\text{O}$ f. $^{238}_{92}?$

2. Using Table 6.3 as a source of information, what general relationship do you note between the total number of protons and total number of neutrons for atoms of

   a. lighter elements with atomic numbers less than 20?

   b. heavier elements with atomic numbers greater than 50?

▍MODELING MATTER

A.7 ISOTOPIC PENNIES

You learned earlier (Unit 1, page 34) that pre-1982 and post-1982 pennies have different compositions. As you might expect, these pennies also have different masses. In this activity, a mixture of pre- and post-1982 pennies will model or represent atoms of a naturally occurring mixture of two isotopes of the imaginary element *coinium*. Using the pennies, you will simulate one way scientists determine the relative amounts of different isotopes in a sample of an element.

You will receive a sealed container of 10 pennies that contains a mixture of pre-1982 and post-1982 pennies (Figure 6.17). Your container might hold any particular atomic mixture of these two isotopes. Your task is to determine the isotopic composition of *coinium* —*without opening the container.*

Figure 6.17 *How many pre- and post-1982 pennies are in your 10-coin sample of coinium?*

- Your teacher will give you some pre-1982 and post-1982 pennies, and a sealed container with a mixture of 10 pre- and post-1982 pennies, and will tell you the mass of the empty container. Record this information and the code number of your sealed container.

- Determine the isotopic composition of the element *coinium*. That is, find the percent pre-1982 and percent post-1982 pennies in your container. There is more than one way to find these answers.

Now answer these questions:

1. Describe the procedure that you followed to find the percent composition of coinium.

2. What property of the element *coinium* is different in its pre- and post-1982 forms?

3. Name at least one other familiar item that could serve as a model for isotopes.

4. You have examined and created models throughout the year. What are the limitations of this model in explaining isotopes?

concept check 2

1. Why was it necessary for Rutherford to know something about radiation before he could make a claim regarding atomic nuclei?
2. Describe the isotope symbolized $^{210}_{82}Pb$.
3. There are two naturally occurring isotopes of copper $^{63}_{26}Cu$ and $^{65}_{26}Cu$. If the average atomic mass of copper is 63.55 g/mol, which isotope occurs more frequently in nature? Explain.

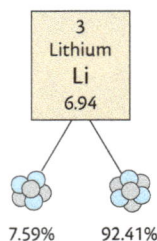

| 3 |
| Lithium |
| Li |
| 6.94 |

7.59%   92.41%

Figure 6.18 *Lithium's average atomic mass is a weighted average of the molar masses of its two isotopes, Li-6 and Li-7. Most lithium atoms (92.41%) are Li-7. Thus, lithium's average atomic mass is closer to 7 than to 6.*

A.8 ISOTOPES IN NATURE

Most elements in nature are mixtures of isotopes. Some isotopes of an element may be radioactive, whereas others are not. All isotopes of an element behave almost exactly the same way chemically, because they have the same electron distribution and differ only slightly in mass. If you consider chemical changes only, knowledge of isotopes is not particularly helpful. The average atomic mass of an element, as shown on the periodic table, is based on the relative natural abundances of isotopes of that element (see Figure 6.18 for lithium).

Marie Curie originally thought that only heavy elements were radioactive. It is true that naturally occurring **radioisotopes** (i.e., radioactive isotopes) are more common among the heavy elements. In fact, all naturally occurring isotopes of elements with atomic numbers greater than 83 (bismuth) are radio-

Some Natural Radioisotopes	
Isotope	**Abundance (%)**
Hydrogen-3	0.00013
Carbon-14	Trace
Potassium-40	0.0012
Rubidium-87	27.8
Indium-115	95.8
Lanthanum-138	0.089
Neodymium-144	23.9
Samarium-147	15.1
Lutetium-176	2.60
Rhenium-187	62.9
Platinum-190	0.012
Thorium-232	100

Table 6.4

active. However, many natural radioisotopes are also found among lighter elements. Modern technology has made it possible to create a radioisotope of any element. Table 6.4 lists some naturally occurring radioisotopes and their isotopic abundances.

What is the relationship between an element's molar mass and the percent abundance of the element's istopes? To calculate the molar mass of an element, it is helpful to use the concept of a weighted average, as illustrated in Developing Skills A.9.

No stable (nonradioactive) isotopes have yet been found for elements with atomic numbers of 83 or greater.

DEVELOPING SKILLS

A.9 MOLAR MASS AND ISOTOPIC ABUNDANCE

Sample Problem 1: *Consider an isotopic mixture of copper. Naturally occurring copper (Cu) consists of 69.1% copper-63 atoms and 30.9% copper-65 atoms. The molar masses of these two isotopes are:*

$$Copper\text{-}63 = 62.93 \ g/mol$$
$$Copper\text{-}65 = 64.93 \ g/mol$$

What is the average molar mass of naturally occurring copper?

The equation for finding average molar masses is as follows:

$$\text{Molar mass} = \left(\begin{array}{c}\text{Fractional} \\ \text{abundance} \\ \text{of isotope 1}\end{array}\right) \times \left(\begin{array}{c}\text{Molar mass} \\ \text{of isotope 1}\end{array}\right) + \left(\begin{array}{c}\text{Fractional} \\ \text{abundance} \\ \text{of isotope 2}\end{array}\right) \times \left(\begin{array}{c}\text{Molar mass} \\ \text{of isotope 2}\end{array}\right) + \dots \begin{array}{c}\text{(for each isotope} \\ \text{involved)}\end{array}$$

Since there are two naturally occurring copper isotopes, the average molar mass of copper can be calculated as follows:

Molar mass of Cu = (0.691)(62.93 g/mol) + (0.309)(64.93 g/mol) = 63.5 g/mol

> The decimal fractions must add up to 1. Why?

Sample Problem 2: For the coinium example, suppose that you found that the composition of the mixture was 0.4 (40%) pre-1982 pennies and 0.6 (60%) post-1982 pennies.

What is the average mass of a penny?

The equation setup is shown below, using the 40/60 coin mixture:

$$\text{Average penny mass} = (0.4) \times (\text{Mass of pre-1982 penny}) + (0.6) \times (\text{Mass of post-1982 penny})$$

1. Calculate the average mass of a penny in your *coinium* mixture.
2. Calculate the average mass of a penny in your mixture another way: Divide the total mass of your entire penny sample by 10.
3. Compare the average masses that you calculated in Questions 1 and 2. These results should convince you that either calculation leads to the same result. If not, consult your teacher.
4. Naturally occurring boron (B) is a mixture of two isotopes. (Refer to the table below.)
   a. Do you expect the molar mass of naturally occurring boron to be closer to 10 or to 11? Why?
   b. Calculate the molar mass of naturally occurring boron.
5. Naturally occurring uranium (U) is a mixture of three isotopes. (Refer to the table below.)
   a. Do you expect the molar mass of naturally occurring uranium will be closest to 238, 235, or 234? Why?
   b. Calculate the molar mass of naturally occurring uranium.

Isotope Molar Mass and Abundance		
Isotope	**Molar Mass (g/mol)**	**% Natural Abundance**
Boron-10	10.0	19.90%
Boron-11	11.0	80.10%
Uranium-234	234.0	0.0054%
Uranium-235	235.0	0.71%
Uranium-238	238.1	99.28%

■ MAKING DECISIONS

A.10 FACT OR FICTION?

Look again at the statements at the start of this unit (page 586). Answer the following questions about the CANE flyer. This will help you start preparing your presentation for Riverwood senior citizens (see Figure 6.19).

1. Identify the specific statements on the flyer that you can now conclusively identify as either true or false? For each statement:

   a. list two pieces of evidence that helped you make your decision.

   b. list two public concerns about the statement that you plan to address in your presentation.

2. Choose one statement from the leaflet that you understand more completely now, but are still unable to confirm or deny. What else do you have to know before you can make a decision about that statement?

3. How helpful do you think it will be to discuss the history of some discoveries that you studied in this section when you talk to the Riverwood senior citizens? Explain your answer.

4. Think about the new terms that were introduced in this section.

   a. Which terms will you explain as part of your presentation?

   b. Select two terms from your answer to Question 4a. Prepare an explanation of each term that your audience will understand. Describe examples or real-world applications that you may use in your explanation.

The history of science is full of discoveries that build on earlier discoveries. The discovery of radioactivity was such an event. The investigations of Roentgen, Becquerel, the Curies, and Rutherford led to new knowledge and a better understanding of atomic structure. As you start to consider some current applications of nuclear radiation, think about the evidence and reasoning that supported these scientific advancements.

Figure 6.19 *What does your research lead you to conclude about the accuracy of statements made in the flyer? (See page 586.)*

SECTION A SUMMARY

Reviewing the Concepts

> Radioactive nuclei are unstable and undergo spontaneous changes in their structure.

1. Describe the sequence of events that led to Becquerel's discovery of radioactivity.
2. What three radioactive elements did the Curies find in pitchblende?
3. Define radioactivity.
4. List three types of nuclear radiation.
5. How did scientists determine that alpha particles have a positive electrical charge?
6. In what ways is gamma radiation different from alpha and beta radiation?
7. How did the idea that the atom is the smallest particle of matter change after the discovery of radioactivity?

> Radiation can be classified as either ionizing or non-ionizing, depending on the type of energy it transmits.

8. Define and give an example of
   a. ionizing radiation.
   b. non-ionizing radiation.
9. Why is ionizing radiation regarded as more dangerous than non-ionizing radiation?
10. Classify each of the following as ionizing or non-ionizing radiation.
    a. visible light
    c. gamma rays
    b. X-rays
    d. radio waves
11. How does ionizing radiation damage living cells?

> Rutherford's gold-foil experiment results led to a new model of the atom.

12. Describe Rutherford's gold-foil experiment.
13. a. What happened to most of the alpha particles observed in the gold-foil experiment?
    b. What did Rutherford conclude from this observation?
14. a. What happened to about 1 in every 20 000 alpha particles in the gold-foil experiment?
    b. What did Rutherford conclude from this observation?
15. What was the general structure of the atom that Rutherford proposed?
16. Sketch models to show the concept of an atom before and after Rutherford's gold-foil experiment.
17. What characteristic of alpha particles made them desirable as the beam in Rutherford's gold-foil experiment?

> Most elements in nature are a mixture of isotopes. Some isotopes are radioactive.

18. Give the correct isotopic notation for copper-65.

19. Calculate the total neutrons in an atom of sulfur-34.

20. Copy the following table and find the value of each coded letter, **a** through **p**:

Symbol	Total protons	Total neutrons	Mass number
$^{2}_{1}\text{H}$	**a**	1	**b**
$^{37}_{c}\text{Cl}$	**d**	**e**	**f**
$^{g}_{h}\text{Tc}$	43	56	**i**
$^{137}_{j}\text{Cs}$	**k**	**l**	**m**
$^{n}_{o}\text{Ag}$	47	60	**p**

21. How do the nuclei of carbon-12, carbon-13, and carbon-14 differ?

22. Consider the symbol $^{190}_{78}\text{Pt}$.

   a. What does the superscript 190 indicate?

   b. What does the subscript 78 indicate?

   c. How many neutrons are in a Pt-190 nucleus?

23. Neon (Ne) is composed of three isotopes with the following molar masses and relative abundances: Ne-20 (19.99 g/mol), 90.51%; Ne-21 (20.99 g/mol), 0.27%; and Ne-22 (21.99 g/mol), 9.22%.

   a. Based on these data, should neon's average atomic mass be closest to 20, 21, or 22? Why?

   b. Calculate the actual average atomic mass of neon. Show your calculations.

What evidence led to a modern understanding of the composition of atoms?

In this section, you have investigated ionizing radiation, read about the discoveries that led to Rutherford's model of the atom, and learned about isotope symbols, naturally occurring isotopes, and percent abundance. Think about what you have learned, then answer the question in your own words in organized paragraphs. Your answer should demonstrate your understanding of the key ideas in this section.

Be sure to consider the following in your response: key discoveries that led to the modern view of the atom (including the gold-foil experiment and the discovery of radioactivity), isotopes, and the structure of the atom.

Connecting the Concepts

24. A local politician proposes to ban all radiation in your community. Explain why this proposal has little chance of success.

25. Why is it possible to receive a suntan from ultraviolet radiation but not from radio waves?

26. How does the gold-foil experiment demonstrate the importance of evidence regarding events that cannot directly be seen?

27. Describe how the development of atomic theory illustrates the way scientific discoveries build on previous scientific knowledge and experiments.

28. In what way is fluorescence different from radioactivity?

29. In what way is a cathode-ray tube similar to a

    a. modern X-ray device?

    b. plasma lamp?

Extending the Concepts

30. Investigate the properties of gold and discuss why Rutherford probably selected that metal as an alpha-particle target.

31. Imagine that Rutherford proposed using beta particles rather than alpha particles in the gold-foil experiment. What result would you predict?

32. The neutron was discovered decades after the proton and electron were discovered. What made the discovering the neutron such a challenge?

33. Compare earlier models of the atom with

    a. Rutherford's model.

    b. the currently accepted model of the atom.

SECTION B NUCLEAR RADIATION

How do we detect and describe the products of nuclear decay?

Of nearly 2000 known isotopes, there are more radioactive (unstable) isotopes than nonradioactive (stable) isotopes. Actually, most isotopes you encounter are not radioactive. However, naturally occurring radio-isotopes expose everyone to low levels of radiation. This radiation is from radioisotopes in building materials (such as brick and stone) in schools and homes (see Figure 6.20); in air, land, and sea; in foods you eat; and even within your own body. Because the human senses cannot detect nuclear radiation, various devices to detect and measure its intensity have been developed.

Figure 6.20 *Bricks are a common source of background ionizing radiation.*

GOALS

- Identify sources of background radiation.
- Describe effects of ionizing radiation on human tissue and identify factors that determine the extent of damage.
- Distinguish among alpha particles, beta particles, and gamma rays and describe the effects of their emission on the composition of the nucleus.
- Write, complete, and balance nuclear equations.
- Describe methods for detecting and measuring ionizing radiation.

concept check 3

1. In a collection of 500 lithium atoms, would all atoms be identical in terms of electrons, neutrons, and protons? If not, describe what would be the same and what would be different.
2. a. What is your current definition of *background radiation*?
   b. Do you think it is possible to reduce your exposure to background radiation? Why or why not?
3. As you know, scientists use units to convey quantitative information such as distance (m), volume (L), concentration (mol/L), and density (g/mL). If you wanted to quantify the effects of ionizing radiation, what units might be useful? Propose one or two units of measurement and explain your reasoning.

B.1 EXPOSURE TO IONIZING RADIATION

When radioisotopes spontaneously decay, they usually emit alpha, beta, or gamma radiation. The type and intensity of radiation emitted helps to determine possible medical and industrial applications of particular radioisotopes. Each type of nuclear radiation also poses distinct hazards to human health.

Recall that a relatively constant level of radioactivity, called **background radiation**, is always present around and within you. Everyone receives background radiation at low levels from natural sources and from sources related to human activity. You will always experience at least some exposure to ionizing radiation.

Natural sources of background ionizing radiation include:

- High-energy particles from outer space that bombard Earth.
- Radioisotopes in rocks, soil, and groundwater: uranium (U-238 and U-235), thorium (Th-232), and the radioactive isotopes that form as they decay.
- Radioisotopes in the atmosphere: radon (Rn-222) and its decay products, including polonium (Po-210).
- Naturally occurring radioisotopes in foods and the environment, such as potassium-40 and carbon-14.

Advances in science and technology have created additional sources of background radiation, such as:

- Residual radioactive fallout from aboveground nuclear-weapon testing. (Most aboveground testing ended after the signing of the Limited Test Ban Treaty of 1963.)
- Increased exposure to radiation during high-altitude airplane flights (see Figure 6.21).

The total quantity of radioisotopes released to the environment from fossil-fuel power plants is greater than the total quantity released from nuclear power plants.

Figure 6.21 *Air travel increases human exposure to ionizing radiation.*

- Radioisotopes released into the environment from both fossil fuel and nuclear power generation as well as other nuclear technologies.
- Radioisotopes released through the disturbance and use of rocks in mining and in making cement, concrete, and drywall.

Because of the effect of ionizing radiation on living tissue, it is important to monitor the quantity to which people are exposed over time. This quantity is referred to as the ionizing-radiation **dose**. The **gray** (Gy) is the SI unit that expresses the quantity of ionizing radiation absorbed by a particular sample, typically human tissue. An absorbed dose of one gray is defined as one joule of energy absorbed per kilogram of body tissue.

Not all forms of ionizing radiation, however, produce the same effect on living organisms. For example, alpha radiation will cause more harm internally to living organisms than will the same quantity of gamma radiation. The **sievert** (Sv) is the SI unit that expresses the ability of radiation—regardless of type or activity—to cause ionization in human tissue. Any exposure to radiation that produces the same detrimental effects as one gray of gamma rays represents one *sievert* of exposure. It is usually most convenient to express exposure in sieverts; this unit eases direct comparison across different types of ionizing radiation.

While the SI units for radiation exposure are the gray and the sievert, two other units have traditionally been used in the United States: the **rad** and the **rem** (see Table 6.5). The *rad* (like the gray) expresses the absorbed dose of radiation, and the *rem* (like the sievert) indicates ionizing effects on living organisms. Both the rad and the rem are one-hundredth of their corresponding SI units.

Table 6.5

Units of Radiation Dosage			
Unit	**Absorbed Dose**	**Ionizing Effects**	**Definition**
sievert (Sv)		X	1 Sv = radiation exposure that causes same effects as 1 Gy of gamma rays
rem		X	1 rem = 10^{-2} Sv
millirem (mrem)		X	1 mrem = 10^{-3} rem = 10^{-5} Sv
gray (Gy)	X		1 Gy = one joule of energy delivered to one kilogram of body tissue
rad	X		1 rad = 10^{-2} Gy

Even though a rem is only one-hundredth as large as a sievert, it is still much larger than typical exposures. Normal human exposures are so small that doses are expressed in units of *millirem* (mrem), where 1 mrem = 0.001 rem. One millirem of any type of radiation produces the same biological effects, whether the radiation is composed of alpha particles, beta particles, or gamma rays.

Some ionizing radiation comes from within your own body, as depicted in Figure 6.22. On average, people living in the U.S. receive about 360 mrem per person each year. About 300 mrem (83%) come from natural sources. Figure 6.23 shows the approximate proportion from each source.

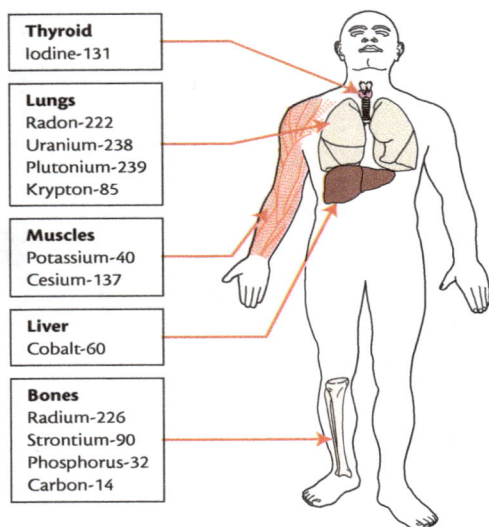

Thyroid
Iodine-131

Lungs
Radon-222
Uranium-238
Plutonium-239
Krypton-85

Muscles
Potassium-40
Cesium-137

Liver
Cobalt-60

Bones
Radium-226
Strontium-90
Phosphorus-32
Carbon-14

Figure 6.22 *All living things contain some radioactive isotopes, including these that are located in particular parts of the human body.*

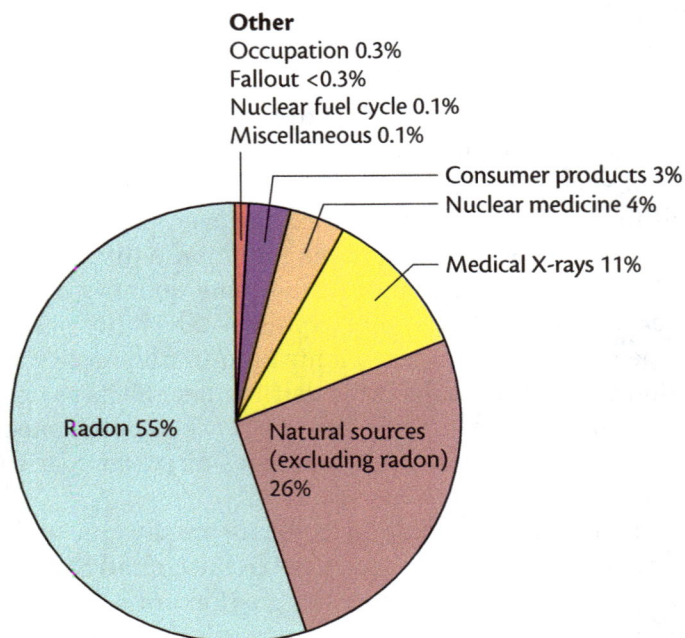

Other
Occupation 0.3%
Fallout <0.3%
Nuclear fuel cycle 0.1%
Miscellaneous 0.1%

Consumer products 3%
Nuclear medicine 4%
Medical X-rays 11%

Radon 55%

Natural sources
(excluding radon)
26%

Figure 6.23 *Sources of ionizing radiation in the United States.*

It is in your best interest to avoid unnecessary ionizing radiation exposure. What ionizing radiation level is considered reasonably safe? The U.S. government's background radiation limit for the general public is 500 mrem (0.5 rem) per year for any individual. The U.S. average exposure value of 360 mrem falls below this. The established U.S. limit for an individual's annual maximum safe exposure in the workplace is 5000 mrem (5 rem).

CHEM**QUANDARY**

RADIATION EXPOSURE STANDARDS

Why might radiation exposure standards for some individuals differ from those for the general public? Why do the standards differ for those who are exposed in their workplace?

◼ MAKING DECISIONS

B.2 YOUR YEARLY IONIZING RADIATION DOSE

How does your yearly ionizing-radiation dose compare to the U.S. average of 360 mrem? Could you make decisions that would decrease your annual dose? On a sheet of paper, list the numbers and letters of each category in the following table. Then fill in the blanks on your sheet with appropriate quantities. Add all these quantities to estimate your annual ionizing-radiation dose.

Your Yearly Ionizing Radiation Dose

Common Sources of Radiation				Your Yearly Dose (mrem)
1. Where You Live				
a. Cosmic Radiation (from outer space)				
Your exposure depends on elevation. These are annual doses.				
sea level	**26 mrem**	4000–5000 ft	**47** mrem	
0–1000 ft	**28 mrem**	5000–6000 ft	**55** mrem	
1000–2000 ft	**31 mrem**	6000–7000 ft	**66** mrem	
2000–3000 ft	**35 mrem**	7000–8000 ft	**79** mrem	
3000–4000 ft	**41 mrem**	8000–9000 ft	**96** mrem	_____ mrem
b. Terrestrial Radiation (from the ground)				
If you live in a state bordering Gulf or Atlantic coasts		**23**	mrem	
If you live in AZ, CO, NM, or UT		**63**	mrem	
If you live anywhere else in continental United States		**46**	mrem	_____ mrem
c. House Construction				
If you live in a stone, adobe, brick, or concrete building		**7**	mrem	_____ mrem
d. Power Plants				
If you live within 50 miles of a nuclear power plant		**0.009**	mrem	
If you live within 50 miles of a coal-fired power plant		**0.03**	mrem	_____ mrem
2. Food, Water, Air				
Internal Radiation (based on average values)				
a. from food (C-14, K-40) and water				
(radon dissolved in water)		**40**	mrem	**40** mrem
b. from air (radon)		**200**	mrem	**200** mrem

(Continued on next page)

Your Yearly Ionizing Radiation Dose

Common Sources of Radiation			Your Yearly Dose (mrem)
3. How You Live			
Smoking cigarettes (1 pack / day)[+]	15	mrem	_____ mrem
Weapons test fallout*	1	mrem	_____ mrem
Travel by jet aircraft (per hour of flight)	0.5	mrem	_____ mrem
If you have porcelain crowns or false teeth	0.07	mrem	_____ mrem
If you wear a luminous (LCD) wristwatch	0.06	mrem	_____ mrem
If you go through airport security (each time)	0.002	mrem	_____ mrem
If you watch TV*	1	mrem	_____ mrem
If you use a video display (computer screen)*	1	mrem	_____ mrem
If you live in a dwelling with a smoke detector	0.008	mrem	_____ mrem
If you use a gas camping lantern with an old mantle	0.2	mrem	_____ mrem
If you wear a plutonium-powered pacemaker	100	mrem	_____ mrem
4. Medical Uses (radiation dose per procedure)			
X-rays: Extremity (arm, hand, foot, or leg)	1	mrem	_____ mrem
Dental	9	mrem	_____ mrem
Chest	6	mrem	_____ mrem
Pelvis/hip	65	mrem	_____ mrem
Skull/neck	20	mrem	_____ mrem
Barium enema	405	mrem	_____ mrem
Upper GI	245	mrem	_____ mrem
CT scan (head and body)	110	mrem	_____ mrem
Nuclear medicine (e.g., thyroid scan)	14	mrem	_____ mrem
Your Estimated Yearly Radiation Dose			_____ mrem

[+]This value represents a conservative estimate based upon information from the U.S. Department of Energy. Other estimates range from 20–1300 mrem.

* The value is less than 1 mrem, but adding that value would be reasonable.

Adapted from *Estimate your personal annual radiation dose*, American Nuclear Society, 2000.

1. Do any of the yearly dose values surprise you? Explain.

2. Compare your annual ionizing-radiation dose to the
   a. U.S. limit of 500 mrem.
   b. average background radiation (360 mrem).

3. Why is it useful to keep track of how many X-rays you receive each year?

4. How do geographic factors affect your annual ionizing-radiation dose?

5. a. What lifestyle changes could reduce a person's exposure to ionizing radiation?
   b. Would you decide to make those changes? Explain.

B.3 IONIZING RADIATION—HOW MUCH IS SAFE?

The two main factors that determine tissue damage due to ionizing radiation are radiation density (the number of ionizations within a given volume) and dose (the quantity of radiation received).

Gamma rays and X-rays are ionizing forms of electromagnetic radiation that penetrate deeply into human tissue. Ionizing radiation causes tissue damage by breaking bonds in molecules. At low levels of ionizing radiation, only a few molecules are damaged. In most low-dose cases, a body's systems can repair the damage. As the dose received increases, the total number of molecules affected by the radiation also increases. Generally, the damage to proteins and nucleic acids is of greatest concern because of their role in body structures and functions. Proteins form much of the body's soft tissue structure and compose enzymes, molecules that control the rates of cellular chemical reactions. If a large number of protein molecules are destroyed within a small region, too few functioning molecules may remain to enable the body to heal itself in a reasonable time (see Figure 6.24).

See Unit 7 for more information on proteins and enzymes.

Figure 6.24 *Mutations are caused by damage to DNA, which can result from exposure to radiation. Compare the normal blood cells (left) to blood cells that are deformed (right) as a result of a mutation.*

Nucleic acids in DNA can be damaged by ionizing radiation. Minor damage causes **mutations**, which change the structure of DNA. Such mutations may result in the production of altered proteins and often kill the cell in which they occur. If the cell is a sperm or ovum, a mutation may lead to birth defects in offspring. Some mutations can lead to cancer, a disease in which cell growth and metabolism are out of control. When the DNA in many body cells is severely damaged, cells cannot synthesize new proteins to replace the damaged ones, and the organism or person dies.

DNA molecules control cell reproduction and the synthesis of proteins.

Biological Damage from Radiation	
Factor	**Effect**
Dose	Most scientists assume that an increase in radiation dose produces a proportional increase in risk.
Exposure time	The more a given dose is spread out over time, the less harm it does.
Area exposed	The larger the body area exposed to a given radiation dose, the greater the damage.
Tissue type	Rapidly dividing cells, such as blood cells and sex cells, are more susceptible to radiation damage than are slowly dividing or non-dividing cells, such as nerve cells. Fetuses and children are more susceptible to radiation damage than are adults.

Table 6.6 *These four factors determine the actual effects of particular ionizing-radiation exposure.*

Table 6.6 lists factors determining the extent of biological damage from ionizing radiation. Table 6.7 summarizes the biological effects of large dosages of ionizing radiation. Because the values in Table 6.7 are so large, they are reported in rems, not millirems.

Becquerel observed a red spot on his chest after carrying a radium sample in his breast pocket. The dangers of ionizing radiation were unknown at that time.

Ionizing-Radiation Effects	
Dose (rem)	**Effect**
0–25	No immediate observable effects.
25–50	Small decreases in white blood cell count, causing lowered resistance to infections.
50–100	Marked decrease in white blood cell count. Development of lesions.
100–200	Radiation sickness—nausea, vomiting, hair loss. Blood cells die.
200–300	Hemorrhaging, ulcers, death.
300–500	Acute radiation sickness. 50% of those exposed die within a few weeks.
>700	100% die.

Table 6.7 *You can see how the consequences of radiation exposure change as dose increases.*

Large ionizing-radiation doses can have drastic effects on humans. Conclusive evidence that such doses produce increased cancer rates has been gathered from uranium miners and nuclear-accident victims. Some of the first cases of exposure to large doses of ionizing radiation occurred among workers who used radium compounds to paint numbers on watch dials that would glow in the dark (see Figure 6.25). The workers used their tongues to smooth the tips of their paintbrushes, and unknowingly swallowed small amounts of radioactive compounds. Later, these workers began to lose hair and became quite weak. Sometimes, this exposure even led to death.

Leukemia, a rapidly developing cancer of white blood cells, is commonly associated with exposure to high doses of ionizing radiation. Exposure also promotes other forms of cancer, anemia, heart problems, and cataracts (opaque spots on an eye lens).

Considerable controversy continues regarding whether very low doses of ionizing radiation, such as those from typical background sources, can cause cancer. Most of the data on cancer incidence have been based on human exposure to high doses of radiation. These data are extrapolated to much lower doses. Few studies have directly linked low radiation doses with cancer development. Most scientists agree that typical background levels of ionizing radiation are safe for most people. Some authorities argue that any increase above normal background levels increases the probability of developing cancer.

Figure 6.25 *These women, working in a factory in Orange, New Jersey, in the mid-1920s, painted radioactive radium onto watch dials so the watches could be read in the dark. Their exposure to ionizing radiation from radium often resulted in illness. Modern glow-in-the-dark watches do not contain radium.*

INVESTIGATING MATTER

B.4 ALPHA, BETA, AND GAMMA RADIATION

Preparing to Investigate

Investigating Matter A.1 provided an opportunity for you to make some initial ionizing radiation measurements and you now know more about its origin and effects on people. This investigation will allow you to learn more about specific properties of three types of ionizing radiation—alpha, beta, and gamma radiation.

To prepare for this investigation, review what you have already learned:

- Background radiation is always present (and can be detected and measured).

- Some objects emit ionizing radiation, which can be measured with the same devices used to measure background radiation.

- Alpha particles, beta particles, and gamma rays are all forms of ionizing radiation, but they have properties that differ from each other. For instance, you learned in Section A.2 (page 590) that alpha particles are positively charged, beta particles are negatively charged, and gamma rays have no electrical charge.

Also consider these two claims from Sections B.1 and B.3: "alpha radiation will cause more harm internally to living organisms than will the same quantity of gamma radiation" and "gamma rays . . . penetrate deeply into human tissue." Given these claims, you may want to know more about the likelihood of alpha particles penetrating human tissue—how does their penetrating ability compare to that of gamma rays? You might also wonder whether beta particles can penetrate skin or cause internal harm. After reading about doses and effects, perhaps you wonder how distance from a radiation source affects the dose you receive or how you can shield yourself from ionizing radiation.

This investigation will allow you to address these questions. Keep in mind that with proper handling, the radioactive materials in this investigation pose no danger to you. Nuclear materials are strictly regulated by state and federal laws. The radioactive sources you will use emit only very small quantities of radiation. Using them requires no special license. Nevertheless, you should handle all radioactive samples with great care, wearing protective gloves. Do not allow the radiation counter to come in direct contact with the radioactive material. Check your hands with a radiation monitor before you leave the laboratory.

Asking Questions

Read through *Gathering Evidence*. Write scientific questions that reflect the goals of Parts I and II. Think about a question that you might like to investigate for Part III. Write a draft of that question now (you will have the opportunity to revise this question after you complete Parts I and II). Construct a data table suitable for recording all relevant data that you will need to collect to answer your questions.

Gathering Evidence

Part I: Comparing the Penetrating Ability of Ionizing Radiation

1. Before you begin, put on your goggles, and wear them properly throughout the investigation.

2. Set up the apparatus shown in Figure 6.26. There should be space between the source and the detector for several sheets of glass or metal.

3. Turn on the counter. Allow it to warm up for at least 3 min. Determine the intensity of background radiation by counting the clicks for 1 min without any radioactive sources present. Record this background radiation value in counts per minute (cpm) in your data table.

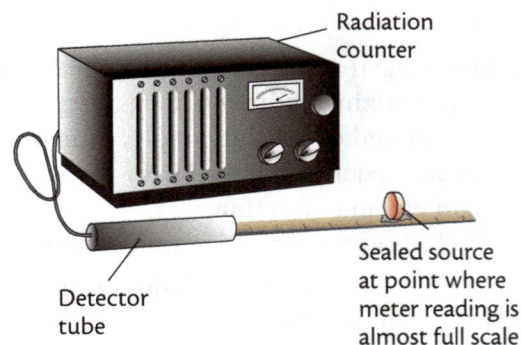

Figure 6.26 *You will need to determine the level of radiation detected in the absence of shielding.*

4. Put on protective gloves. Using forceps, place a gamma-ray source on the ruler at a point where it produces a nearly full-scale reading. Record the distance between the source and the detector.

5. Observe the meter for 30 s and estimate the number of counts per minute detected over this period. Record this approximate value. Subtract the background reading from that value and record the corrected results.

6. Without moving the radiation source, place a piece of cardboard (or an index card) between the detector and the source, as shown in Figure 6.27.

7. Observe the meter for 30 s. Record the typical reading. Then correct the reading for background radiation and record the corrected result in your data table.

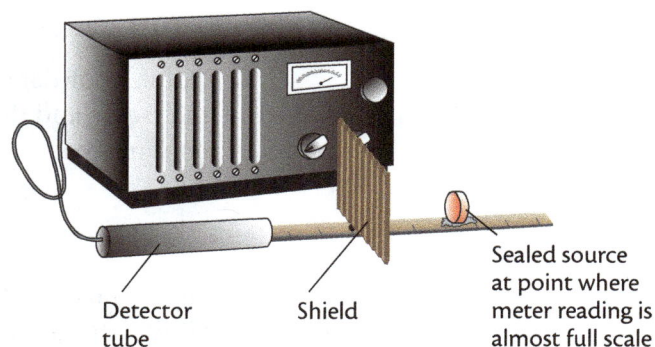

Sealed source at point where meter reading is almost full scale

Detector tube   Shield

Figure 6.27 *What materials are penetrated by gamma rays, beta particles, and alpha particles?*

8. Repeat Steps 6 and 7, replacing the cardboard with a glass or plastic sheet.

9. Repeat Steps 6 and 7, replacing cardboard with a lead sheet.

10. Repeat Steps 4 through 9, using a beta-particle source.

11. Repeat Steps 4 through 9, using an alpha-particle source.

Part II: Effect of Distance on Intensity

In Part I, you measured radiation intensity at a single distance from the source. What relationship would you expect between distance and radiation intensity? You will design your own procedure to investigate this relationship.

12. Write down your prediction about the relationship between distance and intensity. For instance, do you expect doubling the distance to lead to half the intensity reading?

13. Design a procedure to test your prediction. Think about how many different distances you should test and what data you will need to record. Write down your detailed procedure and obtain your teacher's approval before proceeding.

14. Carry out your investigation. Be sure to record all relevant data.

Part III: Further Investigations

15. In this part of the investigation, you will design, and possibly conduct, an additional investigation of your own design.

16. Note the materials available to conduct this investigation.

17. Think about one or two additional questions that relate to, or extend, the investigations you have already conducted using these ionizing radiation sources. Choose one question that you can address using the available equipment and materials. Write down the scientific question that you will investigate.

18. Propose and write down a detailed procedure for your investigation.

19. If you will carry out the investigation, first obtain your teacher's approval, then conduct your investigation. Be sure to record all relevant data.

Analyzing Evidence

1. Graph your data from Part II, plotting corrected cpm values on the *y*-axis and distances from the source to the detector (in cm) on the *x*-axis.

Interpreting Evidence

1. Order the three types of ionizing radiation from "least penetrating ability" to "greatest penetrating ability." Support your answer with data or observations from Part I.

2. Interpret your data from Part II.

   a. By what factor did the intensity of radiation (measured in counts per minute) change when the initial distance was doubled?

   b. Did this same factor apply when the distance was doubled again? Explain.

   c. State the mathematical relationship between distance and intensity.

> A *factor* is a number by which a value is multiplied to give a new value.

3. It has been claimed that "alpha radiation will cause more harm internally to living organisms than will the same quantity of gamma radiation."

   a. Based on your investigations, which type of radiation is more likely to penetrate internal organs if the source of radiation is outside the body? Support your answer with data or observations.

   b. Describe what would need to happen for alpha particles to cause damage to internal organs.

Making Claims

4. Of the shielding materials tested, which do you conclude is the

   a. most effective in blocking radiation? Cite supporting evidence.

   b. least effective in blocking radiation? Cite supporting evidence.

5. Based on your observations, what properties of a material appear to affect its ability to be penetrated by radiation?

6. Figures like the one on the next page (Figure 6.28) are often used to illustrate the relationship between distance from a radiation source and its intensity.

   a. Does the information in this figure fit with the data you collected in Part II?

   b. Write a caption for the figure that would help someone else interpret this diagram.

   c. What would the diagram look like at a total distance of 15 cm from the source? Explain your reasoning.

7. If you conducted an additional investigation in Part III:

   a. State the scientific question that you were addressing in your investigation.

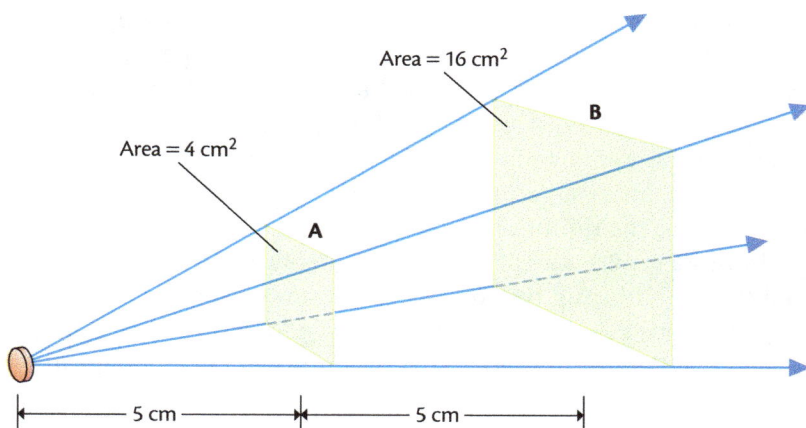

Area = 16 cm²

B

Area = 4 cm²

A

|← 5 cm →|← 5 cm →|

Figure 6.28

b. Summarize how you designed your experiment to address your question.

c. State your conclusions or claims. Support these conclusions or claims using data or observations from your investigation.

Reflecting on the Investigation

8. A patient receiving an X-ray is covered by a protective shield (Figure 6.29).

   a. What material would be a good choice for this apron?

   b. Why?

9. Whether or not you conducted an additional investigation, propose one additional question you have about alpha, beta, and gamma radiation.

You have found, through this investigation, that the three kinds of nuclear radiation differ greatly in their penetrating ability. You may wonder whether penetrating ability has any relationship to the structure of these nuclear emissions. In the next section, you will learn more about the composition of alpha, beta, and gamma radiation.

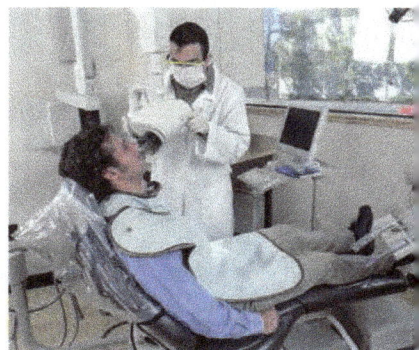

Figure 6.29 *What type of body shielding is used as this dental X-ray is obtained?*

B.5 NATURAL RADIOACTIVE DECAY

You know that alpha particles are positively charged. Why is this so? An alpha particle is composed of two protons and two neutrons. It is identical in composition to the nucleus of a helium-4 atom. Since alpha particles have no electrons, they have a double-positive electrical charge and are often symbolized as a doubly charged helium-4 atom, $^{4}_{2}\text{He}^{2+}$. Alpha radiation (also called alpha emission) is released by many radioisotopes of elements with atomic numbers greater than 83.

Compared to a beta particle, an alpha particle has 5 to 50 times more energy and is more than 7300 times more massive. However, the larger and slower alpha particles are easy to stop when they are outside the body. Once inside the body, however, the electrical charge and energy of alpha particles can cause great damage to tissues. This damage occurs over very short distances (about 0.025 mm).

Because of their relatively large mass, slower velocities, and large (2+) electrical charge, alpha particles lose most of their energy within a small distance.

Because alpha particles are very powerful tissue-damaging agents once inside the body, alpha emitters in air, food, or water are particularly dangerous to human life. Fortunately, outside the body, alpha particles are easy to block. As you noted in Investigating Matter B.4, alpha particles are stopped within a few centimeters by air.

Figure 6.30 illustrates a radium-226 nucleus emitting an alpha particle. During this process, the radium nucleus loses two protons, so its atomic number drops from 88 to 86, and it becomes an isotope of a different element, radon. In addition to losing two protons, radium-226 loses two neutrons, so its mass number drops by 4. The net result is the formation of radon-222. The decay process can be represented by the following nuclear equation:

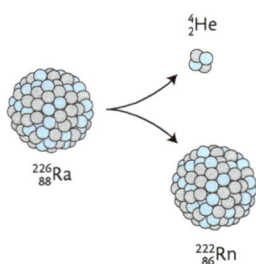

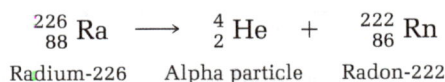

$$ {}^{226}_{88}\text{Ra} \longrightarrow {}^{4}_{2}\text{He} + {}^{222}_{86}\text{Rn} $$

Radium-226   Alpha particle   Radon-222

Figure 6.30 *Alpha-particle emission by radium-226. After emission, its mass number decreases by four (2 p + 2 n) and its atomic number decreases by two (2 p).*

Atoms of two elements—helium and radon—have been formed from one atom of radium. (A radium compound is shown in Figure 6.31.) Note that atoms are *not* necessarily conserved in nuclear reactions, as they are in chemical reactions. Atoms of different elements can appear on both sides of a nuclear equation. Total mass numbers and atomic numbers, however, *are* conserved. In the above equation for radium-226 decay, the reactant mass number equals the sum of mass numbers of products (226 = 4 + 222). Also, the atomic number of radium-226 equals the sum of atomic numbers of products (88 = 2 + 86). Both relations hold true for all nuclear reactions.

Beta particles are fast-moving electrons emitted from a nucleus. Because they are so much lighter than alpha particles and travel at very high velocities, beta particles have much greater penetrating ability than do alpha particles. On the other hand, beta particles are not as damaging to living tissue.

During **beta decay**, a neutron in a nucleus decays to a proton and an electron. The proton remains in the nucleus, but the electron is ejected. The electron emitted from the nucleus is a beta particle. The following equation describes the process:

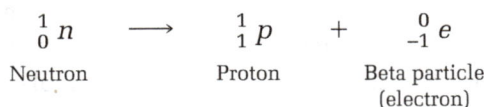

Figure 6.31 *Radium's name comes from the fact that small amounts of radium compounds, such as that on this antique watch face, emit enough radiation to glow in the dark. (See Figure 6.25, page 617.)*

Note that *n*, *p*, and *e* are symbols, respectively, for a neutron, a proton, and a beta particle. An emitted electron (beta particle) can also be symbolized by the Greek letter beta, β.

$$ {}^{1}_{0}n \longrightarrow {}^{1}_{1}p + {}^{0}_{-1}e $$

Neutron   Proton   Beta particle
(electron)

Due to its negligible mass and negative electrical charge, a beta particle is assigned a mass number of 0 and an "atomic number" (nuclear charge) of −1. The net nuclear change due to beta emission is that a neutron is converted to a proton.

The equation below shows a nucleus of lead-210 undergoing beta decay: The nucleus loses one neutron and gains one proton. The mass number remains unchanged at 210, but the atomic number increases from 82 to 83. The new nucleus formed is that of bismuth-210:

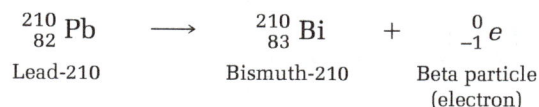

$$^{210}_{82}\text{Pb} \longrightarrow ^{210}_{83}\text{Bi} + ^{0}_{-1}e$$

Lead-210   Bismuth-210   Beta particle
                                                   (electron)

Once again, the sum of all mass numbers remains the same in this nuclear reaction (210 on each side), and the sum of atomic numbers (the nuclear charge) remains unchanged (82 on each side).

Alpha and beta decay often leave nuclei in an energetically excited state. This type of excited state is described as **metastable** and is designated by the symbol m. For example, the symbol ^{99m}Tc represents a technetium isotope in a metastable or excited state. Energy from isotopes in such excited states is released as gamma rays—high-energy electromagnetic radiation that has as much or more energy than X-rays. Because gamma rays have neither mass nor charge, their release does not change the mass balance or charge balance in a nuclear equation. Table 6.8 summarizes general changes involved in natural radioactive decay.

Changes Resulting from Nuclear Decay				
Decay Type	**Symbol**	**Change in Atomic Number**	**Change in Total Neutrons**	**Change in Mass Number**
Alpha	^4_2He, α	Decreased by 2	Decreased by 2	Decreased by 4
Beta	$^0_{-1}e$, β	Increased by 1	Decreased by 1	No change
Gamma	γ	No change	No change	No change

Table 6.8
Summary of the results of alpha, beta, and gamma radioactive decay.

New isotopes produced by radioactive decay may also be radioactive, and therefore undergo further nuclear decay. Such a succession of decays is called a **decay series**. Uranium (U) (see Figure 6.32) and thorium (Th) are *parents* (reactants) in three natural decay series that start with U-238, U-235, and Th-232, respectively. Each decay series ends with formation of a stable isotope of lead (Pb). The decay series starting with uranium-238 contains 14 steps, as shown in Figure 6.33 (page 624).

Figure 6.32 *Uranium ore contains U-238 and U-235 isotopes, each of which participates in a decay series, producing several generations of radioactive decay products.*

Figure 6.33 *The uranium-238 decay series. Diagonal arrows show alpha decay. Horizontal arrows show beta decay. Here is how to interpret this chart: Locate radon-222 (Rn-222). The arrow pointing left shows that this isotope decays to polonium-218 by alpha (α) emission. This nuclear equation applies:*

$$^{222}_{86}Rn \longrightarrow \, ^{218}_{84}Po + \, ^{4}_{2}He$$

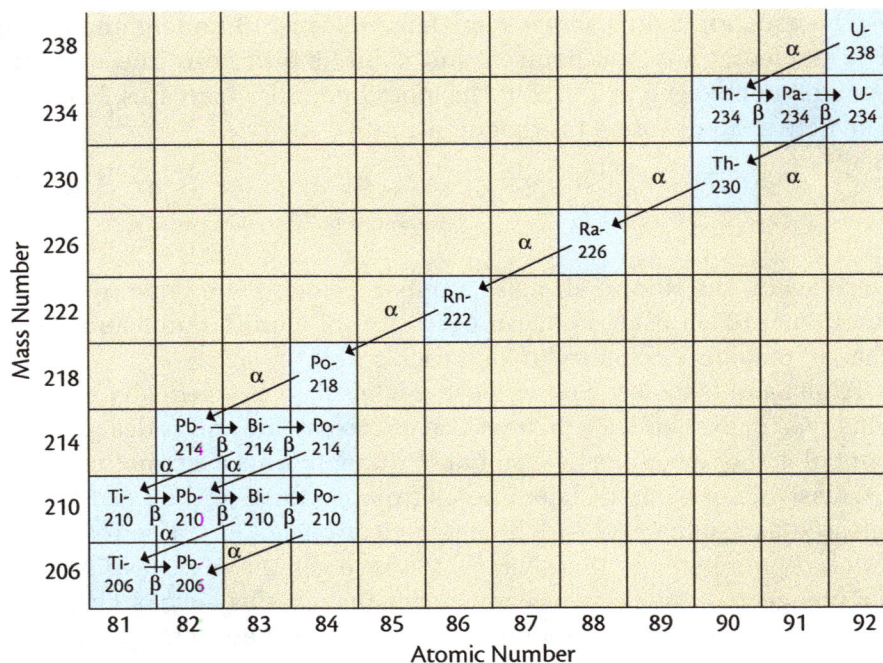

DEVELOPING SKILLS

B.6 NUCLEAR BALANCING ACT

The key to balancing nuclear equations is recognizing that both atomic numbers and mass numbers are conserved. Use the information in Table 6.8 (page 623) to complete the following exercises.

Sample Problem: *Cobalt-60 is one source of ionizing radiation for medical therapy. Complete this equation for the beta decay of cobalt-60:*

$$^{60}_{27}Co \longrightarrow \, ^{0}_{-1}e + \, ?$$

Beta emission causes no change in mass number. Therefore, the new isotope will also have a mass number of 60. Thus, the unknown product can be written as $^{60}?$. Because the atomic number increases by one during beta emission, the new isotope will have atomic number 28, one more than cobalt's atomic number. The periodic table indicates that atomic number 28 is nickel (Ni). The final equation is:

$$^{60}_{27}Co \longrightarrow \, ^{0}_{-1}e + \, ^{60}_{28}Ni$$

Practice writing nuclear equations by completing these questions:

1. Write the appropriate symbol for the type of radiation given off in each reaction.

   a. The following decay process illustrates how archaeologists date the remains of ancient biological materials. Living organisms take in carbon-14 and maintain a relatively constant amount of it over their lifetimes. After death, no more carbon-14 is taken in, so the amount gradually decreases due to decay:

   $$_{6}^{14}\text{C} \longrightarrow \,_{7}^{14}\text{N} + \,?$$

   b. The following decay process takes place in some types of household smoke detectors:

   $$_{95}^{241}\text{Am} \longrightarrow \,_{93}^{237}\text{Np} + \,?$$

2. The two decay series beginning with Th-232 and U-238 are believed responsible for much of the thermal energy generated inside Earth. (Thermal contributions from the U-235 series are negligible. The natural abundance of U-235 is low.) Complete the following equations, which represent the first five steps in the Th-232 decay series. Identify the missing items *A, B, C, D,* and *E.* Each code letter represents a particular isotope or a type of radioactive emission. For example, in the first equation, Th-232 decays by emitting alpha radiation to form *A.* What is *A?*

   > An alpha particle can be symbolized as either $_{2}^{4}\text{He}$ or $_{2}^{4}\alpha$.

   a. $_{90}^{232}\text{Th} \longrightarrow \,_{2}^{4}\text{He} + \,A$

   b. $_{88}^{228}\text{Ra} \longrightarrow \,_{-1}^{0}e + \,B$

   c. $_{89}^{228}\text{Ac} \longrightarrow \,C + \,_{90}^{228}\text{Th}$

   d. $D \longrightarrow \,_{2}^{4}\text{He} + \,_{88}^{224}\text{Ra}$

   e. $_{88}^{224}\text{Ra} \longrightarrow \,E + \,_{86}^{220}\text{Rn}$

concept check 4

1. Is it likely that a gamma ray—and nothing else—would be emitted from a nucleus? Explain.
2. How is writing nuclear equations different from writing chemical equations?
3. Radon is a noble gas, which means it is chemically unreactive. However, there is great concern over high levels of radon in homes. Why?

B.7 RADON

The gaseous element radon, which is the most massive of the noble gases, has always been a component of Earth's atmosphere. It is a radioactive decay product of uranium. In the 1980s, unusually high concentrations of radioactive radon gas found in some U.S. homes became a public health concern.

Radon is produced as the radioisotope uranium-238 decays in soil and building materials. (You can locate this radioactive decay product as Rn-222 in Figure 6.33, page 624.) Once it is emitted, radon is transported throughout the environment in many ways. Some radon produced in soil dissolves in groundwater. In other cases, radon gas seeps into houses through cracks in foundations and basement floors (see Figure 6.34).

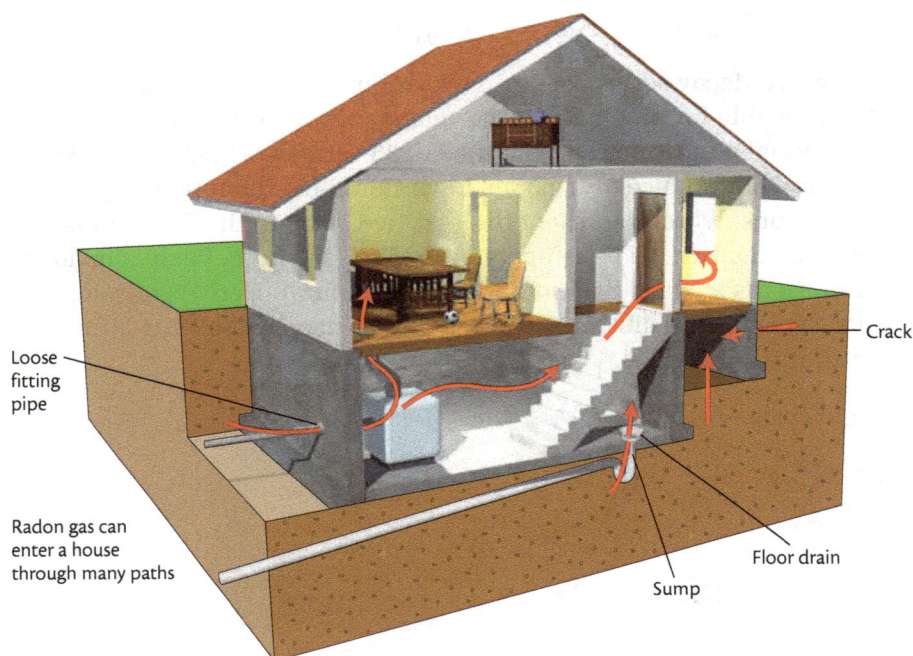

Loose fitting pipe

Radon gas can enter a house through many paths

Crack

Floor drain

Sump

Figure 6.34 *Radon gas is naturally released from some soils containing radioisotopes. Radon gas (red arrows) can enter a house by many paths.*

Figure 6.35 *Radon's high density and the fact that it enters a residence through foundation cracks means this radioactive gas can be removed by installing a venting system that withdraws air from the basement.*

In older houses, outdoor air entering through gaps in doors and windows can dilute the radon gas. However, to conserve energy, newer houses are built more airtight than are older houses. In a tightly sealed house, radon gas cannot mix freely with outdoor air or escape from the house. Consequently, radon gas concentrations may reach higher levels in newer homes. Remedies for high radon levels in houses include increasing ventilation and sealing cracks in floors (see Figure 6.35). Inexpensive radon test kits are available for home use.

The most serious danger of radon gas results from reactions that occur after it is inhaled. Radon decays to produce, in succession, radioactive isotopes of polonium (Po), bismuth (Bi), and lead (Pb). When radon gas is inhaled, it enters the body and is transformed, through radioactive decay, into these toxic heavy-metal ions, which cannot be exhaled as gases. These radioactive heavy-metal ions also emit potentially damaging alpha particles within the body.

Estimates indicate that about 6% of homes in the United States have radon levels higher than the exposure level recommended by the U.S. Environmental Protection Agency (EPA). See Figure 6.36. It is estimated that 10–15% of annual U.S. deaths from lung cancer are linked to the effects of indoor radon gas. These figures, although sobering, should be kept in perspective. About 80% of all U.S. lung-cancer deaths each year are attributed to cigarette smoking.

EPA Map of Radon Zones

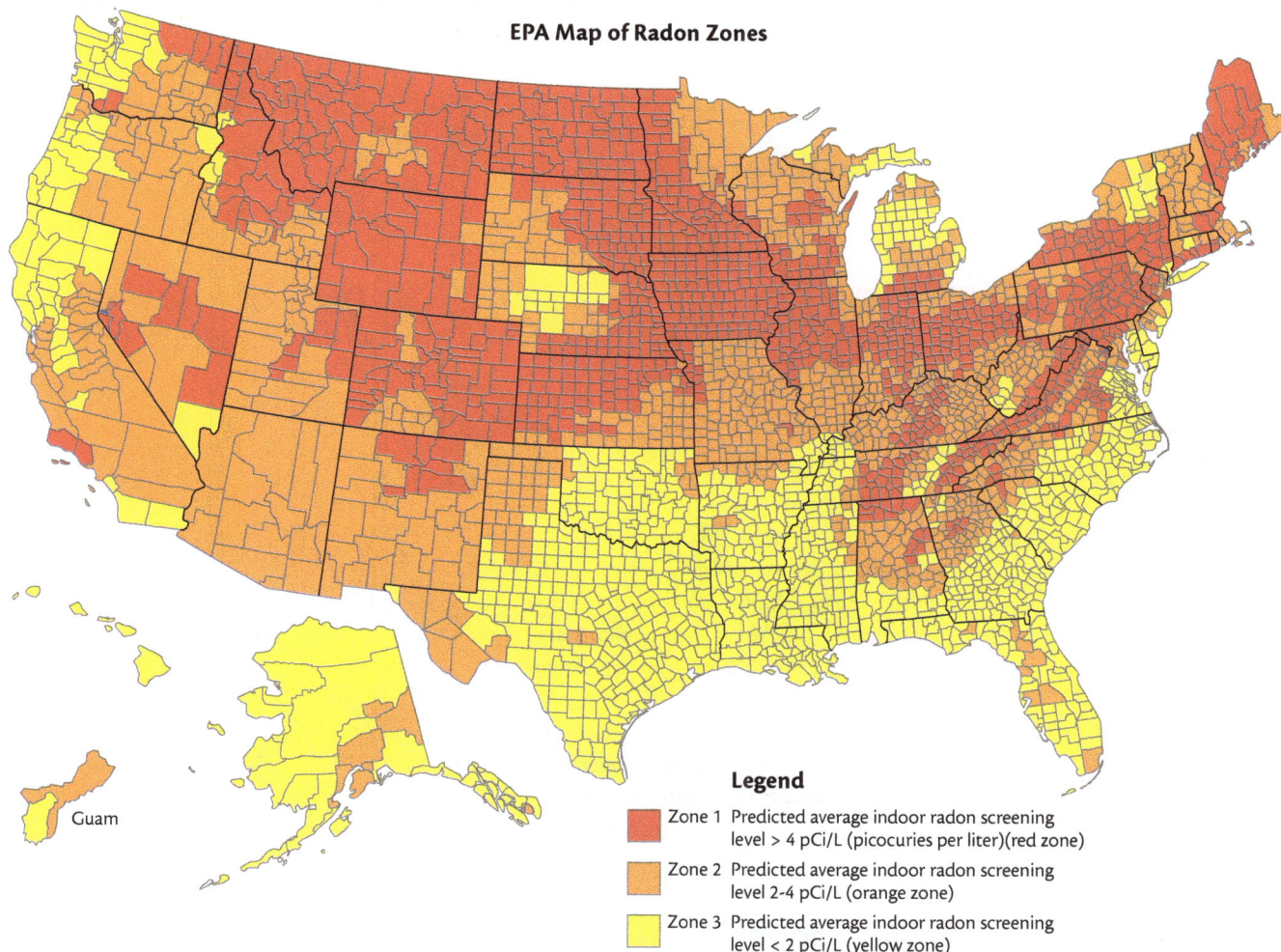

Legend

Zone 1 Predicted average indoor radon screening level > 4 pCi/L (picocuries per liter)(red zone)

Zone 2 Predicted average indoor radon screening level 2-4 pCi/L (orange zone)

Zone 3 Predicted average indoor radon screening level < 2 pCi/L (yellow zone)

Figure 6.36 *Map highlighting U.S. areas by county that have experienced radon problems.*

B.8 NUCLEAR RADIATION DETECTORS

The only way to detect radioactive decay is to observe the results of nuclear radiation interacting with matter. In each of the following detection methods, visible and/or electronically detectable changes in matter enable a technician to determine when such radiation is present.

Geiger–Müller Tubes and Counters

An early device for detecting radioactivity still used today is the Geiger–Müller counter (Figure 6.37). This device contains argon gas that is ionized when radiation enters the tube. As illustrated in Figure 6.38, when the ionized gas strikes the detector, it produces electrical signals. Most radiation counters register these signals as both audible clicks and meter readings. The intensity of the radiation is indicated by the number of electronic signals or counts detected per minute (cpm).

Figure 6.37 *Using a Geiger–Müller counter to measure radiation dose.*

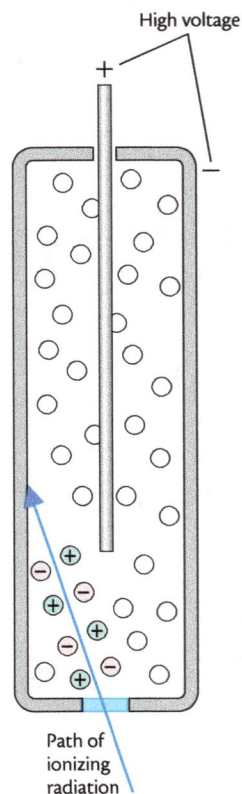

High voltage

Figure 6.38 *As ionizing radiation passes into a Geiger–Müller detector tube, ions form in the gas inside. Positive ions are attracted to the tube's negatively charged outer wall, while negative ions are attracted to the positively charged center. These electrically charged particle movements constitute a pulse of electrical current. Each pulse is detected and counted.*

Path of ionizing radiation

Film Badges

You learned earlier in this unit that emissions from radioactive materials will expose photographic film. Workers who handle radioisotopes wear film badges or other detection devices to monitor their exposure. Film badges are periodically collected and the film is developed and analyzed. The more ionizing radiation workers encounter, the greater the extent of exposure in their film badges. If workers receive doses in excess of federal limits, they are temporarily reassigned to jobs that minimize their future exposure to ionizing radiation.

Scintillation Counters

Another detection approach involves devices called **scintillation counters** that contain solid substances whose atoms emit light when they are excited by ionizing radiation. In modern scintillation counters, the flashes of light are detected electronically. The scintillation counter probe pictured in Figure 6.39 has a sodium iodide (NaI) detector that emits light when ionizing radiation strikes it.

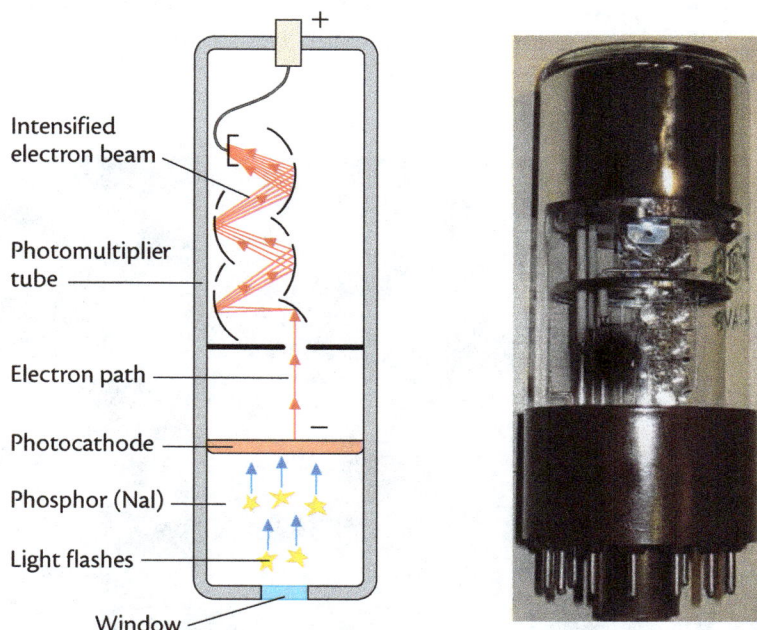

Figure 6.39 *A scintillation counter probe (left). Ionizing radiation causes flashes of light (scintillations) in the detector (NaI crystal). Each light flash is converted to an electron pulse that is increased many times as it moves through the photomultiplier tube (see close-up on right).*

Solid-State Detectors

In another technique, **solid-state detectors** monitor changes in the movement of electrons through silicon-based semiconductors as they are exposed to ionizing radiation. These detectors are often used in research laboratories.

Cloud Chambers

Ionizing radiation can also be detected in a cloud chamber. You will learn about this detection method in the following investigation.

◼ INVESTIGATING MATTER

B.9 CLOUD CHAMBERS

Preparing to Investigate

A **cloud chamber** is a container filled with air saturated with water or another vapor, similar to saturated air on a very humid day. If cooled, this air becomes supersaturated with vapor. This is an unstable condition. When a radioactive source is placed near a cloud chamber filled with supersaturated air, the radiation passes through the chamber, ionizing gas molecules along its path. Vapor condenses to liquid on the ions that are formed, leaving a white trail along each passing radioactive emission, thus revealing the path of the particle or ray. Figure 6.40 is an image of particle tracks under such conditions.

Cloud-chamber trails resemble the vapor trails from high-flying aircraft.

Figure 6.40 *A cloud chamber, showing many particle trails created by ionizing radiation. The radioactive source is located on the left side.*

The cloud chamber you will use consists of a small plastic container and a felt band moistened with 2-propanol (isopropyl alcohol). The alcohol evaporates faster than water, so it saturates the enclosed air more readily. You will chill the cloud chamber with dry ice to promote supersaturation and cloud formation (Figure 6.41).

Figure 6.41
Cloud-chamber setup. Note the positions of the radiation source, dry ice, and flashlight.

Gathering Evidence

1. Before you begin, put on your goggles, and wear them properly throughout the investigation.

2. Fully moisten the felt band inside the cloud chamber with 2-propanol. Also place dark paper on the chamber bottom and moisten it with a small quantity of alcohol.

3. Using gloves and forceps, quickly place the radioactive source in the chamber.

4. To cool the chamber, place it on a flat surface of dry ice, ensuring that the chamber remains level.

5. Leave the chamber on the dry ice for 3–5 min.

6. Your teacher will adjust the room lighting. Focus a light source at an oblique angle (not straight down) through the container to illuminate the chamber base. If you do not observe any vapor trails, shine the light through the side of the chamber.

7. Observe the air inside the chamber near the radioactive source. Record your observations.

Interpreting Evidence

1. How would you describe your observations to someone who had never seen a cloud chamber?

2. What differences, if any, did you observe among the tracks?

3. Which type of radiation do you think would make the most visible tracks? Why?

Reflecting on the Investigation

4. What is the purpose of the dry ice?

5. Could a cloud chamber be used to detect non-ionizing radiation? Explain.

■ MAKING DECISIONS

B.10 ENSURING PUBLIC SAFETY

Consider again the statements found in the CANE flyer (page 586) and also your planned presentation to the senior citizens of Riverwood. With a small group of classmates, discuss the appropriateness of each of the following proposals for protecting the public from radiation hazards. For each proposal, identify statements in the CANE flyer that might encourage public groups to make such a proposal.

1. Because all radiation is bad, government regulators should ban exposure to all forms of radiation.

2. Although there are several different types of ionizing radiation (alpha, beta, and gamma), government standards for protecting the public against possible dangers should treat all radiation exposures identically.

3. Because nuclear radiation can be harmful, doctors should be required by law to inform patients of any medical procedure that involves nuclear radiation. Patients can then reject the treatment if they have concerns about radiation exposure (Figure 6.42).

4. The government should provide a waste-disposal site for permanently storing all wastes that have ever been identified as radioactive (Figure 6.43).

In this section, you studied the origins of ionizing radiation, sources of exposure to ionizing radiation, and some possible consequences of that exposure. Keep these ideas and concerns in mind as you explore uses of nuclear radiation in the next section.

Figure 6.42
Nuclear technology is widely used in medicine.

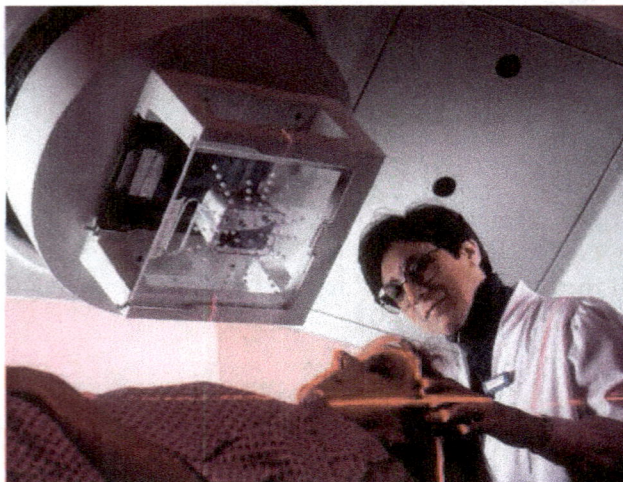

Figure 6.43 *Should the government be responsible for permanently storing all wastes that have ever been identified as radioactive?*

SECTION B SUMMARY

Reviewing the Concepts

> A relatively constant level of radioactivity, called background radiation, is always present and contributes to an individual's ionizing-radiation dose.

1. List five sources of background radiation.

2. Why does background radiation vary from one region to another?

3. Is it possible to eliminate background radiation? Explain.

4. What is the established U.S. background radiation limit for the general public expressed in
   a. mrem?
   b. Sv?

5. In what SI units is exposure to background radiation measured?

6. What radiation dose would you receive
   a. during a five-hour jet aircraft flight?
   b. from a CT scan and a chest X-ray?
   c. if you lived in a brick house 25 miles (40 km) from a coal-fired power plant for one year?

7. How does radon gas get into houses?

8. Why do radon levels tend to be higher in energy-efficient houses?

> Ionizing radiation has sufficient energy to break chemical bonds.

9. Why is the breaking of chemical bonds in living cells by ionizing radiation harmful?

10. What is a *mutation*?

11. How does ionizing radiation lead to an increase in mutation rates?

12. List four factors that determine the extent of biological damage from radiation.

13. What types of human tissue are most susceptible to radiation damage?

14. At what radiation dosage level would a person begin to experience nausea and hair loss? Express your answer in units of
    a. rem.
    b. mrem.

> Alpha, beta, and gamma radiation have different properties that determine their effects on living tissues.

15. Which type or types of radiation would
    a. be stopped by a glass window pane?
    b. penetrate a cardboard box?
    c. penetrate a thin sheet of plastic but not a thin sheet of lead?

16. Explain why materials that emit alpha radiation are more dangerous inside the body than outside the body.

17. What is the relationship between radiation intensity and distance from the radiation source?

18. Suppose a beta source gives a corrected radiation reading of 640 cpm at a distance of 3 cm. Predict the corrected reading at a distance of
    a. 6 cm.
    b. 12 cm.

> The emission of nuclear radiation changes the composition of the nucleus.

19. a. What is the composition of an alpha particle?
    b. List two symbols used to represent an alpha particle.
20. a. What is a beta particle?
    b. List two symbols used to represent a beta particle.
21. How is a beta particle formed?
22. Why does the emission of an alpha or beta particle create an atom of a different element?
23. What is a gamma ray?
24. Copy and complete the following nuclear equations.

    a. $^{6}_{3}\text{Li} + ^{1}_{0}n \longrightarrow ^{4}_{2}\text{He} + \underline{\quad}$

    b. $^{42}_{19}\text{K} \longrightarrow ^{0}_{-1}e + \underline{\quad}$

    c. $^{235}_{92}\text{U} \longrightarrow \underline{\quad} + ^{231}_{90}\text{Th}$

    d. $^{1}_{0}n + \underline{\quad} \longrightarrow ^{142}_{56}\text{Ba} + ^{91}_{36}\text{Kr} + 3\,^{1}_{0}n$

> Ionizing radiation may be detected by its interaction with matter using a variety of methods.

25. Describe how each of the following devices can detect the presence of ionizing radiation:
    a. Geiger–Müller tube and counter
    b. scintillation counter
    c. solid-state detector
    d. cloud chamber
    e. film badge
26. Would you expect alpha, beta, and gamma radiation to produce the same kinds of trails in a cloud chamber? Explain your answer.

How do we detect and describe the products of nuclear decay?

In this section, you investigated properties of alpha, beta, and gamma radiation, learned about sources of exposure to ionizing radiation and its possible health effects, considered nuclear decay, and explored how ionizing radiation can be detected. Think about what you have learned, then answer the question in your own words in organized paragraphs. Your answer should demonstrate your understanding of the key ideas in this section.

Be sure to consider the following in your response: background radiation; alpha, beta, and gamma radiation; penetrating power; dose; radioactive decay; and radiation detection methods.

Connecting the Concepts

27. High-level radioactive wastes are generally stored deep underground. Suggest two ways in which this method serves to keep people safe from excessive exposure to ionizing radiation.

28. In the 1940s and 1950s, some shoe stores invited customers to check the adequacy of shoe fit by X-raying their feet. Why did shoe stores discontinue that practice? (That shoe-store device was called a *fluoroscope*.)

29. Are the effects of shielding and distance the same for both ionizing and non-ionizing radiation? Explain your answer.

30. Radon is an inert noble gas that is fairly harmless to living things. Explain why its presence in homes constitutes a health hazard for occupants.

31. Describe fundamental differences between nuclear and chemical reactions.

32. A student sets up a cloud chamber and sees no white trails. What are some possible explanations for this result?

33. Two students live next door to each other. One receives three times more yearly radiation than the other. Explain how this could be possible.

34. Why do radiation detectors register counts even though no apparent source of radioactivity is near?

35. A heavy apron is provided for a patient who receives a dental X-ray.

   a. What element is probably used in the apron?

   b. What is the purpose of the apron?

   c. Why does the dentist or hygienist leave the room while the X-ray device is turned on?

Extending the Concepts

36. In terms of radiation, how is a sunburn different from a suntan? How do sunscreens work to prevent both?

37. Identify the five fastest-growing metropolitan areas in the United States. Rank them according to the average level of ionizing radiation exposure that inhabitants of similar houses receive.

38. Investigate and explain whether or not beta particles can be distinguished from electrons.

39. Why does ionizing radiation break some chemical bonds but not others?

CHEMISTRY *AT WORK*
Q&A

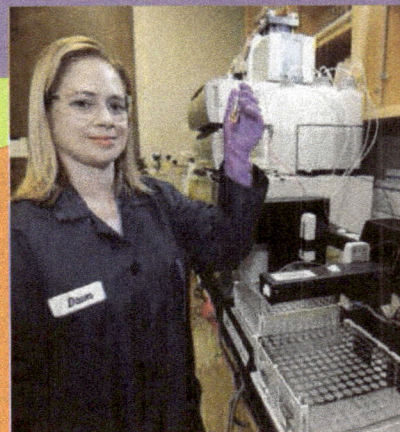

Dawn Shaughnessy, Nuclear Chemist at Lawrence Livermore National Laboratory in Livermore, California

With only a few exceptions, elements in the periodic table after polonium (Po, atomic number 84) are radioactive. Unlike polonium, radium, and a few other naturally occurring radioactive elements, almost all radioactive elements are synthesized, or created in a lab, including the "superheavy" elements—the ones with the largest atomic numbers. For many years, nuclear chemists at Lawrence Livermore National Laboratory have been leaders in synthesizing new superheavy elements. Read on to see how one chemist is focusing her efforts on extending the periodic table.

Q. What is nuclear chemistry, and why is it important for synthesizing new elements?

A. Nuclear chemistry is the study of radioactivity and nuclear reactions, including fission and fusion. Nuclear chemistry is central to creating new elements because we use fusion to synthesize these elements and radioactive signatures from their decay to detect their existence.

Q. How did you get into nuclear chemistry?

A. When I was in junior high, my parents got me a chemistry set. It sounds like a cliché, but it's true! I played with it all the time. Unfortunately, my high school didn't have much equipment, so we couldn't do many chemistry experiments there. But when I got to college, having access to a lab full of interesting tools made me realize that chemistry was what I wanted to do. One of my introductory chemistry professors was a nuclear chemist, and he spent a few weeks teaching us the basics of nuclear chemistry. I had never heard of nuclear chemistry, and I thought it was the coolest thing ever. When I finally got to take a nuclear chemistry class in college, I was hooked.

Q. Why did you end up working at Lawrence Livermore National Laboratory?

A. There are lots of places that chemists can work. But if you want to work in nuclear chemistry, one of the U.S. Department of Energy's National Laboratories is the place to be. National labs do basic research with the specific goal of helping the country, and each one has a different focus. At Lawrence Livermore, one of our areas of focus is national security, which involves protecting the United States from nuclear weapons and disposing of radioactive waste safely. We also do basic nuclear chemistry research, such as discovering new elements. Scientists from Lawrence Livermore, including me, have played a part in discovering all the elements from 113 to 118.

How element 118 was made. (top) Artist's conception of calcium ions traveling down the accelerator at a high velocity toward the rotating californium target. (bottom) The new element 118 travels through the accelerator to the detector.

Q. How do you go about discovering a new element?

A. We start by looking for two elements in the periodic table that have the right combination of protons and neutrons to add together to make the new element's atomic number. For example, if you want to make element 118 (also called ununoctium), you first need to find two elements with atomic numbers that add up to 118. You might try krypton (Kr) and lead (Pb), with atomic numbers 36 and 82. You then accelerate atoms of one of these elements to a very high velocity and slam them into atoms of the other element. In this process, the two elements fuse and a new element is created. This whole process happens inside a device called a particle accelerator.

Q. Lawrence Livermore doesn't have its own particle accelerator. How do you discover new elements without this device?

A. We have many partnerships, or collaborations, with nuclear chemists who work at particle accelerators around the world. We supply them with one of the starter elements. They supply the other starter element and conduct the reaction in their accelerator. We both get to bring something to the table. One of the labs we collaborate with most often is Dubna, a particle accelerator facility in Russia we've been working with since the 1980s. I fly to Russia to work there at least once a year.

The Joint Institute for Nuclear Research is in Dubna, near Moscow. This photo shows its synchrocyclotron.

Q. Superheavy elements can have very short half-lives. Why is it important for us to learn about them?

A. We don't know much about the bottom of the periodic table, including where it ends. Trying to create new elements will tell us if element 118 is truly the last element in the periodic table or if we can extend it even further. We also know very little about how these superheavy elements behave. Right now, we have them placed in groups based on their atomic number. But does element 116 really behave like other group 16 elements, such as oxygen? Creating these elements and trying to measure their properties is the only way to tell.

Experimenting with "atomic energy" at home is not safe, which is why modern chemistry sets do not claim to—unlike this vintage set. Yet a chemistry set is a good way to start investigating activity at the atomic level.

Q. What advice do you have for students interested in nuclear chemistry?

A. Take as many chemistry, physics, and math classes in high school as you can. When you get to college, think about doing a summer internship at a national lab like Lawrence Livermore. An internship is a great way to get hands-on experience in a wide variety of sciences, including nuclear chemistry—it can really help accelerate your career.

USING RADIOACTIVITY

Why and how are radioactive isotopes useful?

Each radioisotope decays and emits ionizing radiation at its own special rate. Scientists have devised convenient ways to measure, analyze, and report how rapidly (or slowly) particular radioisotopes decay. In this section, you will learn about radioactive decay rates, a characteristic that helps determine how useful or hazardous a radioisotope may be.

GOALS

- Define and describe half-life.
- Using its half-life, calculate the amount of a particular isotope that remains undecayed after a specified time.
- Describe how radioisotopes are used as diagnostic tracers in medicine.
- Explain how radioisotopes are used to kill cancerous cells.
- Describe the process of nuclear transmutation.
- Write, complete, and balance nuclear transmutation equations.

concept check 5

1. What does it mean to describe radiation as *ionizing radiation*?
2. What is balanced in a nuclear equation? How does this compare to a chemical equation?
3. What does half-life mean?

C.1 HALF-LIFE: A RADIOACTIVE CLOCK

How long does it take for a sample of radioactive material to decay? There is no simple answer to this question. However, by understanding how radioactive materials decay, scientists can predict how long a radioisotope used in a medical diagnostic test, for example, will remain active within the body; plan the long-term storage of hazardous nuclear wastes; and estimate the ages of old organisms, civilizations, or rocks.

The concept of **half-life** has several interpretations. The most common interpretation is that a half-life is the time it takes for one-half of the total radioactive atoms originally present in a sample to decay. Under normal conditions, the half-life for a particular radioisotope remains constant. This interpretation of half-life is useful when dealing with a sample that contains a large number of radioactive atoms. Because one gram of any element contains well over 10^{21} atoms, the typical sample size involves very large numbers of atoms and this definition of half-life is particularly useful.

Another interpretation, proposed by Ernest Rutherford in 1904, applies to an individual atom in a sample. In this view, half-life expresses the time within which a radioactive atom has a 50–50 chance to undergo radioactive decay. Table 6.9 lists the half-lives and decay reactions of several radioisotopes.

Table 6.9

Decay Equations and Half-Lives for Five Radioactive Isotopes

Radioisotope	Symbol	Half-Life
Hydrogen-3	$^{3}_{1}H \longrightarrow {}^{3}_{2}He + {}^{0}_{-1}e$	12.3 y
Carbon-14	$^{14}_{6}C \longrightarrow {}^{14}_{7}N + {}^{0}_{-1}e$	5.73×10^{3} y
Phosphorus-32	$^{32}_{15}P \longrightarrow {}^{32}_{16}S + {}^{0}_{-1}e$	14.3 d
Potassium-40	$^{40}_{19}K \longrightarrow {}^{40}_{20}Ca + {}^{0}_{-1}e$	1.28×10^{9} y
Radon-222	$^{222}_{86}Rn \longrightarrow {}^{218}_{84}Po + {}^{4}_{2}He$	3.82 d

Although it is not possible to predict when a particular radioactive atom will decay, each way of thinking about half-life is equally valid when working with the large numbers of atoms in typical chemical samples. The decay of carbon-14 is represented by this equation:

$$^{14}_{6}C \longrightarrow {}^{14}_{7}N + {}^{0}_{-1}e$$

All living matter contains carbon and, therefore, a small amount of radioactive C-14. Thus, all living organisms emit a small but constant level of radioactivity. In living matter, decaying C-14 atoms are constantly replaced by new carbon atoms. After death, no new carbon atoms are taken in. The radioactive C-14 atoms continue to decay, so the longer an organism has been dead, the fewer C-14 atoms it contains.

Table 6.9 indicates that C-14's half-life is 5730 years. If an organism contains 50.0 billion atoms of carbon-14 at the time of death, half of those atoms will have decayed after 5730 years pass, leaving 25.0 billion atoms of carbon-14. During the next 5730 years, another one-half of the atoms will decay, leaving 12.5 billion atoms of carbon-14. Therefore, if a sample from a previously living organism contains only one-fourth the number of C-14 atoms expected in a living organism, we can estimate that the sample is about 11 460 years old—that is, two half-lives must have passed since the organism died.

The relative abundance of C-14 has remained reasonably constant over thousands of years. C-14 is constantly being produced in the upper atmosphere by cosmic rays interacting with nitrogen atoms, while C-14 atoms are also undergoing radioactive decay.

Because the decay rate is directly related to the number of C-14 atoms present, scientists usually express the decay rate rather than the number of C-14 atoms.

In 2004, scientists reported that they could decrease the half-life of beryllium-7 by half a day (a 1% change) by trapping it in an electron-rich environment. The half-life of sodium-22 has also been decreased, in that case by implanting it in palladium (Pd) metal cooled to 12 K.

Every radioisotope has a specific half-life that is constant under normal circumstances. Half-lives of radioisotopes can be as short as a fraction of a second or as long as several billion years. For example, the half-life of polonium-212 is 3×10^{-7} seconds, while that of uranium-238 is 4.5 billion years. Thus, in one year, all atoms in a small sample of polonium-212 will probably have decayed, while well over 99% of the original uranium-238 atoms will still be present.

After 10 half-lives, only about 1/1000th or 0.1% of the original radioisotope atoms are still left to decay. (You can verify that statement with your own calculations.) That means that the rate of radioactive decay of the isotope has dropped to 0.1% of its initial level. This reduced level is often considered safe because it roughly approaches the level of normal background radiation.

Because there is no way to change the radioactive decay rate significantly for a particular isotope, radioactive waste disposal (or storage) can pose challenging problems, especially for radioisotopes with very long half-lives. You will examine that issue later in this unit.

In the following activity, you will model and explore the concept of half-life with heads-up and heads-down coins.

MODELING MATTER

C.2 UNDERSTANDING HALF-LIFE

In this activity you will receive 80 pennies and a box. Place all pennies heads up to represent the starting sample of *headsium*. Assume each heads-up penny represents an atom of the radioactive headsium. Its decay produces a tails-up penny—*tailsium*. Each shake of the closed box containing pennies represents one half-life. During this time a certain number of headsium nuclei will decay—flip over—to produce tailsium. You will investigate the relationship between the passage of time and the quantity of radioactive nuclei (heads-up pennies) that decay.

The following steps will lead you through this activity:

1. Prepare a data table for recording the undecayed headsium and decayed tailsium atoms after each of four half-lives. Include initial values for passage of 0 half-lives.

2. Place the 80 pennies heads up in the box.

3. Close the box and shake it vigorously.

4. Open the box. Remove all atoms that decayed into tailsium. Record the number of undecayed (headsium) and decayed (tailsium) atoms after this first half-life.

5. Repeat Steps 3 and 4 three more times. You will now have simulated the passage of four half-lives. Record your results for each half-life.

6. Follow your teacher's instructions to obtain pooled class data for total undecayed headsium atoms remaining after each half-life.

7. Using your own data and class-pooled data, prepare a graph by plotting the number of half-lives on the *x*-axis and the number of undecayed atoms remaining after each half-life on the *y*-axis. Plot and label two graph lines—one representing your own data and the other representing pooled class data.

Now answer these questions, based on your data:

1. a. Describe the appearance of your two graph lines. Are they straight or curved?

    b. Which set of data—yours or the pooled class data—provides a more convincing demonstration of half-life? Why?

2. About how many headsium nuclei would remain after three half-lives, if the initial sample had 600 headsium atoms?

3. If 190 headsium nuclei remain from an original sample of 3000 headsium nuclei, about how many half-lives must have passed?

4. Describe one similarity and one difference between your model based on pennies and actual radioactive decay. (*Hint:* Why did you pool the class data?)

5. How could you modify this model to demonstrate that different isotopes have different half-lives?

6. a. How many half-lives would be needed for one mole of a radioisotope to decay to 6.25% of the original number of atoms?

    b. Is it likely that any of the original radioactive atoms would still remain after

        i. 10 half-lives? Explain your answer.

        ii. 100 half-lives? Explain your answer.

7. a. In this simulation, can you predict when a particular headsium nucleus will "decay"?

    b. If you could follow the fate of an individual atom in a sample of radioactive material, could you predict when it would decay? Why or why not?

8. What other ideas could model the concept of half-life?

DEVELOPING SKILLS

C.3 APPLICATIONS OF HALF-LIVES

Use what you learned about half-life to answer the following questions. Figure 6.44 may help.

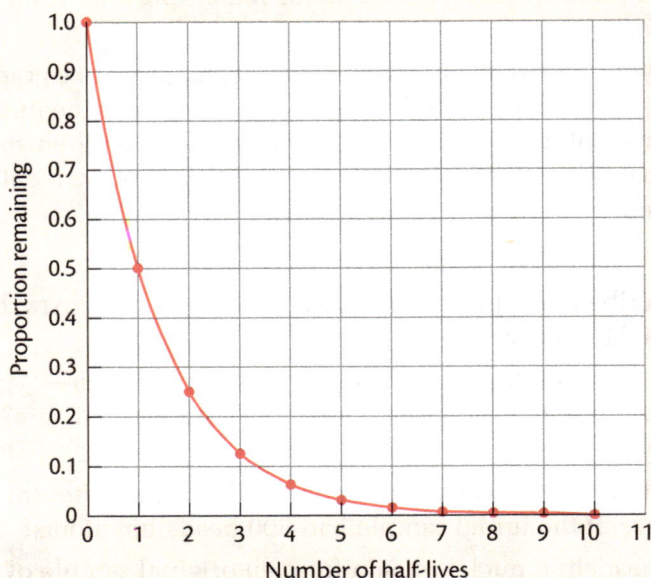

Figure 6.44 *This graph depicts the quantity of radioactive material remaining versus total half-lives that have passed. What proportion of material would remain after four half-lives?*

Sample Problem: *The half-life of O-15 is 2.0 min. Its radioactive decay produces N-15. How much O-15 will remain undecayed after 5.0 min, if the original sample contained 14.0 g O-15?*

First, determine the total half-lives that 5.0 min represents:

$$5.0 \text{ min} \times \frac{1 \text{ half-life}}{2.0 \text{ min}} = 2.5 \text{ half-lives}$$

Because the total number of half-lives is not an integer, use Figure 6.44 to estimate the proportion of O-15 remaining. Locate 2.5 half-lives on the x-axis. From that point, move directly upward on the graph until you touch the curved line. Then move left until you intersect the y-axis. Read this point from the graph: about 0.18. This means that 18% of the original O-15 sample remains:

$$(14.0 \text{ g O-15}) \times (0.18) = 2.5 \text{ g O-15 remain}$$

You can check to see if an answer of 2.5 g makes sense: After one half-life, 7.00 g O-15 would remain in the sample (the remaining mass represents N-15). After two half-lives, 3.50 g remain, and after three half-lives 1.75 g O-15 would remain undecayed. The answer we found, 2.5 g, is between 3.50 g and 1.75 g, which is what we would expect.

1   Suppose you received $1000 (see Figure 6.45) and could spend one-half in the first year, one-half of the balance in the second year, and so on. (*Note:* One year corresponds to one half-life in this analogy.)

   a. If you spent the maximum allowed annually, at the end of which year would you have $31.25 left?

   b. How much money would be left after 10 half-lives?

2. Potassium is a necessary nutrient for all living things and is the seventh most abundant element on Earth's surface, composing about 1.5% of its crust. About 0.01% of natural potassium atoms are the radioisotope potassium-40. K-40 has a half-life of nearly 1.3 billion (1.3×10^9) years.

   a. Assuming Earth is ~4.5 billion years old, how much of the K-40 at Earth's formation remained after one half-life?

   b. Roughly how many times more K-40 was present when Earth formed than is present now?

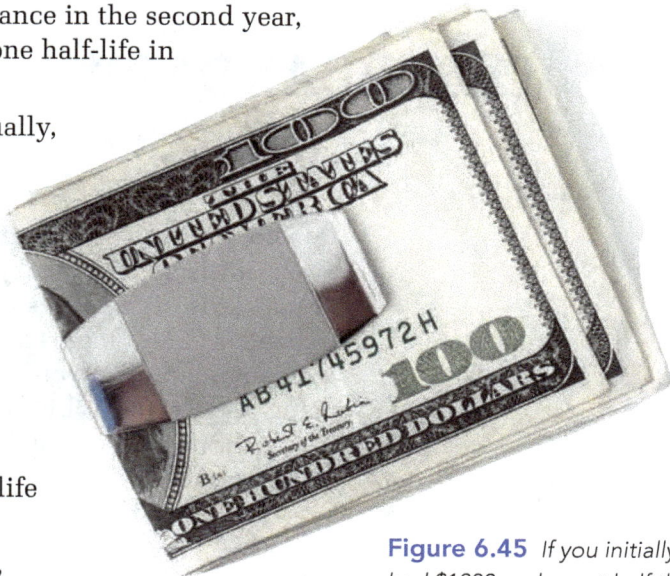

Figure 6.45 *If you initially had $1000 and spent half the money each year, how many years would it take until you had only $125?*

3. Strontium-90 is one of many radioisotopes generated by nuclear-weapon explosions. This isotope is especially dangerous if it enters our food supply. Strontium behaves chemically like calcium, because these elements are members of the same chemical family. Rather than passing through the body, radioactive strontium-90 is incorporated into calcium-based material, such as bone. A nuclear test-ban treaty in 1963 ended most aboveground weapons testing. But some Sr-90 released in previous testing is still present in the environment.

   a. Sr-90 has a half-life of 28.8 years. Track the decay of Sr-90 atoms that were present in the atmosphere in 1963 as follows:

      i. Using 1963 as year zero, when 100% of released Sr-90 was present, identify the years that represent completion of one, two, three, four, and five half-lives.

      ii. Calculate the percent of the original 1963 Sr-90 radioisotope present at the end of each half-life.

   b. Plot the percent of the original 1963 Sr-90 radioactivity level on the *y*-axis and the years 1963 to 2110 on the *x*-axis. Draw a smooth curve through the data points.

      i. What percent of Sr-90 formed in 1963 is present now?

      ii. What percent will remain in 2100?

   c. Compare your graph to that shown in Figure 6.44.

      i. How do the two graphs differ?

      ii. In what ways are the graphs similar?

CHEM**QUANDARY**

CARBON-14 DATING

It is not possible to determine the age of every artifact using carbon-14 dating. What kinds of materials might be good candidates for cabon-14 dating? What are some materials you could not date using carbon-14? Why does carbon-14 dating have a practical limit of about 50 000 years?

A conservator cuts a sample of a document for carbon-14 dating.

C.4 RADIOISOTOPES IN MEDICINE

During the late 19th and early 20th centuries, the rapid advancements in science fascinated the general public. This excitement about the "wonders of science" made people very susceptible to exaggerated claims about the healing powers of radiation. Many products were marketed as medicine with remarkable powers to cure. Some of these medicines advertised that they contained radium, claiming supposed benefits of that substance. Unfortunately, some actually contained radioactive materials and were quite hazardous. Radioactive medicines were sold in the United States as late as the 1930s.

The careful use of ionizing radiation and radioisotopes can be quite effective in medical diagnosis and treatment (see Figures 6.46 and 6.47). Such uses can be classified as either *diagnostic* or *therapeutic*. **Diagnostic** use helps doctors understand what is happening inside the body (see Figure 6.48), while **therapeutic** use involves treating a medical condition.

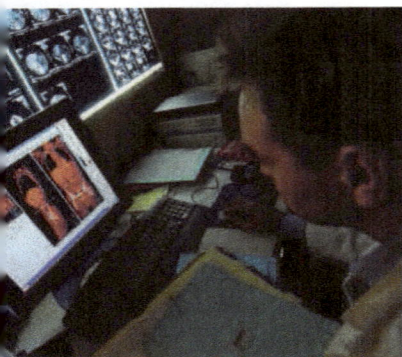

Figure 6.46 *Injection of radioactive tracer molecules allows noninvasive study of the metabolic function of internal organs.*

Figure 6.47 *Radioisotopes are prepared and used with knowledge of their half-lives, thus ensuring appropriate levels of activity as tracers. In this image, a lead-lined enclosure is used for protection as an injection of radioactive substance is prepared.*

A common diagnostic application uses radioisotope-tracer studies, based on detecting a radioisotope in particular parts of the body. Physicians know that certain elements collect in specific parts of the body (e.g., calcium in bones and teeth). They can investigate a given part of the body by using an appropriate radio-isotope as a **tracer**. In a radioisotope-tracer study, the physician places radioisotopes with short half-lives in a patient's body. Such studies can identify cellular abnormalities and help physicians select appropriate therapy.

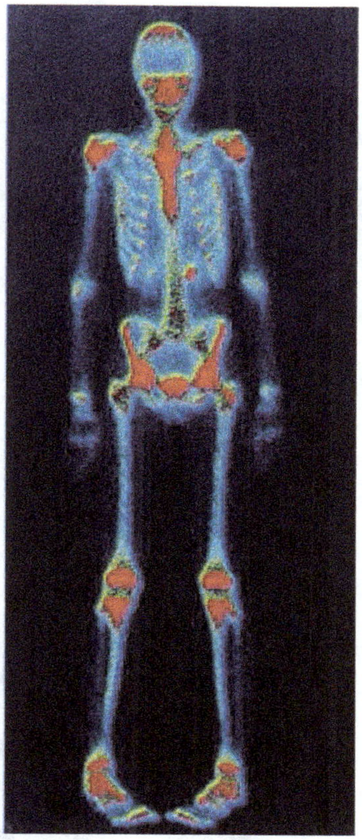

Tracers have properties that make them ideally suited to this task. First, radioisotopes have the same chemical properties as stable (non-radioactive) isotopes of the element. Researchers can apply a solution of an appropriate tracer isotope to the body, or they can feed or inject the patient with a biologically active compound containing the radioactive tracer element. A detection system, as shown in Figure 6.46, then allows a medical technician to track the tracer's location throughout the body.

For example, iodine-123 is used to diagnose problems of the thyroid gland, which is located in the neck. A patient drinks a tracer solution containing sodium iodide (NaI), in which some of the iodide ions are the radioactive isotope

Figure 6.48 *Nuclear-medicine scan of bones of the entire human body.*

I-123. The physician, using a radiation detection system, monitors the rate at which this tracer is taken up by the thyroid. A healthy thyroid will incorporate a known amount of iodine. An overactive or underactive thyroid will take up, respectively, more or less iodine. The physician compares the measured rate of I-123 uptake by the patient to the normal rate for an individual of the same age, gender, and weight, then takes appropriate therapeutic action.

Technetium-99m (Tc-99m), a synthetic radioisotope, is the most widely used diagnostic radioisotope in medicine. It has replaced exploratory surgery as a way to locate tumors in the brain, thyroid, and kidneys. Tumors are areas of runaway cell growth; technetium concentrates where cell growth is fastest. A bank of radiation detectors around the patient's body can pinpoint the Tc-99m at the tumor's precise location.

Physicians use therapeutic radioisotopes because they emit radiation that carries enough energy to destroy living tissue. In some cancer treatments, doctors kill cancerous cells with ionizing radiation. For thyroid cancer, a patient takes a liquid or a pill that contains radioiodine. The radioiodine collects in and destroys the cancerous portion of the thyroid gland. In other cancer treatments, physicians may direct an external beam of ionizing radiation (from cobalt-60) at the cancerous spot. Physicians must administer treatments with great care. High radiation doses also kill normal cells.

Radioiodine is a way of designating the radioactive isotope of iodine.

CHEM**QUANDARY**

USING RADIOISOTOPES IN MEDICINE

One source of ionizing radiation is cobalt-60, which kills rapidly dividing cells. Consider two equally sized Co-60 samples shipped to two hospitals at the same time. At one hospital, the Co-60 is used to treat dozens of individuals, while at the other hospital it is used only once or twice. Why would *both* hospitals dispose of their Co-60 samples after five years?

> Radiosodium is usually administered as a NaCl solution.

Other medical applications include the use of radiosodium (Na-24) to detect circulatory system abnormalities and radioxenon (Xe-133) to help locate lung embolisms (blood clots) and abnormalities. Table 6.10 summarizes medical uses for several radioisotopes.

Selected Medical Radioisotopes

Radioisotope	Half-Life	Use
Used as Tracers		
Technetium-99m	6.01 h	Measure cardiac output; locate strokes and brain and bone tumors.
Gallium-67	78.3 h	Diagnosis of Hodgkin's disease
Iron-59	44.5 d	Determine rate of red blood cell formation (these contain iron); anemia assessment
Chromium-51	27.7 d	Determine blood volume and lifespan of red blood cells
Hydrogen-3 (tritium)	12.3 y	Determine volume of the body's water; assess vitamin D use in body
Thallium-201	72.9 h	Assess coronary artery disease
Iodine-123	13.3 h	Diagnose diseases of the thyroid gland
Used for Radiation Therapy		
Cesium-137	30.1 y	Treat shallow tumors (external source)
Phosphorus-32	14.3 d	Treat leukemia, a bone cancer affecting white blood cells (internal source)
Iodine-131	8.0 d	Treat thyroid cancer (external source)
Cobalt-60	5.3 y	Treat shallow tumors (external source)
Yttrium-90	64.1 h	Treat pituitary gland cancer (internal source)

Table 6.10

You have learned that ionizing radiation, like that observed in a cloud chamber (page 630), is emitted by an unstable radioactive isotope as it decays, eventually to a stable nonradioactive isotope. Do you think it would be possible to reverse that process, converting a stable isotope into an unstable, radioactive isotope? Think about it. This question will be addressed later in this section.

C.5 NUCLEAR MEDICINE TECHNOLOGIES

Computers touch nearly all aspects of modern life. For example, health and medical science now employ two nuclear medicine technologies that rely heavily on computers to make sense of the large quantities of data obtained. These technologies are positron emission tomography (see Figure 6.49, page 648), which involves radioisotopes, and magnetic resonance imaging.

Positron emission tomography (PET) scans are based on a very unusual form of radioactive decay involving a few particular radioisotopes. Although most radioisotopes emit alpha, beta, or gamma radiation, a few radioisotopes emit radiation in the form of positrons. **Positrons** originate in the nucleus and have the same mass as beta particles (electrons). However, positrons differ from electrons in fundamental ways. Positrons have a positive electrical charge, while electrons are negatively charged.

Tomography refers to producing a 3-D image of an internal object of interest.

Positrons are composed of antimatter. When a positron encounters an electron, both particles are annihilated (destroyed) and produce two gamma rays that are emitted in opposite directions. PET detects these gamma-ray pairs and, with the help of computers, determines where they originated. By observing a large number of such events, a computer-generated image gradually emerges.

The radioisotope tracer that emits positrons in PET scans is attached to a sugar molecule. Physicians can accurately determine the movement of each tagged sugar molecule as it progresses through the body. Because cancers grow faster than normal tissues, cancerous tissue metabolizes more sugar in a given time than does normal tissue. The radioisotope trace (sugar tag) eventually becomes more concentrated in regions of the body containing cancerous tissue. PET technology can thus detect and display metabolic activity. Doctors can use this information to investigate brain functioning without invasive surgery.

Magnetic resonance imaging (MRI) does not employ ionizing radiation. This technique relies on the properties of the protons in the nuclei of hydrogen atoms in large biomolecules. Because of its reliance on the properties of nuclear particles, this technique is classified as a nuclear medicine technique. MRI is an application of a laboratory process known as nuclear magnetic resonance (NMR), which was developed in the mid-20th century. NMR is a noninvasive technique that can identify atoms within a sample without altering and affecting the sample itself.

Scientific American Working Knowledge Illustration

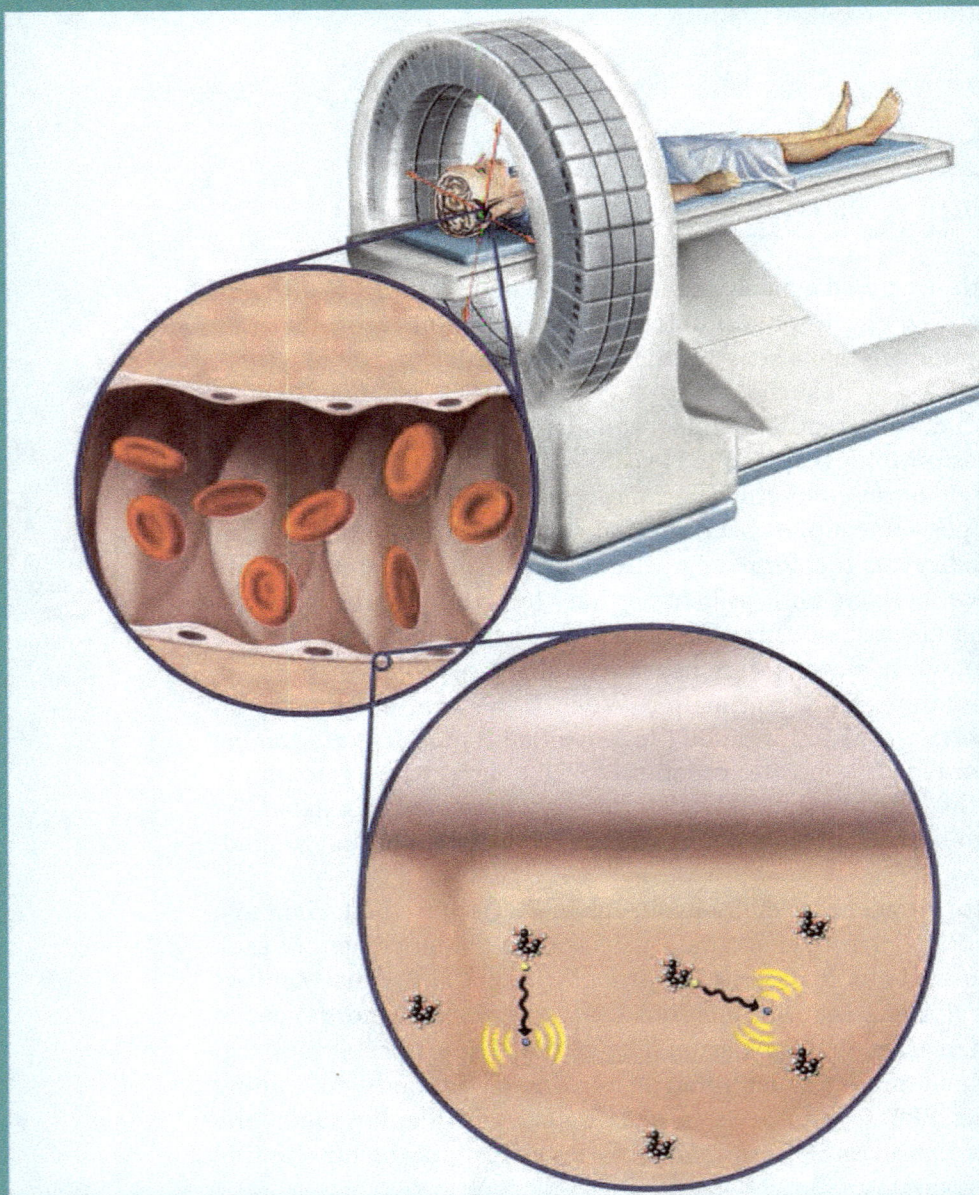

Figure 6.49 *Positron emission tomography (PET). Cancerous tissue can be detected using PET scans. In this illustration, a brain tumor is identified by detecting gamma radiation emitted when positrons collide with electrons. As depicted in the particulate-level representation (bottom image), sugar molecules are tagged with a radioisotope tracer that emits positrons (yellow spheres). When a positron collides with an electron (blue spheres) in the immediate vicinity, both particles are destroyed and produce two gamma rays (depicted as yellow waves). The gamma-ray pair is then detected.*

MRI imagery can produce useful images of soft tissues. A major benefit of MRI is that it does not rely on ionizing radiation. Unlike most other nuclear-medicine technologies, MRI uses radio waves of very low energies and involves no known health risks. Some patients were hesitant to undergo the procedure when it was called by its original name, nuclear magnetic resonance, due to fear evoked by the term nuclear. This unfounded fear prompted the name change to magnetic resonance imaging.

concept check 6

1. Would carbon-14 dating be useful for dating artifacts if the half-life of carbon-14 was not constant? Explain.
2. What specific characteristics of radioisotopes make them useful in
   a. medical diagnosis?
   b. treatment of cancer?
3. You are familiar with the synthesis of compounds. What does it mean to synthesize an element?

C.6 ARTIFICIAL RADIOACTIVITY

In 1919, Ernest Rutherford enclosed nitrogen gas in a glass tube and bombarded the sample with alpha particles. After analyzing the gas remaining in the tube, he found that some nitrogen atoms had been converted to an isotope of oxygen, according to the following equation:

$$^{4}_{2}\text{He} + ^{14}_{7}\text{N} \longrightarrow ^{17}_{8}\text{O} + ^{1}_{1}\text{H}$$

| Helium-4 (alpha particle) | Nitrogen-14 | Oxygen-17 | Hydrogen-1 |

Rutherford had produced the first synthetic or artificial **transmutation** of an element, the first documented conversion of one element to another. He continued this work but was limited by the moderate energies of alpha particles then available. By 1930, scientists had developed particle accelerators that could produce highly energetic particles for bombardment reactions. Using these higher-energy particles, scientists created many other synthetic atoms, some of which were radioactive.

The first synthetic radioisotope (one not occurring in nature) was produced in 1934 by French physicists Irène and Frédéric Joliot-Curie (the daughter and son-in-law of Marie and Pierre Curie; see Figure 6.50). They bombarded aluminum atoms with alpha particles, producing radioactive phosphorus-30 and neutrons:

$$^{27}_{13}\text{Al} + ^{4}_{2}\text{He} \longrightarrow ^{30}_{15}\text{P} + ^{1}_{0}n$$

Figure 6.50 *Irène and Frédéric Joliot-Curie produced the first synthetic radioisotope in 1934.*

Since then, researchers have accomplished many transformations of one element to another and have synthesized new radioactive isotopes of various elements. Many of the diagnostic radioisotopes noted in Table 6.10 (page 646) are synthetic. Technetium-99m, for example, is both a synthetic element and a radioisotope.

Most synthetic radioisotopes are produced by bombarding elements with neutrons, which are captured by target nuclei. This process requires less energy than many other bombardment reactions, because neutrons have no electrical charge and are not repelled by the positive charge of the nucleus. Such reactions produce radioactive nuclei that tend to emit beta particles, thus changing the atomic number and producing a different element.

The following examples show the formation of two synthetic radioisotopes often used as medical tracers, calcium-45 and iron-59.

$$^{44}_{20}\text{Ca} \ + \ ^{1}_{0}n \ \longrightarrow \ ^{45}_{20}\text{Ca}$$

$$^{58}_{26}\text{Fe} \ + \ ^{1}_{0}\text{n} \ \longrightarrow \ ^{59}_{26}\text{Fe}$$

Nuclear-bombardment reactions generally involve four particles:

- *Target nucleus:* the stable isotope that is bombarded.
- *Projectile particle (bullet):* the particle fired at the target nucleus.
- *Product nucleus:* the isotope produced in the reaction.
- *Ejected particle:* the lighter nucleus or particle emitted from the reaction.

For example, consider how the Joliot-Curies produced the first synthetic radioactive isotope, phosphorus-30. The four types of particles involved are identified as follows:

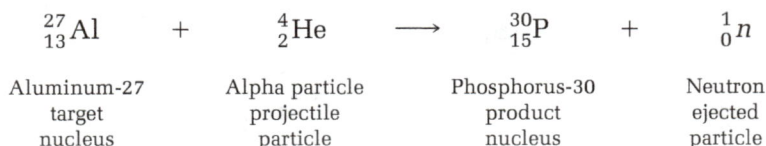

$$^{27}_{13}\text{Al} \ + \ ^{4}_{2}\text{He} \ \longrightarrow \ ^{30}_{15}\text{P} \ + \ ^{1}_{0}n$$

| Aluminum-27 target nucleus | Alpha particle projectile particle | Phosphorus-30 product nucleus | Neutron ejected particle |

> Reactions may release more than one ejected particle. For example, see the Cm-246 reaction on the next page.

CHEM**QUANDARY**

TRANSMUTATION OF ELEMENTS

Alchemists searched in vain for ways to transform (transmute) lead or iron into gold. Has such transmutation now become a reality? From what you know about nuclear reactions, do you think that lead, iron, or mercury atoms could be changed to gold? If so, try to write equations for the possible reactions.

DEVELOPING SKILLS

C.7 NUCLEAR-BOMBARDMENT REACTIONS

Sample Problem: Nobelium (No) can be produced by bombarding curium (Cm) atoms with nuclei of a low-mass element. What element serves as the projectile particle in this reaction?

$$^{246}_{96}\text{Cm} \ + \ ? \ \longrightarrow \ ^{254}_{102}\text{No} \ + \ 4\,^{1}_{0}n$$

Because the sum of product atomic numbers is 102, the projectile must have the atomic number 6, to make the sum of reactant atomic numbers also 102. Therefore, the projectile must be carbon (atomic number = 6). The total mass numbers of products is 258 (254 + 4), indicating that the projectile must have been carbon-12 (258 − 246 = 12).

The completed equation is as follows:

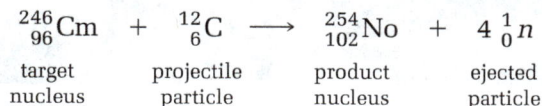

$$^{246}_{96}\text{Cm} \ + \ ^{12}_{6}\text{C} \ \longrightarrow \ ^{254}_{102}\text{No} \ + \ 4\,^{1}_{0}n$$

| target nucleus | projectile particle | product nucleus | ejected particle |

As you learned earlier, completing nuclear equations involves balancing atomic numbers and mass numbers.

Complete the following equations by supplying the missing numbers or symbols. Name each particle. Then identify the target nucleus, projectile particle, product nucleus, and ejected particle.

1. $^{59}_{27}? \ + \ ^{?}_{?}n \longrightarrow \ ^{60}_{?}?$ (Scientists produce most medically useful isotopes by bombarding stable isotopes with neutrons. This process converts the original nuclei to radioactive forms of the same element.)

2. $^{96}_{42}? \ + \ ^{?}_{?}\text{H} \longrightarrow \ ^{97}_{43}? \ + \ ^{1}_{0}?$ (Until it was synthesized in 1937, technetium was only an unfilled gap in the periodic table. All of its isotopes are radioactive. Any technetium originally on Earth has decayed. Technetium, the first element artificially produced, is now used in industry and medicine.)

3. $^{58}_{?}? \ + \ ^{209}_{?}\text{Bi} \longrightarrow \ ^{?}_{109}\text{Mt} \ + \ ^{1}_{0}?$ (In 1992, a research group in Darmstadt, Germany, created element 109 by bombarding bismuth-209 nuclei. The name meitnerium (Mt) honors Lise Meitner, the Austrian physicist who first proposed the idea of nuclear fission. (See page 659.)

Not only does the ability to transform one element into another provide new and powerful technological capabilities, it also has changed our view of elements.

C.8 EXTENDING THE PERIODIC TABLE

Seaborg and coworkers also identified over 100 new isotopes of various elements.

Since 1940, nearly 20 **transuranium** elements—with atomic numbers greater than the atomic number of uranium (92)—have been added to the periodic table. These elements have been synthesized in nuclear reactions, usually conducted in accelerators known as **cyclotrons** (see Figure 6.51). From 1940 to 1961, Glenn Seaborg and coworkers at the University of California–Berkeley synthesized and identified 10 new elements with atomic numbers 94 to 103, a prodigious feat.

Figure 6.51 *Cyclotrons, pioneered by Ernest O. Lawrence, allow scientists to investigate high-energy bombardment of heavy nuclei with various particles. The images here illustrate the development of the cyclotron over several decades: (top left) an 11-inch chamber from 1932, (bottom left) the Berkeley Radiation Laboratory's 60-inch cyclotron, which was used by Glenn Seaborg, Albert Ghiorso, and their coworkers to discover several new elements, and (right) the world's largest cyclotron (~18 m in diameter) at Canada's National Laboratory for Particle and Nuclear Physics.*

None of those 10 elements occurs naturally. All were made by high-energy bombardment of heavy nuclei with various particles. For example, alpha-particle bombardment of plutonium-239 produced curium-242:

$$^{239}_{94}\text{Pu} + ^{4}_{2}\text{He} \longrightarrow ^{242}_{96}\text{Cm} + ^{1}_{0}n$$

Bombarding Pu-239 with neutrons yielded americium-241, a radioisotope now used in home smoke detectors:

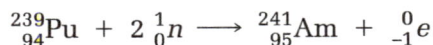

$$^{239}_{94}\text{Pu} + 2\,^{1}_{0}n \longrightarrow ^{241}_{95}\text{Am} + ^{0}_{-1}e$$

Seaborg was awarded the 1951 Nobel prize in chemistry for his work. Albert Ghiorso, a colleague of Seaborg, led the way in producing several new elements beyond lawrencium (element 103). One is element 106, produced by

Figure 6.52 *Glenn Seaborg and coworkers at the University of California–Berkeley synthesized and identified ten new elements beyond uranium. Can you identify any transuranium elements that they may have created?*

bombarding a californium-249 target with a beam of oxygen-18 nuclei, producing an isotope of element 106. To honor Seaborg's pioneering work, element 106 was named *seaborgium* (Sg). Glenn Seaborg, shown in Figure 6.52, has been called the father of the modern periodic table.

Traditionally, the discoverer of an element selects its name. For example, when Marie Curie first discovered element number 84 she named it *polonium* (Po) in honor of Poland, her home country.

Several scientific laboratories have claimed to have synthesized elements with atomic numbers greater than 92. For example, laboratories in both the former Soviet Union and the United States claimed the discovery of elements 104 and 105. Soviet scientists proposed naming them *kurchatovium* (Ku) and *dubnium* (Db), while U.S. scientists proposed the names *rutherfordium* (Ru) and *hahnium* (Ha). The International Union of Pure and Applied Chemistry (IUPAC) examined the claims for element discovery before recommending the official names. In 1997, the IUPAC approved the element names *rutherfordium* and *dubnium*.

At present, the IUPAC recognizes official names and symbols for the first 112 elements. Although claims for the discovery of other elements, 113 to 118, have been reported, the IUPAC has not yet recognized these discoveries or their official names. Scientists temporarily identify such unnamed elements by Latin prefixes indicating their atomic numbers. For example, element 113 is temporarily named ununtrium (un = 1, un = 1, tri = 3), which is symbolized as *Uut* until IUPAC officially recognizes the original discovery of that new element.

The name proposed by discoverers of element 111—roentgenium (Rg)—was approved by IUPAC in 2004. It honors W. K. Roentgen, who discovered X-rays. Roentgen was the first scientist you learned about in this unit. The process of confirmation and naming of new elements is very thorough and rigorous and takes a great deal of time. Evidence for element 112 was first reported in 1996. In 2010, the IUPAC officially recognized the discovery of and the name copernicium (Cn) for element 112.

The production of transuranium elements has enriched our understanding of the atomic nucleus. With the synthesis and identification of elements beyond atomic number 92, the periodic table has expanded to fill the actinide series as well as nearly all of Period 7.

Scientific research, like other human endeavors, can involve strong personalities, competition, and controversy.

IUPAC is a federation of national organizations that represent chemists around the world. It plays an important role in the international standardization of chemical nomenclature and terminology.

■ MAKING DECISIONS

C.9 OPINIONS ABOUT RADIOACTIVITY

Some older people tend to associate nuclear technologies with the use of atomic weapons at the close of World War II and with the atomic-weapon threats of the Cold War in the 1950s through the 1980s. Some of these people are likely to be in the audience when you speak to the Riverwood senior citizens.

The following opinions might be expressed by such community members. Decide how you would respond to each opinion, using knowledge you have gained in this unit (see Figure 6.53).

1. "I'm against having any isotopes in Riverwood. They're too dangerous."

2. "I don't want to live near anything that is radioactive."

3. "I don't know why scientists keep trying to make new elements. All the new ones are radioactive."

4. "I don't understand how cancer can be treated with radiation. I thought radiation *caused* cancer."

5. "We must outlaw radioactive material to eliminate the possibility of a nation using radioactive material to harm others."

Join with a classmate and share your responses to each opinion. How are your responses similar? How are they different?

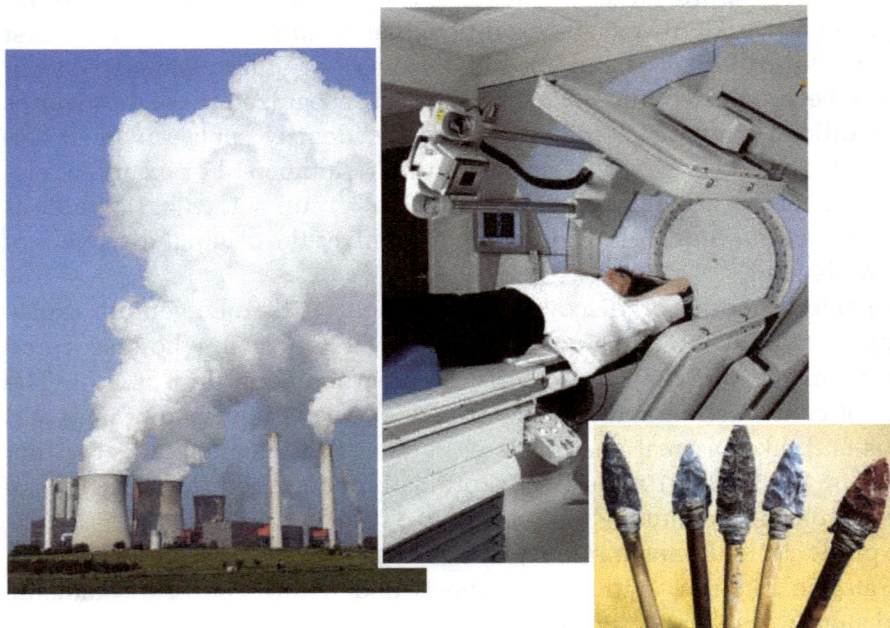

Figure 6.53 *How can each of these images suggest ways to respond to the concerns of Riverwood's citizens?*

SECTION C SUMMARY

Reviewing the Concepts

> Half-life can be defined as the time required for half of the radioactive atoms in a sample to decay.

1. How can the concept of half-life be used to determine a material's age?

2. The half-life of carbon-14 is 5730 years. Provide a rough estimate of the percent of C-14 radioisotope that would be left after

   a. 24 hours.

   b. 100 years.

3. The half-life of astatine-209 is 5.4 h. Estimate the percent of At-209 that would be left after

   a. 24 hours.

   b. 100 years.

4. Phosphorus-32 has a half-life of 14.3 d. How long would it take for 1/8 of a sample of P-32 to remain unchanged?

5. Given a sample of 4.5 mol radon-222 (see Table 6.9, page 639), how many moles of Rn-222 would remain undecayed after

   a. 3.82 d?

   b. 15.3 d?

   c. 28.0 d?

> Radioisotopes have properties that make them useful as tracers for diagnostic purposes.

6. How is metastable Tc-99m useful in medical diagnosis?

7. How is iron-59 used for diagnostic purposes?

8. Why do doctors use a radioisotope of iodine to detect thyroid problems?

9. Radioactive sodium chloride is appropriate for diagnosing circulatory problems, while radioactive xenon is helpful in searching for lung problems. Explain why each is used for its specific applications.

10. Medical personnel are selecting a radioisotope for diagnostic use. Why is each of the following considerations important in selecting the most suitable radioisotope?

   a. half-life

   b. mode of decay

   c. chemical properties of the element

> Ionizing radiation emitted by some radioisotopes can be used for medical treatment.

11. How are cancer cells different from normal cells?

12. How do physicians use ionizing radiation to treat cancer?

13. What could happen if the source of ionizing radiation used to treat cancerous growth were

   a. too weak?

   b. too strong?

14. Name three radioisotopes that radiologists can use externally to treat cancer.

The conversion (transmutation) of one element to another can be accomplished by the high-energy bombardment of atomic nuclei with subatomic particles or other nuclei.

15. Define *transmutation*.

16. Why was Rutherford's ability to complete transmutations limited compared to the Curies' transmutation abilities?

17. Why are neutrons frequently used as projectiles in transmutation reactions?

18. Copy and complete each of the following transmutation equations:

a. $\underline{\hspace{1cm}} \longrightarrow {}^{4}_{2}\text{He} + {}^{259}_{104}\text{Rf}$

b. ${}^{238}_{92}\text{U} \longrightarrow \underline{\hspace{1cm}} + {}^{234}_{90}\text{Th}$

c. ${}^{210}_{83}\text{Bi} \longrightarrow {}^{0}_{-1}\text{e} + \underline{\hspace{1cm}}$

d. ${}^{95}_{40}\text{Zr} \longrightarrow \underline{\hspace{1cm}} + {}^{95}_{41}\text{Nb}$

e. $\underline{\hspace{1cm}} \longrightarrow {}^{4}_{2}\text{He} + {}^{214}_{80}\text{Hg}$

Why and how are radio-active isotopes useful?

In this section, you explored the concept of half-life, learned about applications of radioisotopes in medicine, and considered transmutation and how it relates to the extension of the periodic table. Think about what you have learned, then answer the question in your own words in organized paragraphs. Your answer should demonstrate your understanding of the key ideas in this section.

Be sure to consider the following in your response: half-life, carbon dating, diagnostic and therapeutic uses of nuclear medicine, bombardment, and transmutation.

Connecting the Concepts

19. A student wrote this statement in a homework assignment: "After one half-life, half of the mass of a material has disappeared." Do you agree or disagree? Explain.

20. Would carbon dating help scientists determine the age of dinosaur remains? Explain.

21. Explain why an externally applied alpha emitter is an ineffective treatment for tumors deep within the body.

22. Graph the data in the following table and determine the

a. half-life of thorium-234.

b. minimum number of days required for thorium-234 to decay to 20% of its original activity.

Elapsed time (days)	% thorium-234 activity remaining
0	100
7	82
16	60
28	45
42	30
62	17
94	7

23. Scientists using carbon-14 dating generally do not use it to go back further than seven half-lives.

a. Explain why.

b. What is the maximum number of years a substance can be dated using carbon-14 dating?

24. A radiologist injects a patient's bloodstream with a radioisotope tracer sample registering 10 000 cpm (counts per minute) of radioactivity. Soon after, a technician draws 6.0 mL of blood, and the sample shows an activity of 10 cpm. What is the patient's total blood volume, assuming essentially no decrease in the activity of the radioisotope occurred during this clinical procedure?

Extending the Concepts

25. Compare PET and MRI in terms of
    a. their method of operation.
    b. data they provide for diagnosis.
    c. radiation exposure for a patient.

26. Scientists originally were uncertain whether the oxygen gas produced during photosynthesis came from CO_2, H_2O, or both. How could radioisotope tracers be used to help settle that uncertainty?

27. A newly discovered element with an extremely short half-life is detected by analyzing its decay products. Explain how scientists can "work backward" to identify the original element.

28. Many gemstones are irradiated. Research and report on the reasons for this process.

29. Gold and other precious metals can be created from other metals by nuclear transformations.
    a. So why don't commercial firms do this as a source of profit?
    b. Is the synthetic precious metal distinguishable from the naturally occurring metal? Explain.

30. Research how the proton–neutron ratio is related to the stability of an atomic nucleus. Describe how you could use this ratio to predict the type of radioactive decay a particular atomic nucleus undergoes.

31. List foods that are currently irradiated with nuclear radiation before they are marketed for human consumption. Describe the process and evaluate its risks and benefits.

SECTION D

NUCLEAR ENERGY: BENEFITS AND BURDENS

What burdens and benefits accompany uses of nuclear energy?

In the 1930s, a bombardment reaction involving uranium unlocked a new energy source and led to development of both nuclear power and nuclear weapons. This event marked the start of the nuclear age. How did scientists first unleash the enormous energy of the atom, and how have nuclear engineers harnessed atomic energy for both useful and destructive purposes?

GOALS

- Define and describe nuclear fission.
- Write and balance nuclear equations for fission reactions.
- Describe how nuclear energy is used to generate electricity in a nuclear power plant.
- Define and describe nuclear fusion.
- List and describe types of radioactive waste, including their sources.
- Describe and evaluate methods for disposal of radioactive wastes, including burdens and benefits of each.

concept check 7

1. List at least three beneficial uses of nuclear technology or nuclear energy.
2. Describe how synthetic elements are created.
3. What is nuclear fission?

D.1 UNLEASHING NUCLEAR FORCES

Just before the start of World War II, German scientists Otto Hahn and Fritz Strassmann bombarded uranium with neutrons in the hope of creating a more massive nucleus and, thus, a new element. Much to their surprise, they found that one reaction product was atoms of barium, with only about half the atomic mass of the original target uranium atoms.

The first to understand what had happened was the Austrian physicist Lise Meitner (Figure 6.54), then living in Sweden, who had previously worked with Strassmann and Hahn. Meitner and her nephew Otto Frisch suggested that neutron bombardment had split the uranium atom into two parts of nearly equal mass. Other scientists quickly verified Meitner's explanation.

Hahn and Strassmann had actually triggered an array of related reactions. One of the reactions that produced barium is:

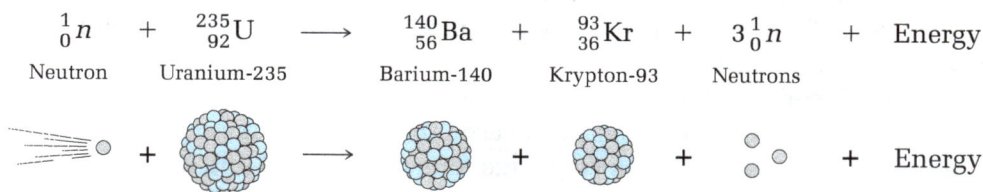

$$\underset{\text{Neutron}}{^{1}_{0}n} \;+\; \underset{\text{Uranium-235}}{^{235}_{92}U} \;\longrightarrow\; \underset{\text{Barium-140}}{^{140}_{56}Ba} \;+\; \underset{\text{Krypton-93}}{^{93}_{36}Kr} \;+\; \underset{\text{Neutrons}}{3\,^{1}_{0}n} \;+\; \text{Energy}$$

Splitting an atom into two smaller atoms is called **nuclear fission**. Scientists soon found that the uranium-235 nucleus can fission (split) into numerous pairs of smaller nuclei. The uranium usually did not split into two equal halves but into one element accounting for about 60% of uranium's mass (such as barium) and another element equivalent to about 40% of uranium's mass (such as krypton). Here is another example of a nuclear fission reaction involving uranium-235:

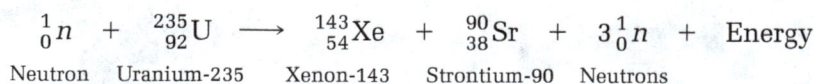

$$\underset{\text{Neutron}}{^{1}_{0}n} \;+\; \underset{\text{Uranium-235}}{^{235}_{92}U} \;\longrightarrow\; \underset{\text{Xenon-143}}{^{143}_{54}Xe} \;+\; \underset{\text{Strontium-90}}{^{90}_{38}Sr} \;+\; \underset{\text{Neutrons}}{3\,^{1}_{0}n} \;+\; \text{Energy}$$

The nuclear fission of heavy atoms such as uranium releases a huge quantity of energy. Gram for gram, the released energy is at least a million times more than the energy of any chemical reaction. This is what makes nuclear explosions so devastating and nuclear energy so powerful.

Why does a nuclear reaction release much more energy than a chemical reaction? Recall what you know about chemical reactions, such as burning petroleum. Chemical reactions involve breaking chemical bonds in reactants and making new chemical bonds in products. When bonds are stronger in products than in reactants, energy is released, often as thermal energy. Thus, chemical energy is converted into thermal energy. There is no overall energy loss or gain. Similarly, mass is conserved in a chemical reaction. The nucleus of each atom, and thus its identity, remains intact in all chemical reactions. As a result, the number of atoms of each element remains unchanged. The atoms simply become rearranged. Balanced chemical equations illustrate this conservation of atoms and mass.

Figure 6.54 *Lise Meitner was first to suggest that nuclei might split due to neutron bombardment.*

Lise Meitner fled to Sweden when Nazis assumed control of Austria and revoked her citizenship.

Fission of U-235 produces many other pairs of nuclei, such as Te-137 and Zr-97.

Not all nuclei are fissionable. U-235 is the only naturally occurring isotope that undergoes fission with lower-energy (thermal) neutrons. However, many synthetic nuclei (such as U-233, Pu-239, and Cf-252) also fission under neutron bombardment.

Nuclear reactions are also based on conserving energy and mass. However, during nuclear fission, very small quantities of mass are converted into measurable quantities of energy. Where does this energy originate?

The origin of nuclear energy lies in the force that holds protons and neutrons together in the nucleus. This force, called the **strong force**, is fundamentally different from the electrical forces that hold atoms and ions together in chemical bonds. It is also a thousand times stronger than the forces that result in chemical bonds. The strong force operates over very short distances, extending only across an atom's nucleus.

The forces holding nuclear particles together in the two atomic nuclei produced during U-235 fission are stronger than those in the nucleus of the uranium atom that was split. A small loss of mass results from forming two new nuclei and is converted into large quantities of released energy.

How much mass and energy are involved? The mass loss is very small, often less than 0.1% of the total mass of the fissioning atom. Even so, the conversion of these small quantities of mass into energy accounts for the vast power of nuclear reactions.

Albert Einstein's famous equation relates mass and energy: $E = mc^2$. This equation indicates that the energy released (E) equals the mass lost (m) multiplied by the speed of light (a very large number) squared (c^2). If one gram of matter were fully converted to energy, the energy released would equal that produced by burning 700 000 gallons of high-octane gasoline!

Such nuclear energy release has been harnessed by engineers to generate electricity (see Figure 6.55) and to create atomic weapons. However, the fission of one nucleus does not produce enough energy for practical use. How are fission reactions sustained to involve much larger quantities of nuclei?

> If one kilogram of U-235 fissions, a mass of about one gram is converted into energy.

> One gram of mass loss (1×10^{-3} kg) times the speed of light (3×10^8 m/s) squared equals 9×10^{13} J of energy.

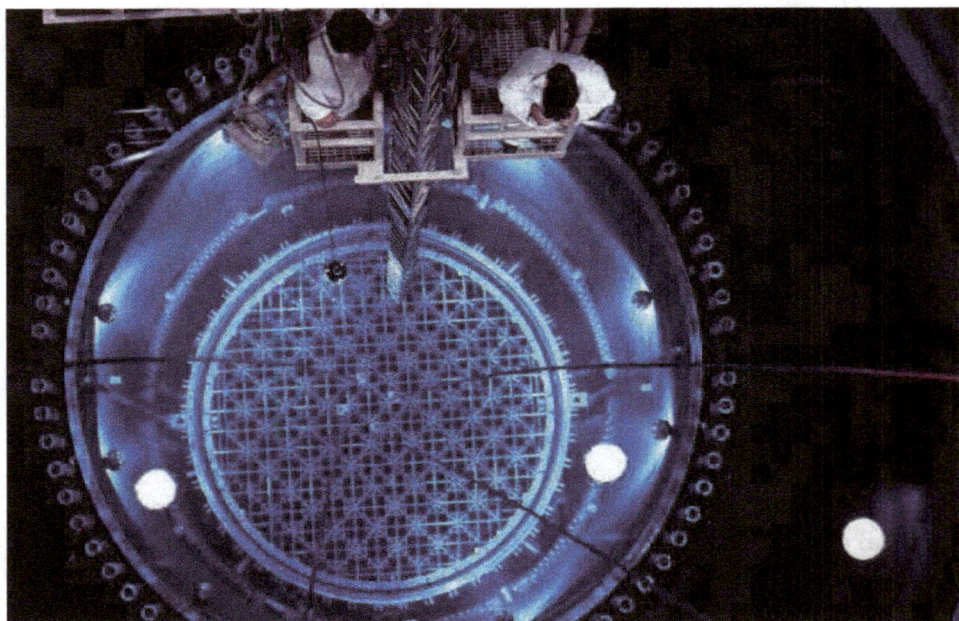

Figure 6.55 *The core of a fission reactor, based on a nuclear chain reaction, emits visible light due to ionizing radiation released.*

Note from the equations on page 659 that another product of nuclear fission is the release of neutrons. These emitted neutrons can sustain the fission reaction by serving as reactants to split additional fissionable nuclei, which produce additional neutrons, which can split additional fissionable nuclei, and so on. The result is a **chain reaction** (see Figure 6.56).

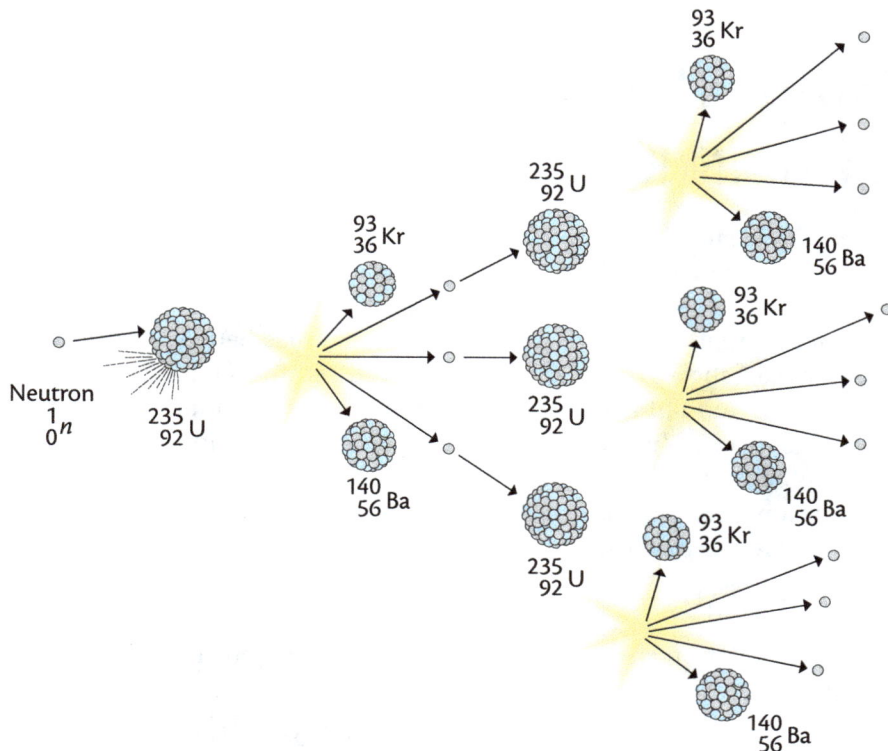

Figure 6.56 *A nuclear chain reaction. A neutron colliding with a uranium-235 nucleus initiates the reaction (left). The reaction continues and grows, as emitted neutrons encounter and split the nuclei of other fissionable atoms.*

Recall, however, that most of an atom is empty space. The probability that a neutron from a fission reaction will hit and split another fissionable nucleus depends on how much fissionable material is available. Unless a certain **critical mass** (minimum quantity) of fissionable material is present, the neutrons are not likely to encounter enough fissionable nuclei to sustain the reaction. However, if a critical mass of fissionable material is present, a chain reaction can occur, as depicted in Figure 6.56.

Shortly after the first fission reactions were explained in 1939, scientists recognized that they could employ large-scale nuclear reactions in military weapons. Germany and the United States soon initiated projects to build atomic bombs during World War II. In 1945, U.S. planes dropped two such bombs on Hiroshima and Nagasaki in Japan, which led quickly to the end of the war.

More recently, nuclear engineers have used the energy produced by nuclear fission chain reactions to generate electricity. They carefully monitor and control the rate of fission for such uses. Nuclear power plants harness the enormous energy produced by nuclear fission reactions, while also minimizing the risks of an uncontrolled chain reaction. You will soon learn more about those design features.

MODELING MATTER

D.2 CHAIN REACTIONS

Chain reactions sustain nuclear fission reactions in applications such as electrical power generation and atomic weapons. In this activity, dominoes will model some aspects of a chain reaction.

Each domino that falls represents a nucleus that has been split during fission. Figure 6.57 shows one way that you could set up the dominoes so that making one domino fall causes all other dominoes also to fall. Because you will model a specific fission reaction, your models will not match the one depicted in Figure 6.57.

The uranium-235 nucleus can fission into more than 100 different pairs of atoms. One way, for example, produces tellurium-137 and zirconium-97:

$$\ _{0}^{1}n \ + \ _{92}^{235}U \ \longrightarrow \ _{52}^{137}Te \ + \ _{40}^{97}Zr \ + \ 2_{0}^{1}n \ + \ Energy$$

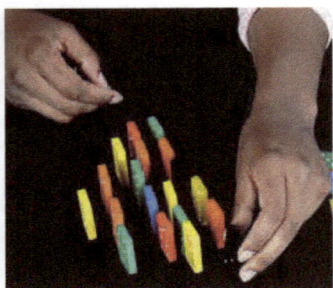

Figure 6.57 *Domino behavior can model key characteristics of a chain reaction.*

1. As this equation shows, splitting one U-235 nucleus releases two neutrons.

   a. Set up all the dominoes you receive from your teacher so that each falling domino will make two more erect dominoes fall.

   b. Sketch your setup.

   c. Push over the first domino and record what happens.

   d. Explain how this models the release of neutrons during the fission of U-235 as in the equation above.

   e. What aspects of the U-235 fission reaction are not modeled by the behavior of your domino setup?

2. Now you will assemble a critical mass of atoms (dominoes) to initiate a chain reaction.

   a. Set up the dominoes as in Step 1a.

   b. Assume that only dominoes (atoms) with seven total dots are fissionable.

   c. Remove all dominoes from your setup that do not have seven dots.

   d. Sketch your new setup.

   e. Push over the first domino and record what happens.

   f. In what way does this model help clarify the idea of critical mass?

   g. How can you ensure that you have a critical mass of "fissionable" dominoes in a particular setup?

3. Suppose you need to control the total neutrons emitted so that fission is just sustained.

   a. Set up the dominoes as in Step 1a.

   b. Devise a plan so that only half the dominoes will fall when you push over one domino. You should not remove any of the dominoes that are already set up.

   c. Describe and sketch your strategy.

   d. Try your plan and record what happens.

   e. What stopped the dominoes from falling?

   f. Use a domino model to describe how to control fission chain reactions.

4. Compare the domino arrangements in Question 1 and 3.

   a. Which is a better model of an atomic-weapon explosion? Explain.

   b. Which is a better model of fission in a nuclear power plant? Explain.

5. Propose another way to model a nuclear chain reaction.

   a. Explain how your model illustrates features of a nuclear chain reaction.

   b. Explain some limitations of your model.

D.3 NUCLEAR POWER PLANTS

The first nuclear reactors were designed and built during World War II. Since then, commercial companies have built many nuclear reactors to generate electricity, such as the nuclear power plant shown in Figure 6.58. In 2010, slightly more than 100 commercial nuclear reactors were generating electricity in the United States. Figure 6.59 (page 664) shows where these reactors are located. Globally, an estimated 438 nuclear reactors in 29 nations produce about 14% of the world's electricity.

Figure 6.58 *A nuclear power plant. Notice the reactor containment building with its domed roof.*

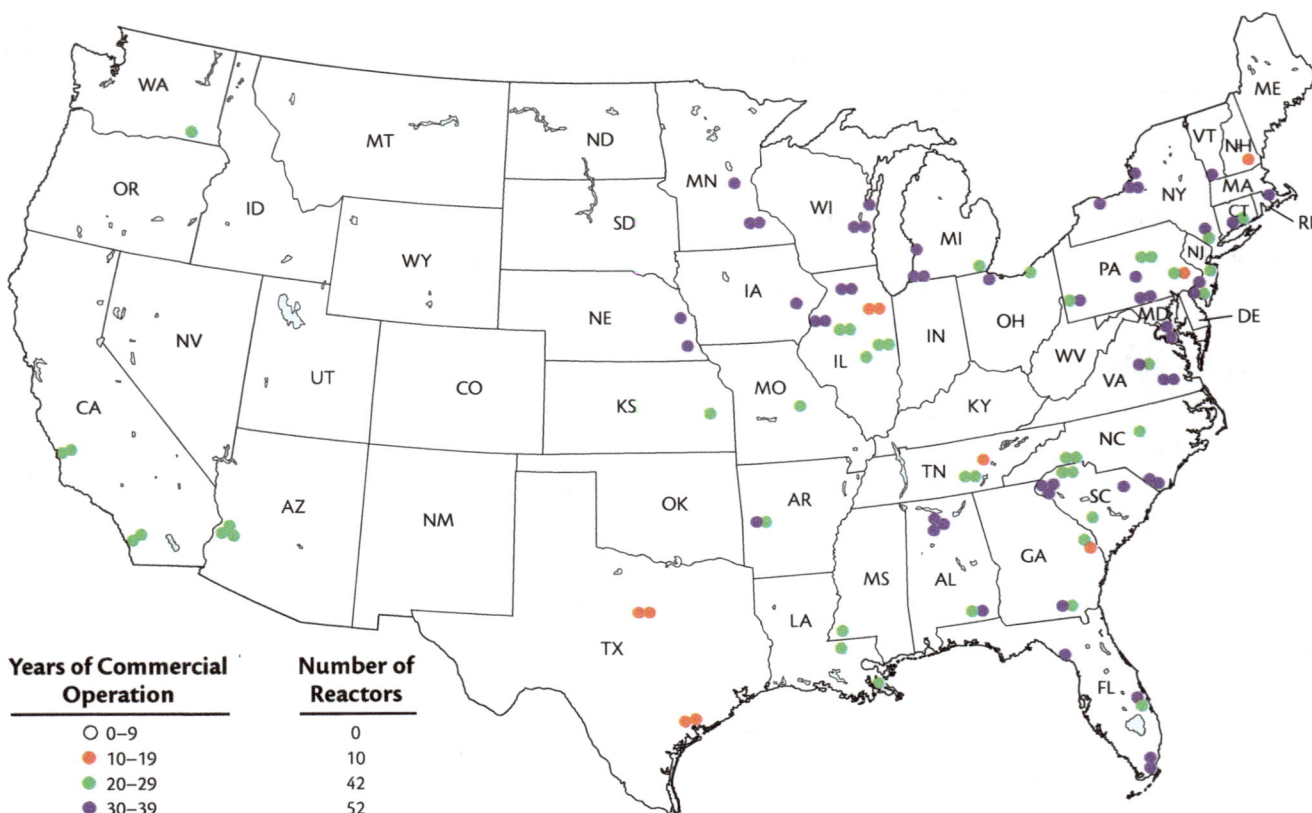

Figure 6.59 *Locations (dots) of all licensed U.S. nuclear power plants in 2010.*

Most conventional power plants generate electricity by burning fossil fuels to boil water and produce steam. A nuclear power plant operates in much the same way. However, instead of using fossil-fuel combustion to boil water, **nuclear power plants** use the thermal energy released from nuclear-fission reactions to heat water and produce steam. The steam spins turbines of giant generators, producing electricity.

The essential parts of a nuclear power plant, diagrammed in Figure 6.60, include the fuel rods, control rods, moderator, generator, and cooling system.

Fuel Rods

Coal-fired power plants burn thousands of tons of coal daily. By contrast, nuclear reactor fuel occupies a fraction of the volume needed for coal and is replenished only about once a year. Nuclear reactor fuel is small uranium dioxide (UO_2) pellets about the size and shape of short pieces of chalk. The energy in one uranium fuel pellet equals the energy in one ton of coal or 126 gal of gasoline. One nuclear power plant uses as many as 10 million fuel pellets. These pellets are arranged inside long, narrow steel cylinders—the fuel rods. The fission chain reaction occurs inside these rods (see Figure 6.60).

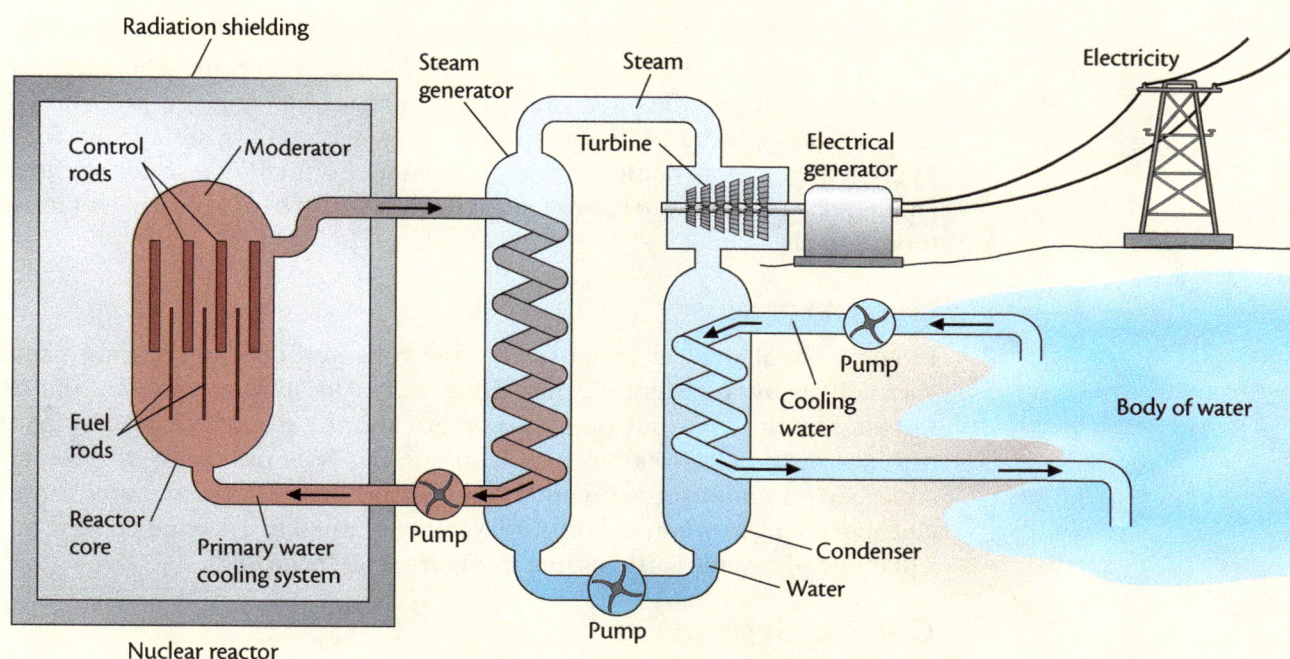

Figure 6.60 *Components of a nuclear power plant. Note the water flowing between the reactor and steam generator (red) and between the steam generator and outside body of water (blue). This water flow is important in controlling the power-plant operating temperature.*

Most uranium dioxide in fuel pellets is composed of the nonfissionable uranium-238 isotope. Only 0.7% of natural uranium is U-235, the fissionable isotope. In reactor fuel rods, U-235 composition has been enriched to about 3%, which is only a small fraction of total fuel-rod material. It is sufficient to sustain a chain reaction, but far less than enough to cause a nuclear explosion.

> Weapons-grade uranium usually contains 90% or more of the fissionable U-235 isotope.

Control Rods

The nuclei of some elements, such as boron or cadmium, can absorb neutrons very efficiently. Such materials are placed in control rods, which regulate the number of neutrons available for causing fission. Moving the control rods up or down (see Figure 6.61) between the fuel rods regulates the rate of the nuclear chain reaction. The reaction can be completely stopped by dropping the control rods all the way down between the fuel rods and absorbing nearly all neutrons released as U-235 fissions.

Figure 6.61 *A nuclear reactor core.*

Moderator

In addition to fuel rods and control rods, the core of a nuclear reactor contains a moderator, which slows down high-speed neutrons. This allows the fuel rods to more efficiently absorb neutrons, enhancing the probability of fission. Heavy water (with the formula D_2O, where D is the symbol for the hydrogen-2 isotope, deuterium), regular water (termed *light water* by nuclear engineers), and graphite (carbon) are the three most commonly used moderator materials.

Generator

In commercial nuclear reactors, the fuel rods and control rods are usually surrounded by a system of circulating water. In simpler reactors, the heat released by the fuel rods boils this water, and the resulting steam spins the turbines of the electrical generator. In another type of reactor, this water is superheated under pressure and does not boil. Instead, it circulates through a heat exchanger, where it boils the water contained in a second cooling loop. This type of reactor is illustrated in Figure 6.60 (page 665).

Cooling System

Steam moves past the turbines and travels through pipes where it is cooled by water drawn from a nearby ocean, lake, or river. The cooled steam condenses to liquid water and circulates inside the generator. So much thermal energy is generated that some steam must also be released into the air. The largest and most prominent features of most nuclear power plants are tall, gracefully curved concrete cylinders, called cooling towers, where excess thermal energy is released. Some observers mistakenly assume that the cooling tower is the nuclear reactor.

A nuclear reactor is designed to prevent the full escape of radioactive material if a malfunction causes the release of radioactive material, including cooling water, within the reactor itself. The core of a nuclear reactor is surrounded by concrete walls two to four meters thick. Further protection is provided by enclosing the reactor in a building with thick walls of steel-reinforced concrete designed to withstand a chemical explosion or an earthquake. The reactor building is also capped by a domed roof that can withstand high internal pressure.

The well-known nuclear accident in 1986 at one of the four reactors at Ukraine's Chernobyl power plant occurred because too many control rods were withdrawn from the reactor and were not replaced fast enough. There was little control of the fission process, resulting in the buildup of much steam. The resulting explosion was not nuclear but was due to the buildup of high-temperature steam and to the chemical reactions that it triggered. Unfortunately, the plant had been built without a surrounding concrete containment building, unlike currently operating U.S. reactors. A large quantity of nuclear material from the reactor was also released directly into the environment.

Nuclear fission is not the only way to liberate nuclear energy. Soon you will learn about the kind of nuclear reaction that fuels the stars (and, indirectly, fuels all living matter).

The white plumes seen rising from generating-plant cooling towers are condensed steam, not smoke.

In the Chernobyl accident, one chemical reaction that caused trouble was high-temperature steam reacting with carbon from the moderator, producing CO and H_2. When H_2 mixed with air, it became explosive.

D.4 NUCLEAR FUSION

In addition to releasing energy by splitting massive nuclei (fission), large quantities of nuclear energy can be generated by fusing, or combining, small nuclei. **Nuclear fusion** involves forcing two relatively small nuclei to combine into a new, more massive nucleus. As with fission, the energy released by nuclear fusion can be enormous, again due to the conversion of mass into energy. Gram for gram, nuclear fusion liberates even more energy than nuclear fission—between 3 and 10 times more energy (Figure 6.62).

Scientific American Conceptual Illustration

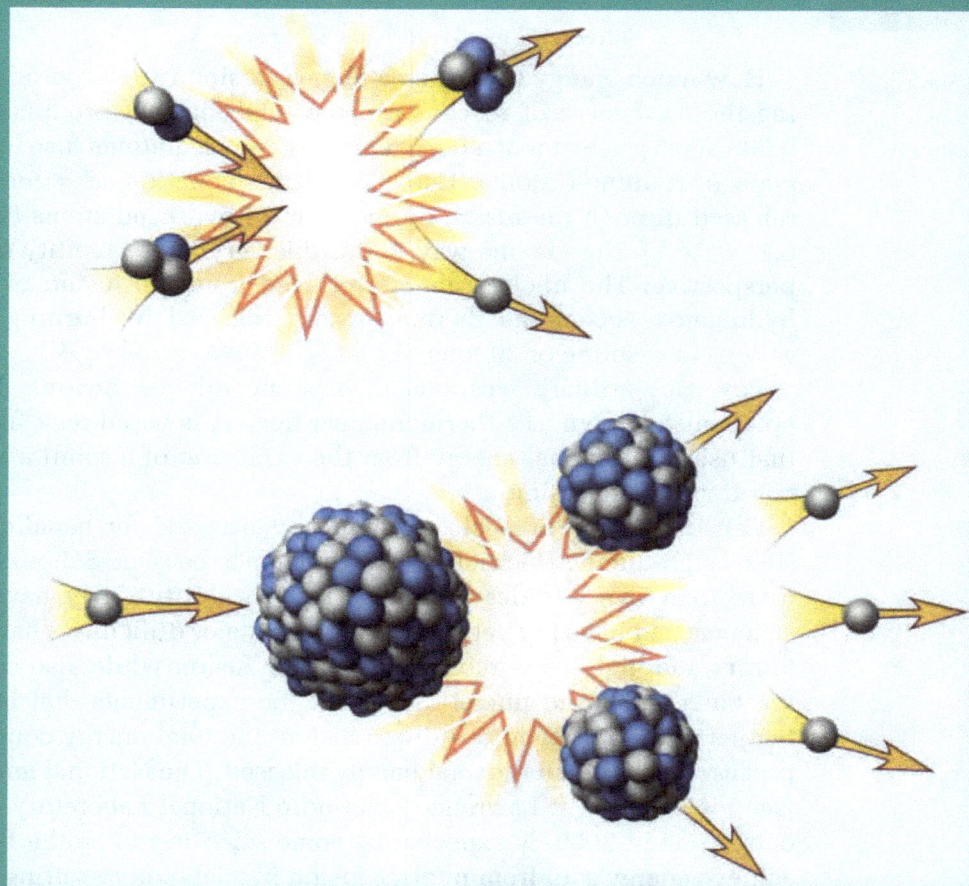

Figure 6.62 *Nuclear fission and fusion. Fusion (top) occurs when smaller nuclei (protons in blue, neutrons in gray) combine to form larger nuclei. Fission (bottom) occurs when a neutron of appropriate energy collides with particular large nuclei, creating two smaller nuclei. Both processes can release large quantities of energy, according to Einstein's famous equation, $E = mc^2$, where m is the mass converted into energy (E) and c is the speed of light. However, the total energy released decreases as the nuclei involved approach iron-56, the most stable nucleus.*

Nuclear fusion powers the Sun and other stars (Figure 6.63). Scientists believe that the Sun formed when a huge quantity of interstellar gas, mostly hydrogen, condensed under the force of gravity. As the volume of gas decreased, its temperature increased to about 15 million °C, and hydrogen atoms began fusing into helium. The nuclei that fused together were all positively charged and tended to repel one another. The high temperature gave each nucleus considerable kinetic energy, which helped overcome the repulsions.

Once fusion was started, the Sun began to shine, converting nuclear energy into radiant energy. Scientists estimate that the Sun, believed to be about 4.5 billion years old, is about halfway through its life.

The nuclear-fusion reactions occurring in the Sun are rather complicated, but the result is the conversion of hydrogen nuclei into helium nuclei. The overall result can be summarized by this equation:

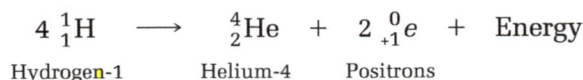

$$4\ {}_{1}^{1}\text{H} \longrightarrow {}_{2}^{4}\text{He} + 2\ {}_{+1}^{0}e + \text{Energy}$$

Hydrogen-1   Helium-4   Positrons

Figure 6.63 *The Sun's energy is produced by continuous nuclear-fusion reactions.*

How much energy does such a nuclear fusion reaction produce? Comparing the total mass of reactants to the total mass of products reveals that 0.006 900 5 g is lost when one gram of hydrogen-1 atoms fuse to produce one gram of helium-4 atoms. Using Einstein's equation, $E = mc^2$, the energy released through the fusion of one gram of hydrogen atoms (one mol H) is 6.2×10^8 kJ. Here is one way to put this very large quantity of energy into perspective: The nuclear energy released from the fusion of one gram of hydrogen-1 equals the thermal energy released by burning nearly 5000 gallons of gasoline or 20 tons of coal.

Powerful military weapons incorporate nuclear fusion. The hydrogen bomb, also known as a thermonuclear device, is based on a fusion reaction that uses the thermal energy from the explosion of a small atomic (fission) bomb to initiate fusion.

Can the energy of nuclear fusion be harnessed for beneficial purposes, such as producing electricity? This remains to be seen. Scientists have spent more than five decades pursuing this possibility. They have tried many schemes, but have not yet succeeded. The major difficulties have been maintaining the high temperatures needed for fusion while also containing the reactants and fused nuclei. So far, in the experiments that have achieved temperatures sufficient to initiate fusion, the total energy consumed by the process is more than the total energy released. The National Ignition Facility (see Figure 6.64) at Lawrence Livermore National Laboratory, which began operations in 2009, is expected by some scientists to be the first facility to achieve energy gain from nuclear fusion in a laboratory setting.

If scientists finally succeed in controlling nuclear fusion in the laboratory, there is still no guarantee that fusion reactions will become a practical source of energy. Low-mass isotopes needed to fuel such reactors are plentiful and inexpensive, but confinement of the reaction could be very costly. Further, although the fusion reaction itself produces less radioactive waste than

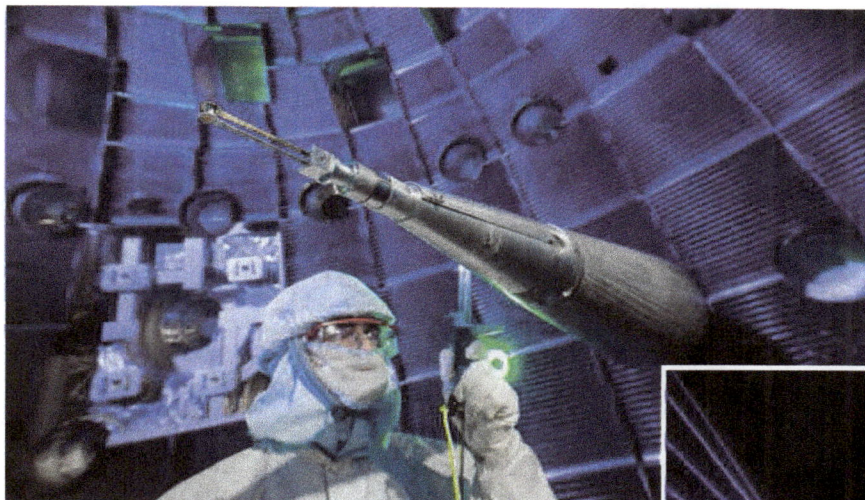

Figure 6.64 *Scientists at the National Ignition Facility (NIF) aim to initiate nuclear fusion in the laboratory. This artist's rendering (right) shows a NIF target pellet (the white ball) inside a hollow capsule. Laser beams enter through openings on each end and compress and heat the target to conditions necessary for nuclear fusion. The target must be precisely centered inside the chamber before each experiment. A NIF technician (above) checks the target positioner, which serves as a reference to align the laser beams.*

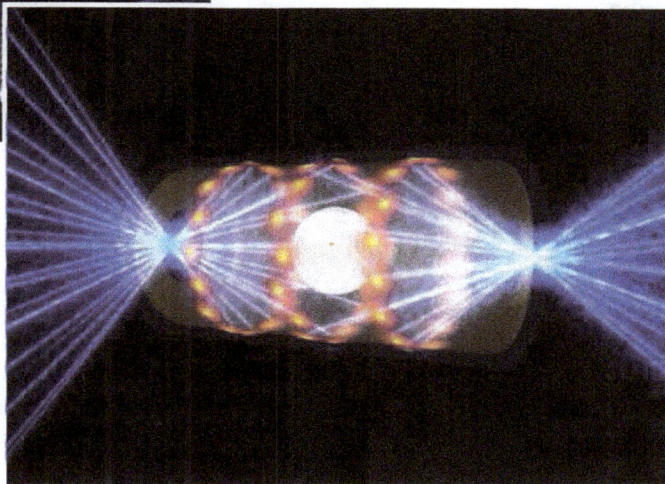

nuclear fission, capturing the positrons and shielding the heat of the reaction could generate nearly as much radioactive waste as that produced now by fission-based power plants.

In splitting and fusing atoms, the nuclear energy that fuels the universe has been unleashed. Much good has arisen from it, but so have scientific, social, and ethical questions. Along with great benefits come great risks. How much risk is worth any potential benefits? In the next activity, you will analyze risks and benefits associated with deciding how to travel to see a friend.

Electing not to make a decision is, in fact, also a decision—one with its own risks and benefits.

concept check 8

1. Consider the nuclear processes you have just studied.
   a. What is fission?
   b. What is fusion?
   c. How are they different?
   d. How are they similar?
2. What are some drawbacks to the use of nuclear energy?

MAKING DECISIONS

D.5 THE SAFEST JOURNEY

Suppose you want to visit a friend who lives 500 miles (800 km) away, using your safest means of transportation. Insurance companies publish reliable statistics on the safety of different methods of travel. Using Table 6.11, answer these questions:

Table 6.11

Risk of Travel	
Mode of Travel	**Distance (Miles) Traveled at Which One Person in a Million Will Suffer Accidental Death**
Bicycle	10
Automobile	100
Train	120
Bus	500
Scheduled Airline	1900

1. Assume there is a direct relationship between distance traveled and chance of accidental death (i.e., assume that doubling the distance doubles your risk of accidental death.)

    a. What is the risk factor (chance of accidental death) for traveling 500 miles by each mode of travel listed in the table? For example, Table 6.11 shows that the risk factor for biking increases by 0.000 001 for each 10 miles. Therefore, the bike-riding risk factor in visiting your friend would be

    $$\frac{500 \text{ miles} \times 0.000\ 001 \text{ risk factor}}{10 \text{ miles}} = 0.000\ 05 \text{ risk factor}$$

    b. What is the safest mode of travel (the one with the smallest risk factor)?

    c. Which is the riskiest mode of travel (the one with the largest risk factor)?

    d. Did your results surprise you? Explain.

2. The risk of accidental death is not the only factor to consider when choosing a mode of travel.

    a. List some benefits associated with each mode of transportation.

    b. List some risks associated with each mode of transportation.

    c. In your view, do benefits of riskier ways to travel outweigh their increased risks? Explain your reasoning.

    d. Identify some situations in which assumptions made in Question 1 would be invalid.

3. Do you think the statistics in Table 6.11 will apply 25 years from now? Explain.

4. What factor(s), beyond the risk to personal safety, would you include in a risk–benefit analysis before you decided how to travel?

5. Now use your results and reasoning to draw conclusions.

   a. Which mode of travel would you choose? Why?

   b. Would another person's risk–benefit analysis always lead to your decision? Why or why not?

CHEM**QUANDARY**

RISK-FREE TRAVEL?

Is there any way to travel to visit your friends or relatives that would be completely risk-free? Would it actually be safer not to visit them at all? Why?

D.6 RADIOACTIVE WASTE

Imagine that you live in a home that was once clean and comfortable, but now you have a major problem: You cannot throw away your garbage. The city forbids garbage removal because it has not decided what to do with the garbage. For about 50 years, your family has compacted, wrapped, and saved the garbage as efficiently as possible, but you are running out of room. Some bundles leak and are a health hazard. What can be done?

The U.S. nuclear power industry, the nuclear weapons industry, and medical and research facilities have a similar problem. Spent (used) nuclear fuel and radioactive waste products have been accumulating for about 50 years (see Figure 6.65). Some of these materials are still highly radioactive, while other materials—even initially—have low levels of radioactivity. It is uncertain where these materials, regardless of radioactivity levels, will be permanently stored or how soon.

Figure 6.65 *Low-level radioactive waste does not require the same disposal methods as spent fuel rods from nuclear reactors do. Yet improper disposal of such wastes poses serious hazards.*

Radioactive Wastes

Two broad categories of radioactive waste are high-level and low-level. **High-level radioactive wastes** are either (a) products of nuclear fission, such as those generated in a nuclear reactor or (b) transuranics, products formed when the original uranium-235 fuel absorbs neutrons. For example, plutonium-239 is a transuranic material. **Low-level radioactive wastes** have much lower levels of radioactivity. These wastes include used nuclear laboratory protective clothing, diagnostic radioisotopes, and air filters from nuclear power plants. Figure 6.66 illustrates the composition of radioactive wastes in the United States.

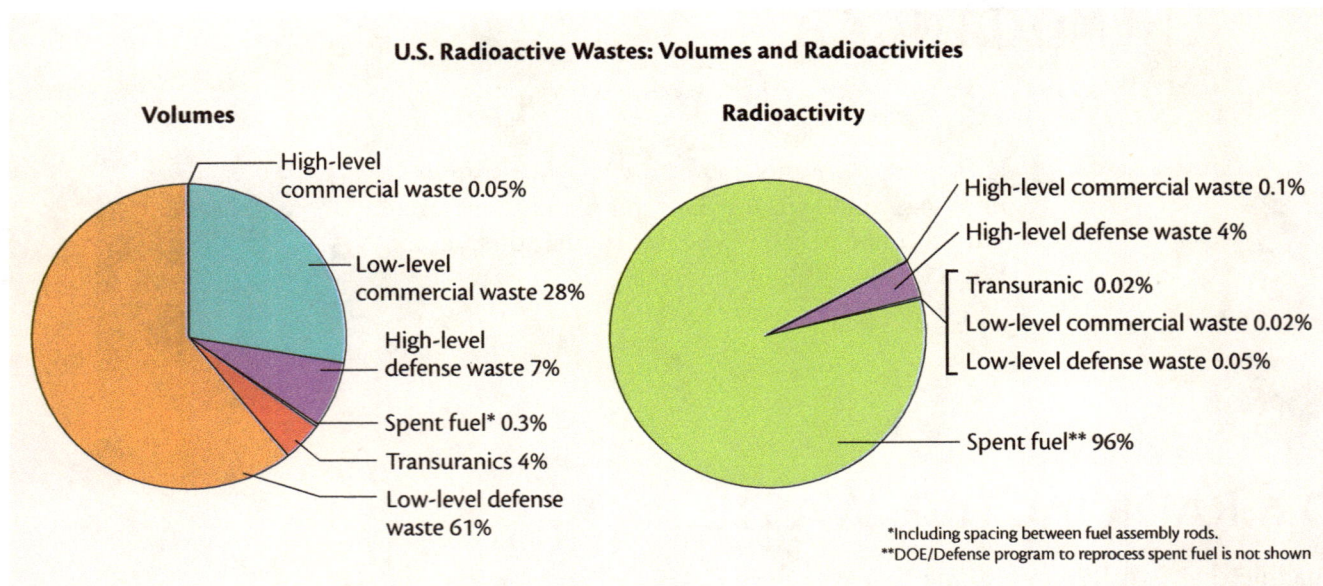

U.S. Radioactive Wastes: Volumes and Radioactivities

Volumes

- High-level commercial waste 0.05%
- Low-level commercial waste 28%
- High-level defense waste 7%
- Spent fuel* 0.3%
- Transuranics 4%
- Low-level defense waste 61%

Radioactivity

- High-level commercial waste 0.1%
- High-level defense waste 4%
- Transuranic 0.02%
- Low-level commercial waste 0.02%
- Low-level defense waste 0.05%
- Spent fuel** 96%

*Including spacing between fuel assembly rods.
**DOE/Defense program to reprocess spent fuel is not shown

Figure 6.66 *Volume and radioactivity of U.S. radioactive wastes.*

Figure 6.66 emphasizes distinguishing waste volume from radioactivity level when considering radioactive wastes. High volume does not necessarily mean high radioactivity. For example, Figure 6.66 shows that military defense efforts produce the largest volume of radioactive wastes (61%), but this waste contributes slightly more than 4% of the total radioactivity in U.S. radioactive wastes. On the other hand, spent fuel from nuclear reactors occupies less than 1% of the volume of all radioactive wastes, yet it contributes 96% of the total radioactivity. Moreover, high-level radioactive wastes also can have extended half-lives, some as long as thousands of years.

Current Disposal Methods

Because high-level and low-level radioactive wastes have different charateristics, people dispose of them differently. We can put low-level wastes into sealed containers and bury the containers in lined trenches 20 feet deep.

Low-level military radioactive waste is disposed of at federal sites maintained by the U.S. Department of Energy (DOE). Since 1993, each state is responsible for disposal of its own commercial low-level radioactive wastes. Groups of states in several regions have formed compacts in which all members use a disposal site in one state for low-level wastes from all compact members.

High-level radioactive waste disposal requires a very different approach. Spent nuclear fuel presents the greatest challenge. A commercial nuclear reactor (power plant) typically produces about 30 tons of spent fuel annually. That means that the commercial nuclear power plants now operating in the United States generate roughly 3100 tons of waste annually. Approximately 60 000 tons of spent fuel are already stored in 34 states. Eventually, of course, the components of nuclear power-plant reactors will also become radioactive waste.

The U.S. nuclear weapons program contributed over 300 million liters of additional stored waste. The volume of military radioactive waste is much greater than the volume of commercial radioactive waste. This waste is in the form of a sludge—the waste product of extracting plutonium from spent fuel rods in military reactors. (Nuclear weapons are created from the plutonium.) As radioisotopes in the waste gradually decay, they emit radiation and thermal energy. In fact, without external cooling, such waste can become hot enough to boil. This makes military waste containment a continual challenge.

Operators must replace up to one-third of a nuclear reactor's fuel rods annually. This is necessary because the uranium-235 fuel becomes depleted as fissions occur and because accumulated fission products interfere with the fission process. The spent fuel rods are highly radioactive, with some isotopes continuing to decay for many thousands of years. Table 6.12 lists the half-lives of three radioisotopes produced by nuclear fission. All of these and many more are contained in spent fuel rods.

By federal law, nuclear reactor waste must be stored on-site, usually in waste storage tanks, until a permanent repository is created. Available storage space on-site is limited, however. The federal government is responsible for the final storage of high-level radioactive waste, but it has not yet opened permanent long-term disposal sites for high-level wastes. A site at Yucca Mountain, Nevada, was selected for this purpose and some construction to prepare for storage of high-level radioactive wastes was completed (see Figure 6.67). However, ongoing legal, political, and regulatory debate has made the future of the Yucca Mountain site uncertain.

Table 6.12

Some Radioisotopes Found in Spent Fuel Rods

Isotope	Half-Life
Plutonium-239	24 110 y
Strontium-90	28.8 y
Barium-140	12.8 y

Figure 6.67 *Yucca Mountain, Nevada, was approved by the U.S. Congress as the site of a national repository for spent nuclear fuel and high-level radioactive waste. Ongoing controversy makes the future of this repository uncertain.*

Investigating Long-Term Disposal Methods

The method of long-term radioactive waste disposal favored by the U.S. government (and by many other nations) is mined geologic disposal. The radioactive waste would be buried at least one kilometer below Earth's surface and at least one kilometer from any water table in vaults that would presumably remain undisturbed.

To prepare radioactive waste for burial, spent fuel rods would first be allowed to cool for several decades in very large tanks of water. Over time, many of the radioisotopes would decay, lowering the radioactivity level to a point where the materials could be transferred to storage containers. These containers are currently stored on-site at many nuclear reactor facilities for transportation to a long-term disposal site when it becomes available.

Another option is to transform cooled radioactive wastes into glassy solids by application of heat, a process known as **vitrification** (see Figure 6.68). Although the encased material would still be highly radioactive, the waste would be much less likely to leak or leach into the environment because of the glasslike envelope encasing it. Technicians could seal the vitrified radioactive wastes in containers made of glass, stainless steel, or concrete. The Savannah River Site vitrification plant in South Carolina is the world's largest facility of its kind. France has used commercial vitrification for more than two decades (nuclear reactors produce about 78% of France's electrical energy).

Nowhere in the world, however, has radioactive waste been permanently buried. The challenge is to find technically, politically, and socially acceptable sites. The Japanese government has even considered deep-ocean burial.

Some geologic sites formerly assumed to be stable enough for radioactive waste disposal were later discovered to be unsafe. For example, technicians buried some plutonium at Maxey Flats, Kentucky, in a rocky formation that geologists believed would remain stable for thousands of years. Within a decade, however, some buried plutonium had moved dozens of meters away.

What are current plans to resolve our long-term radioactive waste disposal problems? By U.S. law, Congress selected two sites for permanent radioactive waste disposal from options provided by the DOE. These sites, located in regions of presumed geologic stability (see Figure 6.69), were the Waste Isolation Pilot Plant (WIPP) near Carlsbad, New Mexico, and Yucca Mountain, Nevada, an extinct volcanic ridge 100 miles northwest of Las Vegas.

Each nuclear reactor site already uses this cooling process for wastes.

Figure 6.68 *One strategy to prevent radioactive waste from entering water supplies is to vitrify the waste—that is, convert it into a glassy solid.*

The Hanford Vitrification Plant, which will be the world's largest when construction is completed in 2019, is designed to treat the 53 million gallons of radioactive and chemical waste stored near its site in eastern Washington State.

Figure 6.69 *Storing radioactive waste requires an underground site that will minimize contact with local water resources and remain geologically stable over many centuries. Siting poses great challenges. The waste will remain dangerously radioactive for thousands of years.*

Both proposals generated considerable debate and controversy about site locations and the means of transporting radioactive wastes to them. Although WIPP has been operating since 1999, the Yucca Mountain Repository is no longer being funded as of 2011, leaving future disposal options for U.S. radioactive waste uncertain.

DEVELOPING SKILLS

D.7 DISPOSING OF HIGH- AND LOW-LEVEL WASTES

Use Figure 6.66 (page 672) to answer the following questions:

1. Approximately what percent of radioactive waste is
   a. low-level waste?
   b. high-level (including transuranic) waste?

2. Which waste source accounts for the greatest volume of high-level radioactive waste?

3. Which two waste sources account for most of the low-level radioactive waste?

4. Should high-level or low-level radioactive wastes receive greater attention? Explain your answer.

5. For each of the following disposal strategies, identify at least one benefit and one risk:
   a. on-site storage
   b. vitrification (see Figure 6.70)
   c. mined geologic disposal

Figure 6.70 *A nuclear technician at La Hague, France, checks vitrified radioactive waste storage containers for radiation emissions.*

▮MAKING DECISIONS

D.8 DISCUSSING NUCLEAR ENERGY

Read the statements in the CANE flyer on page 586 once again, then complete the following activities and questions.

1. Identify the five statements on the flyer that are most closely connected to the ideas you studied in Section D.

2. Choose one of these statements that you now know to be false. Write a short rebuttal of the statement that would be clear to a fellow student who has not studied chemistry.

3. Select one of the statements that you now know to be true. Find a parent, grandparent, or other adult who is willing to discuss this statement with you. Ask the following questions, write down their answers, and respond to their questions to the best of your knowledge.

   a. Do you think this statement is true? Why?

   b. Does this statement cause you to be concerned? Why or why not?

   c. Would the ideas in this statement affect your opinion about whether a nuclear power plant should be located in your community? Why?

   d. What questions do you have about this statement?

4. Do you think that the person that you spoke to would support CANE's mission? Explain.

Whenever you prepare a written or oral presentation, it is important to be aware of your audience's knowledge, interests, and goals. Keep this in mind as you prepare for your presentation to Riverwood's senior citizens.

SECTION D SUMMARY

Reviewing the Concepts

> Some large nuclei, when bombarded by neutrons, undergo nuclear fission.

1. What is nuclear fission?

2. Name three isotopes that can undergo nuclear fission.

3. Write a balanced nuclear equation for
   a. the fission of U-235 by a neutron, producing Br-87, La-146, and several neutrons.
   b. the fission of U-235 by a neutron, producing Ba-144, Kr-90, and neutrons.

4. Why does a nuclear reaction release more energy than a chemical reaction?

5. State Einstein's mass–energy relationship and explain the meaning of each symbol.

6. Name the force that holds nuclear particles together and describe its characteristics.

7. Describe characteristics of a nuclear chain reaction.

8. Why is a critical mass of fissionable material needed to sustain a nuclear chain reaction?

> The electricity produced by nuclear power plants originates from the energy released by fission in controlled chain reactions.

9. Describe how most conventional (non-nuclear) power plants generate electricity.

10. State the equivalent quantities of coal and petroleum needed to produce the same total energy contained in one nuclear fuel pellet.

11. Why is U-235 used in nuclear power plants?

12. How does each of the following affect neutrons in a nuclear power plant?
    a. control rods
    b. moderator

13. Why is it impossible for the fuel in a nuclear power plant to cause a nuclear explosion?

14. List the three common moderators used in nuclear power plants.

15. The core of a nuclear reactor is surrounded by thick concrete walls. Give at least three reasons for these walls.

16. What is the composition of the white plumes often seen rising from nuclear power-plant towers?

> Nuclear fusion is the combination of two relatively small nuclei into a new, more massive nucleus.

17. Why are high pressures and temperatures needed to initiate fusion reactions?

18. How much more energy can nuclear fusion produce than nuclear fission?

19. Why isn't nuclear fusion currently practical for generating electricity in power plants?

20. State the equivalent quantities of coal and gasoline needed to produce the energy released by the fusion of one gram of hydrogen-1.

21. What is a thermonuclear weapon?

22. Explain why both of these statements are true:

    a. Nuclear fusion has not been used as an energy source on Earth.

    b. Nuclear fusion is Earth's main energy source.

> The permanent disposal of radioactive waste poses challenging problems and issues.

23. Radioactive waste is grouped into two major categories.

    a. List and describe these two categories.

    b. What is the major difference between the two categories?

24. In the United States, what are the two largest sources of radioactive waste? Refer to Figure 6.66 (page 672).

25. Refer to Figure 6.66 (page 672). Spent fuel makes up what percent of total U.S. radioactive wastes

    a. by volume?

    b. by radioactivity?

26. Why do technicians have to regularly replace nuclear fuel pellets even if the pellets are still radioactive?

27. Compare current methods for disposing of high and low-level radioactive wastes.

28. Why is vitrification a preferred method for handling and storing radioactive wastes?

29. a. Where are two potential U.S. sites for permanent radioactive waste disposal?

    b. Why were these particular sites selected?

What burdens and benefits accompany uses of nuclear energy?

In this section, you have considered the processes of nuclear fission and fusion and the energy released in these processes, learned how fission energy is harnessed in nuclear power plants, weighed risks of your personal decisions, and evaluated some options for disposal of radioactive waste. Think about what you have learned, then answer the question in your own words in organized paragraphs. Your answer should demonstrate your understanding of the key ideas in this section.

Be sure to consider the following in your response: nuclear fission, nuclear fusion, nuclear power plants, risk, and radioactive waste.

Connecting the Concepts

30. Explain the difference between nuclear fission and nuclear fusion.

31. Sometimes burning is a good way to dispose of some types of extremely toxic material. Why would burning be an unacceptable plan for destroying radioactive waste?

32. Explain why the disposal of radioactive waste is a challenging issue in many nations. Use the concepts of radioactive decay, half-life, and radiation shielding in your answer.

33. How does nuclear fusion compare to a chemical reaction in which two hydrogen atoms combine to form a hydrogen molecule, H_2?

34. Construct a diagram showing the energy transformations involved in producing electricity in a nuclear power plant. How would your diagram differ for a coal-fired power plant?

35. A simple way to minimize the need for long-term storage of radioactive waste might seem to be speeding up all radioactive decay rates involved. Why won't this plan work?

36. What are some factors complicating the cleanup of abandoned or improperly stored radioactive waste?

Extending the Concepts

37. Research how other countries dispose of their high-level radioactive waste and evaluate their methods in terms of risks and benefits.

38. When confronting problems surrounding high-level radioactive waste disposal, students sometimes propose loading the waste in a rocket and shooting it into the Sun. What are some risks and benefits of that plan?

39. Fusion reactions that produce iron and lighter elements can take place in the core of an ordinary star. However, elements with higher atomic numbers generally result from violent stellar explosions. Explain the difference.

40. What would be some advantages of nuclear fusion over nuclear fission for producing electricity?

41. Investigate conditions needed to sustain a critical-mass chain reaction in

a. a nuclear reactor.

b. a nuclear weapon.

PUTTING IT ALL TOGETHER

COMMUNICATING SCIENTIFIC AND TECHNOLOGICAL INFORMATION

Throughout your chemistry studies up to this point, you have used scientific ideas to evaluate and draw your own conclusions regarding published claims in the CANE flyer. Your final task in this unit is to prepare and present what you have learned to help some Riverwood residents draw their own conclusions about nuclear concerns.

As you already know, your audience is composed of senior citizens at a local community center. They are concerned that regulations proposed by CANE will affect their ability to receive a full range of medical care. They are also alarmed by some CANE claims about radioactive waste disposal.

The senior citizens have invited your class to address the scientific aspects of statements made in the flyer from CANE so that they can evaluate the group's message more completely.

INFORMING OTHERS ABOUT NUCLEAR SCIENCE AND TECHNOLOGY

Responding to the CANE Flyer

Each student group will be responsible for responding to one or more CANE statements. In preparing your presentation, coordinate with other groups working on statements that involve similar ideas.

For each statement, follow these steps:

1. Decide whether the flyer statement is true, false, or partially true.
2. Explain the science or technology that the statement involves. For example, if the statement is about nuclear power plants, you should review how such plants work. For statements about radiation, you should explain different types of radiation.
3. Prepare a replacement statement, if necessary, that more accurately and evenhandedly addresses the same topic/issue.

Targeting the Audience

Design your presentation to be understandable by adults with at least a high school education that most likely was completed many years ago. Review the concerns that were mentioned in the unit introduction and address these concerns whenever they are relevant to your topic.

Supporting the Presentation

When possible, include visual aids and everyday examples and applications. Also devise a way to evaluate the effect of your presentation. Try to assess whether or not the audience, after your presentation, is better able to make informed decisions about nuclear policies for Riverwood.

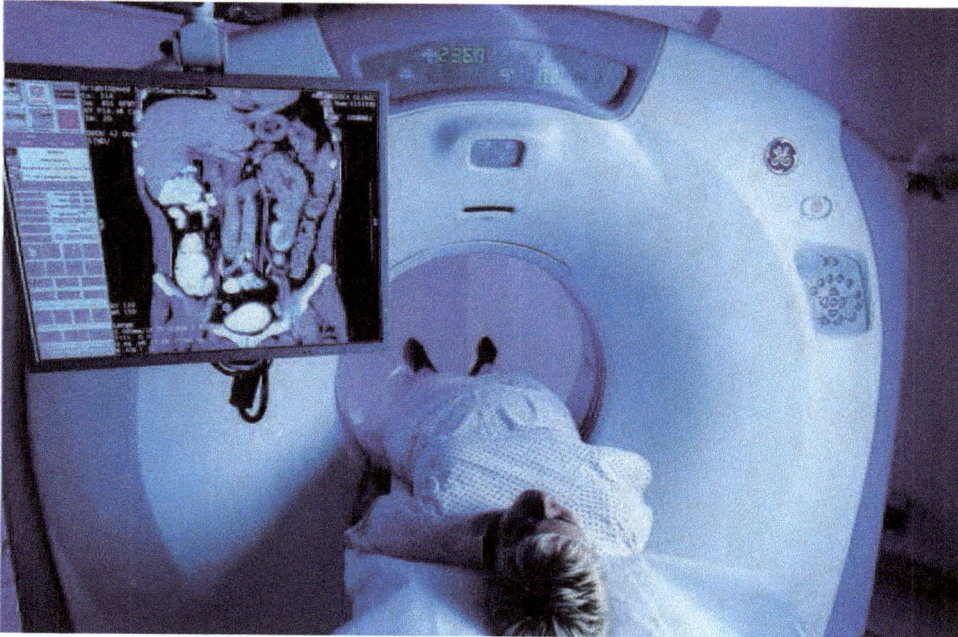

FOOD: MATTER AND ENERGY FOR LIFE

How is food energy stored, transferred, and released?

SECTION A
Food as Energy (page 686)

What role does molecular structure play in metabolism of carbohydrates and fats?

SECTION B
Carbohydrates and Fats
(page 704)

Why are protein molecules essential to living organisms?

SECTION C
Proteins (page 728)

What roles do vitamins, minerals, and additives play in foods we eat?

SECTION D
Vitamins, Minerals, and Additives (page 748)

?

The Parent-Teacher-Student Association (PTSA) at Riverwood High School requests your help in establishing new policies regarding use of the school's food and beverage vending machines. *Turn the page to learn more about nutrition and the role it should play in school vending policies.*

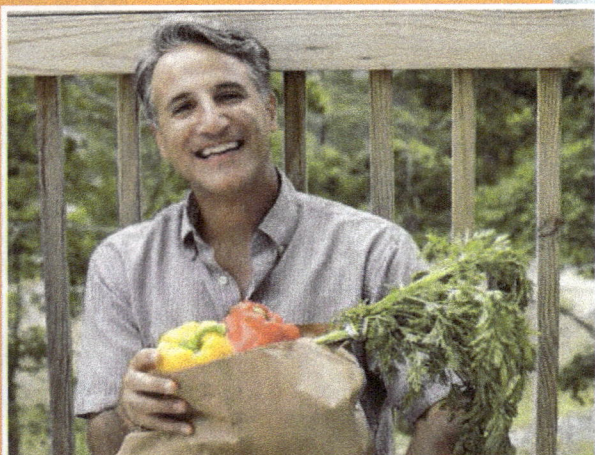

Ervin Kostecky, a parent member of the Riverwood High School Parent-Teacher-Student Association (PTSA), recently read a report about the quality of food sold in U.S. middle school and high school vending machines. Mr. Kostecky is concerned that Riverwood High School's food and beverage vending machines do not meet energy and nutritional needs of students who use them. He has requested that a committee be appointed to recommend school policies regarding these vending machines.

In this unit, you will learn about the chemistry of foods. You will learn how energy contained in food is stored and released and how substances in food promote growth and repair in the body. You will also investigate the chemistry and the nutritional roles of fats, carbohydrates, proteins, vitamins, and minerals. This information will help you make sound policy recommendations to be implemented next year for the school's vending machines.

As a member of the Vending Machine Policy Planning Committee, you receive an inventory that documents a typical Riverwood High student's diet over three days. Together with other committee members (your classmates), you will conduct several analyses of this three-day food inventory. By the end of the unit, your results will provide insight into items that would complement a typical Riverwood High student's diet. You will then write a report summarizing your results and proposing recommended guidelines for selecting new items for the vending machines. This report will be used in the committee's discussions and, ultimately, in selecting future food and beverage items for school vending machines.

To aid your investigation, the PTSA has highlighted characteristics of a healthful diet found in the *Dietary Guidelines for Americans 2010*, issued by the U.S. Department of Health and Human Services.

Keep the following dietary advice as well as chemistry principles you have learned in this course in mind throughout this unit:

Characteristics of a healthful diet:

- Consume a variety of nutrient-dense foods and beverages within and among the basic food groups.

- Balance portion-controlled calories from the foods and beverages ingested with calories expended in physical activity.

- Choose foods that limit the intake of solid fats, added sugars, salt, and alcohol.

- Consume a sufficient quantity and variety of fruits and vegetables, while staying within energy needs.

Is this the vending machine of the future?

SECTION A FOOD AS ENERGY

How is food energy stored, transferred, and released?

The term *diet* does not necessarily imply a weight-loss plan; it refers to the pattern of food and drink that one regularly consumes.

Where does the energy required for walking, running, playing sports, and even sleeping and studying come from? The answer is obvious: It comes from the food you eat. As you begin to analyze the three-day food inventory, you will learn the chemistry that will allow you to understand where that energy originates, how food energy is stored and used, and how decisions about eating can affect well being.

GOALS

- Use calorimetry to determine the energy stored in a particular food.
- Trace the flow of energy in food back to its sources and forward to its uses and dispositions.
- Analyze how the balance between energy intake and expenditure affects body mass and health.

concept check 1

1. How can the amount of energy produced during a chemical reaction be determined?
2. What is a Calorie?
3. What are some components necessary for a healthful diet?

A.1 FOOD GROUPS

When considering foods, people often focus on particular categories, or **food groups**. A healthful diet includes foods from a variety of groups, such as those shown in Figure 7.1. Why? Figure 7.2 highlights the groups. What do you already know about the nutrient and energy contents of foods in each group?

Although it is healthful to consume foods from a variety of groups, the *Dietary Guidelines for Americans 2010* advise that people should increase their daily intake of fruits and vegetables, whole grains, and nonfat or low-fat milk and milk products. In addition, the *Dietary Guidelines* call for decreasing intake of foods rich in fats and sugars.

Some foods listed in the three-day inventory may be represented by categories in Figure 7.2, but others are not. Where should you classify foods not included in Figure 7.2? What range of food groups is represented in the diet you are analyzing? Find out in the next activity.

Figure 7.1 *An assortment of healthful foods.*

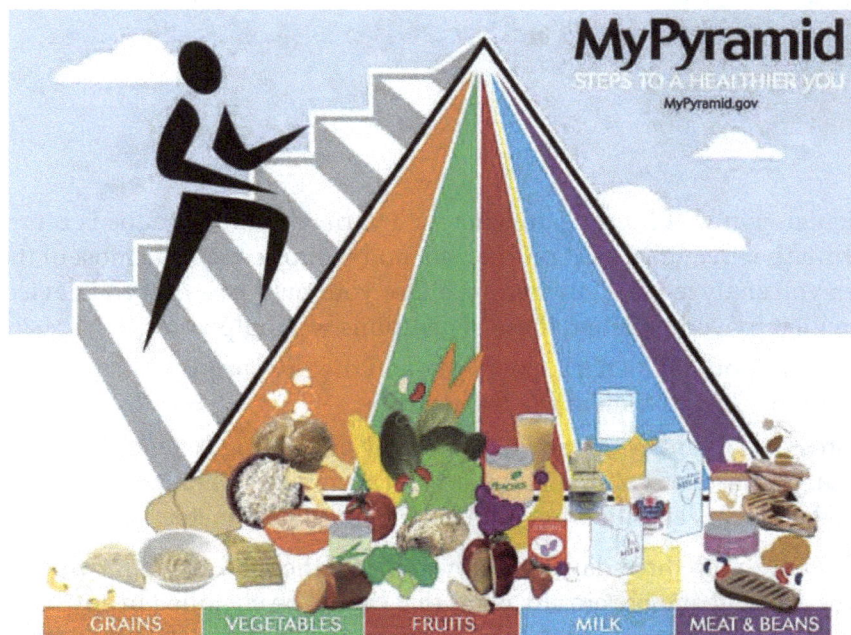

Figure 7.2 *The U.S. Department of Agriculture's food pyramid shows relative quantities of foods that should be consumed from each group, according to the Dietary Guidelines for Americans 2010. The small yellow "slice" represents fats, sugars, and salt, which should be limited in a healthful diet. See MyPyramid.gov for detailed individual recommendations.*

■ MAKING DECISIONS

A.2 DIET AND FOOD GROUPS

Your three-day food inventory analysis begins with the "big picture." That is, before analyzing the foods in terms of their characteristics—such as energy, fat, carbohydrate, protein, mineral, or vitamin content, you will verify what proportions of each food group the food inventory includes.

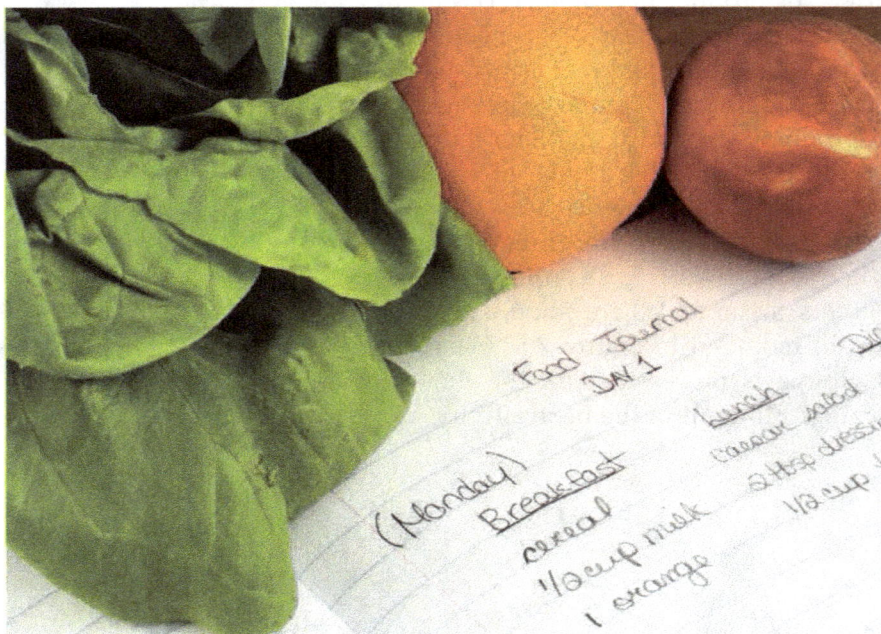

In particular, you will investigate whether a variety of food groups is represented within the inventory, or if one or two food groups constitute most of the food. When you analyze a food inventory, either your own, one that you devise, or one provided by your teacher, these suggestions will help you:

- List each food item, snack, beverage, and dietary supplement consumed during each of the three days.
- Express the quantity of each food item by estimating the number of servings, mass, or volume of the food or beverage. When possible, use food labels on each item to guide your quantity estimates.

For each item in the three-day inventory, indicate the food group to which it belongs. Also indicate if there are any items that cannot be classified in any listed category (such as snacks or soft drinks). You may need to investigate food items about which you are unsure. Then complete the following steps:

1. Construct a data table so that the total number of servings within each food group can be easily recorded. Include four columns, one for each day, and one for the daily average over the three days.

2. Record the quantity consumed within each food group for each day. Then calculate and enter the average number of daily servings for each group over the three days.

3. Using your data from Step 2, which food group or groups

   a. represent most of the foods within the diet?

   b. are minimally represented in the diet or not at all?

4. Overall, is there a reasonable balance of food groups in the inventory? If not, what dietary substitutions would lead to greater variety among food groups?

5. Complete the following steps for food items that were difficult to assign to a particular food group:

   a. List the item and the food group to which the item was assigned.

   b. Explain why it was difficult to assign this food item to a food group.

   c. Explain how you decided to which group to assign the item.

6. Is it better to analyze food consumed over three consecutive days or for only one day? Explain.

7. Under what circumstances would basing your food analysis on a particular three-day average be misleading?

INVESTIGATING MATTER

A.3 ENERGY CONTAINED IN A SNACK

Preparing to Investigate

Figure 7.3 shows part of a food label from a common snack food. Note the number of grams per serving. The label also shows the number of Calories (150). This is the quantity of food energy contained in one serving. In this investigation you will determine the energy contained in a sample of a snack food.

The quantity of energy contained in a particular food can be determined by carefully burning a known mass of the item under controlled conditions and measuring how much thermal energy is released. You completed similar investigations in Unit 3 when you determined the energy released during the combustion of several types of fuel, as well as a candle's heat of combustion (see Figure 3.41, page 331). This procedure is known as **calorimetry** and the measuring device is called a **calorimeter**.

Use the fuel-burning procedure (page 331) to guide you as you design a procedure to determine the energy contained in this particular snack sample. (**Caution:** *Do not use any nut-based snack food for this investigation; some students may be allergic to such items.*)

Figure 7.3 *Part of the label from a popular snack food. Note the mass and total Calories contained in one serving.*

Before starting the investigation, complete the following steps:

- Write a detailed procedure for determining the energy contained in the assigned snack food.
- Sketch your laboratory setup.
- Construct a data table to record all necessary measurements and observations for two separate trials.
- Have your teacher check and approve your procedure and data table.

Gathering Evidence

Before you begin, put on your goggles and wear them properly throughout the investigation. After your teacher approves your procedure, you may start the investigation. Conduct at least two trials.

Analyzing Evidence

If you know the mass of water and its temperature change, you can calculate the quantity of thermal energy that caused the temperature change. In this laboratory investigation, we assume that

Energy to heat water = Energy released by burning snack item

This assumes, of course, that no thermal energy is lost during the process.

Recall from Unit 3 that the specific heat capacity of liquid water is about 4.2 J/(g • °C). Thus, it takes about 4.2 J to raise the temperature of 1 g of water by 1 °C. If you know the mass of water heated and its temperature change, you can calculate the thermal energy absorbed by the water. If necessary, refer to page 338 to remind yourself how to complete this calculation.

Use data collected in this investigation to complete the following calculations:

1. Determine the mass (in grams) of the heated water.
2. Calculate the water's overall temperature change.
3. Calculate the total energy (in joules) required to heat the water.
4. The food **Calorie**, or *Cal* (written with an uppercase C), listed on food labels, is a much larger energy unit than the joule. One Calorie equals 4184 J, which can be conveniently rounded to 1 Cal = 4200 J.

    a. Calculate the total Calories used to heat the water.

    b. How many Calories were released by burning the item?

5. a. Calculate the energy released, expressed as Calories per gram (Cal/g) of the snack item burned, for each trial.

    b. Calculate the average Calories per gram for the snack item.

    c. Use reported values from the package label to calculate the declared Calories per gram for the snack item.

    d. Calculate the percent difference between the declared value found on the label and your average experimental value:

$$\% \text{ Difference} = \frac{|\text{Experimental value} - \text{Label value}|}{\text{Label value}} \times 100\%$$

The precise specific heat capacity of water is 4.184 J/(g •°C).

The *calorie* (cal), another common energy unit, is about four times the size of a joule (1 cal = 4.184 J). One Calorie (Cal), which is a unit still used in the U.S. for food-energy values, equals one kilocalorie (1 kcal) or 1000 cal.

Making Claims

1. Would you describe the snack food that you tested as energy-dense? Why or why not?
2. Compare your results with those of your classmates.
   a. Which snack food had the greatest variance among groups in calculated Cal/g?
   b. What do you think caused these differences?

Reflecting on the Investigation

3. Which aspects of your laboratory setup and procedure might account for any difference between your Cal/g value and the corresponding label value?
4. How could your laboratory setup and procedure be improved to increase the accuracy of your results?
5. Consider the snack item shown in Figure 7.4. What changes in setup or procedure would be necessary to find the Calories per gram in the snack item shown?

Figure 7.4 *Describe how you would determine the quantity of energy contained in these potato chips.*

concept check 2

1. How was the measurement of energy released in the Combustion investigation (Unit 3, page 337) similar to and different from the measurement of energy in snack foods?
2. Where does the energy in food originate?
3. How does your body convert food into usable energy?

A.4 ENERGY FLOW—FROM THE SUN TO YOU

The snack item you burned released enough energy to raise the temperature of a sample of water by several degrees Celsius. Where did that energy come from?

The ultimate answer is easy: All food energy originates from sunlight. Through **photosynthesis**, green plants capture and use solar energy to make large molecules from smaller, simpler ones (see Figure 7.5). Recall from Unit 3 (page 345) that green plants, through photosynthesis, use solar energy to convert water and carbon dioxide into carbohydrates and oxygen gas. Although a variety of carbohydrates are produced, an equation for photosynthesis usually depicts the production of glucose:

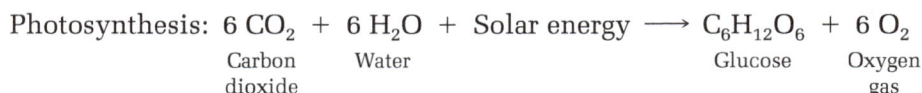

Photosynthesis: $6\,CO_2 \ + \ 6\,H_2O \ + \ $ Solar energy $\longrightarrow \ C_6H_{12}O_6 \ + \ 6\,O_2$

Carbon dioxide   Water   Glucose   Oxygen gas

Figure 7.5 *Energy from the Sun was used to convert water and carbon dioxide into the molecules that constitute these onions. What type of energy conversion may happen next?*

For a discussion of energy involved in chemical reactions, see pages 333–336.

For this reaction to occur, bonds between carbon and oxygen atoms in carbon dioxide molecules and between oxygen and hydrogen atoms in water molecules must be broken. The atoms must then recombine in a different arrangement to form glucose and oxygen molecules. Recall from Unit 3 that breaking bonds always requires energy, whereas bond formation releases energy. In photosynthesis, the bonds in carbon dioxide and water molecules require more energy to break than is released when chemical bonds in glucose and oxygen molecules form. The energy needed to drive this endothermic reaction, as the photosynthesis equation indicates, comes from the Sun. The potential energy diagram in Figure 7.6 shows the energy relationships in this process.

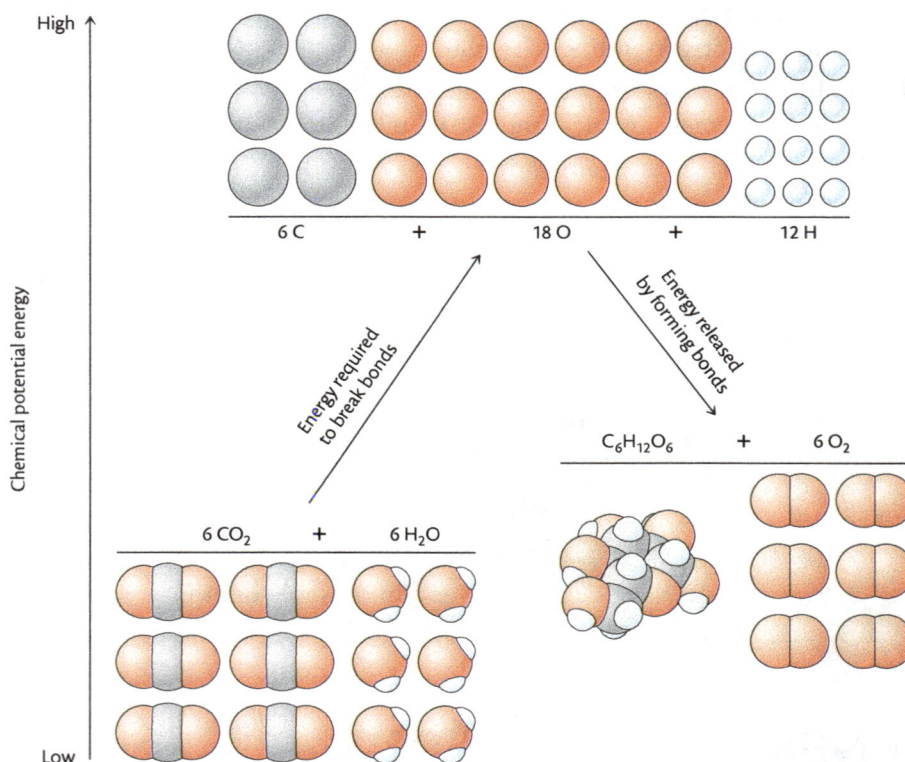

Figure 7.6 *Photosynthesis. Breaking bonds within carbon dioxide and water molecules requires energy. Energy is released when atoms combine to form glucose and oxygen molecules. The first step (bond breaking) requires more energy than the second step (bond forming) releases, so the overall process of photosynthesis is endothermic.*

How would a potential energy diagram for the process of metabolism look?

The Sun's radiant energy is converted to chemical energy stored within bonds of carbohydrate molecules. Living organisms release this chemical energy when they consume and metabolize carbohydrate molecules, converting them into lower-energy CO_2 and H_2O molecules. This chemical energy can be measured via calorimetry, as you just observed in Investigating Matter A.3.

Energy originally delivered as sunlight continues to flow through ecosystems as carnivores (meat-eating animals) consume plant-eating animals. Eventually, plants and animals die and decay; organisms that aid in decomposition use remaining stored-up energy. Thus, energy flows from the Sun to photosynthetic plants to herbivores and then to carnivores and decomposers.

The Sun's captured energy is dispersed and becomes less available as it is transferred from organism to organism. For example, as energy flows from one organism to another, some energy is dispersed into the environment as thermal energy. You may be surprised to learn that only a small fraction (about 10–15%) of food energy consumed by organisms is used for growth—for converting smaller chemical molecules to larger molecules that become part of an animal's structure. Over half the energy contained in consumed food is used to digest food molecules. The supply of useful energy declines as energy continues to transfer away from its original source—the Sun. Figure 7.7 depicts this decline in available energy.

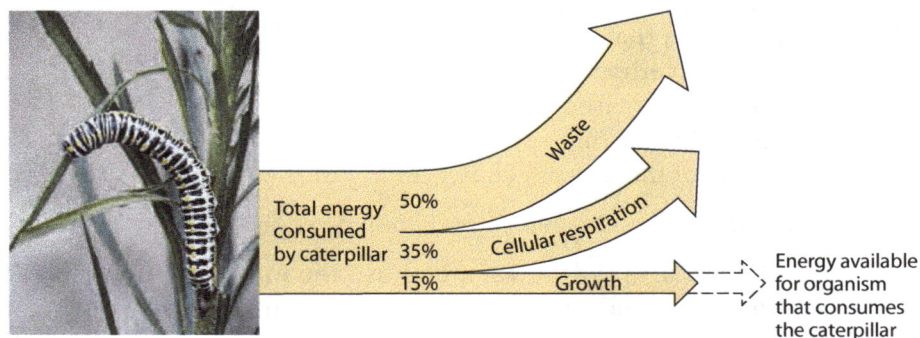

Figure 7.7 *The energy used by this caterpillar originated as solar energy. However, not all incident solar radiation is converted into usable energy. First, only a portion of the Sun's energy is actually stored as chemical energy in molecules within this plant. Next, only a fraction of that stored chemical energy is actually used by the caterpillar.*

CHEM**QUANDARY**

HOW DOES YOUR GARDEN GROW?

From a chemical viewpoint, consider this question: *As garden plants grow, what is the source of the material that causes their mass to increase?*

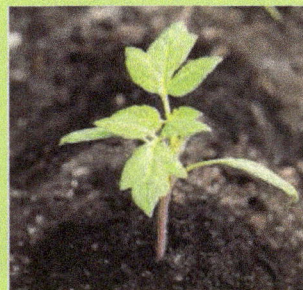

A.5 ENERGY RELEASE AND STORAGE

Some energy in the food you eat is used soon after you digest it (Figure 7.8). The rest of the energy may be stored for later needs. These available reserves of bodily energy are mainly in fats and, to a much smaller extent, in carbohydrates. Figure 7.9 shows some foods rich in carbohydrates and fats.

Whether your body uses energy from food recently ingested or from stored fat, the release of energy from these molecules depends on a series of chemical reactions inside your cells. In **cellular respiration,** plants and animals use oxygen to break down complex organic molecules into carbon dioxide and water molecules. The energy required to break bonds in the reactant molecules is less than the energy released when the bonds in carbon dioxide and water form. Thus energy is released in cellular respiration. You can think of that exothermic process as the reverse of the process shown in Figure 7.6 (page 692).

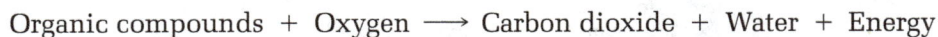

Figure 7.8 *A portion of the energy in food is expended soon after eating, particularly when engaging in energy-demanding activities such as bicycling. The remaining energy is then stored for later activity.*

Carbon dioxide and water are also produced from burning petroleum, coal, and other fossil fuels. See page 345.

Organic compounds + Oxygen ⟶ Carbon dioxide + Water + Energy

Specialized structures within each cell use the energy released to carry out a variety of tasks such as energizing reactions, transporting molecules, disposing waste, storing genetic material, and synthesizing new molecules. That energy ultimately allows you to walk, talk, run, work, and think; it also provides energy to power your heart, lungs, brain, and other organs.

Carbohydrates, fats, and proteins can all be processed as fuel for your body. We will now examine cellular respiration through the oxidation of glucose, a carbohydrate. The overall equation, based on the oxidation of one mole of glucose, is as follows:

This equation is the reverse of the equation for photosynthesis. See Figure 3.47 page 345.

$$C_6H_{12}O_6(aq) \ + \ 6 \ O_2(g) \longrightarrow 6 \ CO_2(g) \ + \ 6 \ H_2O(l) \ + \ 686 \ \text{Calories}$$

Figure 7.9 *Breads, grains, bagels, and pasta (left) contain significant portions of carbohydrates, whereas cooking oil, nuts, and avocado (right) are rich in fats.*

This equation actually summarizes a sequence of more than 20 integrated chemical reactions catalyzed by more than 20 different enzymes. Glucose (see Figure 7.10) is oxidized within cells throughout your body.

The energy needed to perform individual cellular functions is much less than the energy released by "burning" glucose molecules. Between its release from these energy-rich molecules and its later use in cells, energy is stored in biomolecules of ATP (adenosine triphosphate), shown in Figure 7.11.

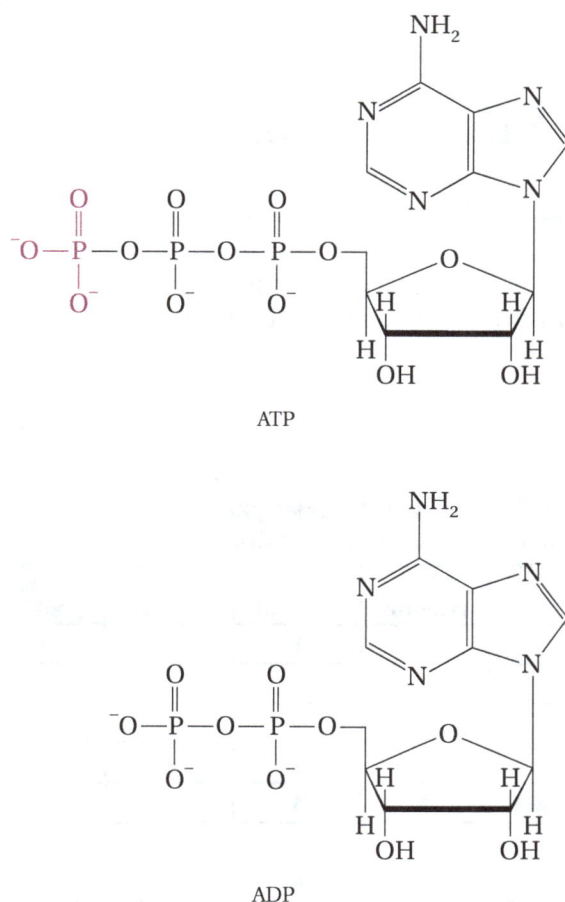

ATP

Figure 7.10 *The icing and decorations on these cupcakes likely contain glucose, which is burned to provide energy or stored as fat in the body.*

ADP

Figure 7.11 *The ionic forms of ATP and ADP. The colored atoms on ATP's left side are removed when ADP forms.*

As indicated by the following equation, more energy is required to break chemical bonds in adenosine diphosphate (ADP) and HPO_4^{2-} than is released when water and ATP form. The difference in those bond energies equals 7.3 Cal, which suggests that an ATP molecule can be regarded as an energy-storage site because energy is released when the reaction proceeds in the opposite direction.

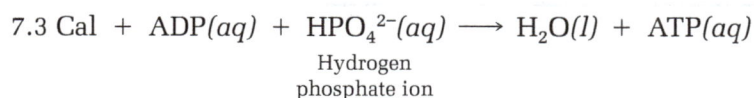

$$7.3 \text{ Cal} + \text{ADP}(aq) + \text{HPO}_4^{2-}(aq) \longrightarrow \text{H}_2\text{O}(l) + \text{ATP}(aq)$$

Hydrogen
phosphate ion

When this reaction is reversed, each mole of ATP releases 7.3 Cal. This conveniently small quantity of energy, compared to the 686 Calories produced as one mole of glucose is oxidized, is used to energize particular cellular reactions. Some of these reactions require less energy than that supplied by a single ATP molecule, whereas other steps require the total energy released by several ATP molecules.

Oxidation of one mole of glucose produces enough energy to add 38 mol of ATP to short-term cellular energy storage. Each day, your body stores and later releases energy from at least 6.02×10^{25} molecules of ATP (100 mol ATP).

DEVELOPING SKILLS
A.6 ENERGY IN ACTION

Table 7.1 summarizes the average energy expended during various activities. Use the information provided to answer questions appearing on the following page.

Table 7.1

Energy Expended in Various Activities

| Activity | Energy Expended (in Calories per Minute) for Individuals of Different Body Mass | | | |
	46 kg (100 lb)	55 kg (120 lb)	68 kg (150 lb)	82 kg (180 lb)
Sleeping	1	1	1	1
Sitting	1	1	1	2
Playing volleyball	2	3	4	4
Weight lifting	2	3	4	4
Walking (17 min/mile)	3	4	5	6
Skateboarding	4	5	6	7
Swimming	5	6	7	9
Wrestling	5	6	7	9
Bicycling (moderate)	6	7	8	10
Playing soccer	6	7	8	10
Doing step aerobics	6	7	8	10
Playing tennis	6	7	8	10
Playing basketball	6	8	10	12
Playing football	6	8	10	12
Playing ice hockey	6	8	10	12
Doing martial arts	8	10	12	14
Running (8 min/mile)	10	12	15	18

Sample Problem: Table 7.1 indicates that a 55-kg (120-lb) person burns 5 Cal/min while skateboarding (Figure 7.12).

   a. *This energy is supplied by ATP. If the oxidation of each mole of glucose ($C_6H_{12}O_6$) produces 38 mol ATP, how many moles of glucose would be needed to provide the energy for that one hour of skateboarding?*

   b. *What mass of glucose does your answer to Part a represent? The mass of one mole of glucose is 180 g.*

To begin, find moles of ATP needed for one hour of skateboarding:

$$\frac{5\text{ Cal}}{1\text{ min}} \times \frac{60\text{ min}}{1\text{ h}} = 300\text{ Cal/h}$$

$$\frac{300\text{ Cal}}{1\text{ h}} \times \frac{1\text{ mol ATP}}{7.3\text{ Cal}} = 41\text{ mol ATP/h}$$

Because 38 mol ATP can be obtained from each mole of glucose, the skateboarder's need for 41 mol ATP can be met by slightly more than one mole of glucose:

$$41\text{ mol ATP} \times \frac{1\text{ mol glucose}}{38\text{ mol ATP}} = 1.1\text{ mol glucose}$$

What mass of glucose does that represent? Because the skateboarder needs more than one mole of glucose, that mass must be greater than 180 g:

$$1.1\text{ mol glucose} \times \frac{180\text{ g glucose}}{1\text{ mol glucose}} = 200\text{ g glucose}$$

Figure 7.12 *How much energy is expended in one hour of skateboarding?*

More than 200 g glucose would actually be needed, since the energy-transfer steps are not 100% efficient.

Now answer the following questions:

1. Assume your body produces approximately 100 mol ATP daily.
   a. How many moles of glucose are needed to produce that amount of ATP?
   b. What mass of glucose does your answer to Question 1a represent?

2. One minute of muscle activity requires about 0.0010 mol ATP for each gram of muscle mass. How many moles of glucose must be oxidized to energize 454 g (1 lb) of muscle to dribble a basketball for one minute?

3. a. How many hours do you typically sleep each night?
   b. How many Calories do you use during one night of sleep? Refer to Table 7.1.
   c. How many moles of ATP does this require?
   d. What kinds of activity require energy while you are sleeping?

A.7 ENERGY IN—ENERGY OUT

People follow various diets for different reasons. Some people may want to lose weight. Others may want to gain weight. Still others put little thought into what they eat. Instead, they merely eat for convenience or pleasure.

Some foods are needed to deliver important molecules to the body, regardless of their energy value. By contrast, other foods provide only energy. This latter category, which includes sugar-sweetened soft drinks, is sometimes described as furnishing "empty Calories."

The *Physical Activity Guidelines for Americans* encourages citizens of all ages to become more active. The report, issued in 2008 by the U.S. Department of Health and Human Services, cites research findings on the benefits of physical activity, including:

- improved cardiac, respiratory, and muscular fitness
- improved bone health
- improved cardiovascular and metabolic health biomarkers
- favorable body composition, and
- reduced symptoms of depression.

These benefits increase as you age. Research also shows that the total amount of physical activity is more important than its intensity, duration, or frequency. The guidelines measure physical activity in a unit called metabolic equivalent (MET). By contrast, the activities in this unit use Calories, so that you can compare energy consumed in your food to energy burned through your activities.

If you want to lose weight, you must consume less energy than you expend. If your goal is to gain weight, you must do just the opposite. If losing weight or gaining muscle mass is desired, then regular physical activity, such as that shown in Figure 7.13, is needed in addition to the particular diet you follow. Now you will explore the interplay between personal activity and diet.

Figure 7.13 *Technology is available for monitoring total energy expenditure.*

> If you eat 100 Cal per day more than you burn, you will gain about a pound each month.

DEVELOPING SKILLS

A.8 GAIN SOME, LOSE SOME

Look again at Table 7.1 on page 696. The quantity of energy expended depends on the duration of a given activity and the person's body mass. For example, during one hour of soccer playing, a 55-kg (120-lb) person expends 420 Cal, whereas a 68-kg (150-lb) person would expend 480 Cal in that same activity.

Use information in Table 7.1 (page 696) to answer the following questions.

Sample Problem: *A 46-kg female usually takes 10.0 min to bike home from school. Would she burn more Calories by running home instead (at 8 min/mile for 15.0 minutes)?*

Bicycling requires: Running requires:

6 Cal/~~min~~ × 10.0 ~~min~~ = 60 Cal   10 Cal/~~min~~ × 15.0 ~~min~~ = 150 Cal

Thus, running home burns more energy by 90 Cal per trip.

1. Consider eating an ice-cream sundae (Figure 7.14). Assume that two scoops of your favorite ice cream contain 250 Cal; the topping adds 125 Cal more.

   a. Assume that your regular diet (without that ice-cream sundae) just maintains your current body weight. If you eat the ice-cream sundae and wish to "burn off" those extra Calories,

      i. for how many hours would you need to lift weights?

      ii. how far would you need to walk at 17 min/mile?

      iii. for how many hours would you need to swim?

   b. One pound of weight gain is equivalent to 3500 Cal. If you choose not to exercise, how much weight will you gain from eating the sundae?

   c. Now assume that you consumed a similar sundae once per week for 16 weeks. If you do not exercise to burn off the added Calories, how much weight will you gain?

2. Question 1 implies that eating an ice-cream sundae will cause weight gain unless you complete additional exercise.

   a. Can you think of a plan that would allow you to consume the sundae, do no additional exercise, and still *not* gain weight?

   b. Explain your answer.

   c. What concerns would you have about this plan?

3. Suppose you drank six glasses (250 g each) of ice water (0 °C) on a hot, summer day.

   a. Assume your body temperature is 37 °C. How many joules of thermal energy would your body use in heating that ice water to body temperature? Recall that the specific heat capacity of liquid water is about 4.2 J/(g • °C).

   b. How many Calories is this? Recall that 1 Calorie equals about 4200 J.

   c. A serving of french fries contains 240 Cal. How many glasses of ice water would you need to drink to "burn off" the Calories in one serving of french fries?

   d. Given your answer to Question 3c, does drinking large quantities of ice water seem like a reasonable strategy to lose weight? Explain.

4. Identify your favorite activity from Table 7.1 (page 696). How many minutes would you need to engage in that activity to burn off the Calories in one serving of french fries (see Question 3c)?

Figure 7.14 *How does an extra helping of ice cream affect your daily balance between energy intake and expenditure?*

■ MAKING DECISIONS

A.9 ENERGY INTAKE AND EXPENDITURE

Based on the chemical knowledge you have gained, you can find the total energy consumed in the three-day food inventory and estimate the energy expended based on typical activity levels. Using appropriate resources suggested by your teacher, calculate and record the food energy (Calories) contained in each item in the three-day inventory.

1. Calculate the total Calories consumed each day.

2. Calculate the average Calories consumed per day.

3. Using results from Making Decisions A.2 (pages 688–689), how many Calories per day (on average) are supplied by each Food Group?

4. Using Table 7.1 (page 696) and other resources provided by your teacher, develop a list of typical daily activities for a high school student and record how much energy is expended during each activity listed.

5. Based on the list you prepared in Question 4, calculate the total Calories expended over an average day. Calories are not only expended during the activities listed, but also by normal, resting-state body processes. Be sure to include this expenditure in your calculation.

6. a. Based only on Calories consumed and expended, decide whether the three-day inventory you are evaluating would be appropriate for weight maintenance (see Figure 7.15). Explain your answer.

   b. Identify and describe any limitations in your analysis that may affect your decision in Question 6a.

7. Based on your analysis thus far, what types of items would you recommend for the Riverwood High School vending machines in terms of the energy they provide?

Figure 7.15 *The balance between energy intake and expenditure affects one's body mass.*

SECTION **A** SUMMARY

Reviewing the Concepts

> Calorimetry can be used to determine the quantity of energy contained in a particular food sample.

1. List three common units used to express the energy content of food.

2. Sketch a simple calorimeter and label the purpose of each component.

3. Why is it important to know accurately the mass of water used in a calorimeter?

4. What is the mathematical relationship between a Calorie and a calorie?

5. Convert these energy quantities:

   a. 4375 cal to Cal

   b. 76 932 J to Cal

   c. 289 Cal to J

   d. 12 226 000 cal to kJ

6. a. How much thermal energy, in joules, is required to increase the temperature of a 115-g water sample by 10.0 °C?

   b. Suppose that a 2.0-g food sample was burned in a calorimeter to provide the thermal energy involved in Question 6a. Assume all thermal energy released was used to heat the water.

      i. How many calories per gram were present in the food sample?

      ii. How many Calories per gram were present in the food sample?

> All food energy originates from sunlight and is stored and released through a series of chemical reactions.

7. What process captures sunlight and transforms it into chemical energy?

8. Write a chemical equation for the production of glucose through photosynthesis.

9. Is photosynthesis an endothermic or exothermic chemical change? Explain.

10. Keeping in mind the law of conservation of energy, describe what happens to the food energy consumed by living creatures.

11. What is meant by an "energy rich" molecule?

12. What is the relationship between photosynthesis and cellular respiration? Write chemical equations that support your answer.

13. What is the difference, in terms of molecular composition and stored energy, between ADP and ATP?

14. How do ADP and ATP permit the controlled use of energy from glucose?

> The energy required for a physical activity depends on the particular activity, the total time involved, and the mass of the person engaged in the activity.

15. How much total energy (in Calories) is expended by a 68-kg (150-lb) person swimming for 35 minutes?

16. How much total energy (in joules) is expended by a 46-kg (100-lb) person walking for 56 minutes?

17. People expend energy even when they sit perfectly still. Why?

18. How many moles of ATP are required to provide energy for an 82-kg (180-lb) student to sit in class for 45 min?

19. Explain why two people performing exactly the same physical exercise may not burn the same total Calories.

20. Suppose a 55-kg (120-lb) man, whose regular diet and exercise just maintains his body weight, adds one chocolate candy bar containing 354 Cal to his diet each day for 30 days without increasing his level of exercise. Predict how much weight the person will gain in 30 days.

How is food energy stored, transferred, and released?

In this section, you began to analyze a three-day food and activity log in terms of quantity of Calories consumed and expended. You applied ideas that you learned in Unit 3 to food and food energy. Think about what you have learned in this unit and in previous units, and how these ideas are related, then answer the question in your own words in organized paragraphs. Your answer should demonstrate your understanding of the key ideas in this section.

Be sure to consider the following in your response: Calories, calorimetry, diet and exercise, ATP, ADP, and energy sources and pathways.

Connecting the Concepts

21. Compare the process of cellular respiration to combustion.

22. Does calorimetry directly measure the quantity of thermal energy liberated by a food or fuel? Explain.

23. Describe problems that could arise in calorimetry by using a food sample that is:

    a. very small.

    b. very large.

24. A student argues that eating a certain mass of chocolate or eating the same mass of apples results in the same gain in body weight. Explain why this is incorrect.

25. Figure 7.6 (page 692) shows an energy diagram for photosynthesis.

    a. Construct a similar energy diagram for cellular respiration.

    b. Compare the energy diagrams for photosynthesis and cellular respiration.

26. The term *respiration* is sometimes used to mean breathing. What is the relationship, if any, between cellular respiration and breathing?

27. In reference to a local seafood restaurant, someone remarked, "When you eat a pound of fish, you're eating ten pounds of flies." Aside from its questionable value as a meal-promoting strategy, how accurate is this message? Why?

28. Explain why low-Calorie food is sometimes described as *lite* or *light*, even though Calories are not a unit of mass.

29. You used similar calorimetry procedures in Unit 3 and in this unit to find the energy content of candles and food products, respectively. Does that imply that anything that can be burned to heat water can be used to fuel your body's metabolism? Explain.

Extending the Concepts

30. a. In what sense does the problem of world hunger involve an "energy crisis"?

    b. In what sense is it a "resource crisis"?

31. Research and report on the difference between *under*nourishment and *mal*nourishment. Could either term ever apply to an overweight person?

32. From an energy standpoint, are there advantages to eating low on the food chain? For example, is it energetically more favorable to use 100 lbs of grain or 100 lbs of beef as a food source?

33. Why is it not possible just to consume only pure ATP and thus eliminate some steps in metabolism?

34. Investigate the characteristics of a professionally-designed calorimeter. Make a sketch of its essential parts and explain its operation.

35. For hibernating animals, the storage of fat is critical for survival. Investigate a particular animal species to find out how it stores optimum quantities of fat for hibernation.

36. A student decides to lose some weight by not wearing a coat in cold winter weather. What knowledge of food energy might have inspired this idea? Does this plan have merit? Explain.

SECTION **B**
CARBOHYDRATES AND FATS

What role does molecular structure play in metabolism of carbohydrates and fats?

You are now ready to explore how the chemical energy contained in foods, studied in Section A, is stored and consumed. Keep in mind the three-day inventory you are evaluating, and think about how this new knowledge may apply to its analysis.

GOALS

- Describe the general structure of a carbohydrate molecule.
- Describe the general structure and components of a fat molecule.
- Distinguish between saturated and unsaturated fats.
- Explain how structural differences in fat and carbohydrate molecules account for their different properties and energy content.
- Identify a limiting reactant, given the equation and reaction conditions.

concept check 3

1. How is the process of cellular respiration similar to the process of combustion?
2. What is the source of the thermal energy released in an exothermic reaction?
3. What is the difference between unsaturated and saturated fat molecules?

B.1 CARBOHYDRATES: ONE WAY TO COMBINE C, H, AND O ATOMS

All **carbohydrate** molecules are composed of carbon, hydrogen, and oxygen. Glucose, which is the key energy-releasing carbohydrate in biological systems, has the molecular formula $C_6H_{12}O_6$. When such formulas were first established, chemists noted a 2:1 ratio of hydrogen atoms to oxygen atoms in carbohydrates, the same as in water. They were tempted to write the glucose formula as $C(H_2O)_6$, implying a chemical combination of carbon with six water molecules. Chemists even invented the term "carbohydrates" (water-containing carbon substances) for glucose and related compounds, although chemists later determined that carbohydrates contain no water molecules. However, like water, carbohydrate molecules *do* contain O—H bonds in their structures.

> Sugars, starch, and cellulose are examples of carbohydrates.

Carbohydrate molecules may be either simple sugars, such as glucose or fructose (see Figures 7.16 and 7.17), or chemical combinations of two or more simple sugar molecules. Simple sugars, called **monosaccharides**, are molecules containing five or six carbon atoms. Glucose (like most other monosaccharides) exists principally in a ring form: however, glucose can also exist in a chain form, as shown in Figure 7.17. Do both forms have the same molecular formula?

Figure 7.16 *Many fruits and vegetables contain simple sugars such as glucose and fructose.*

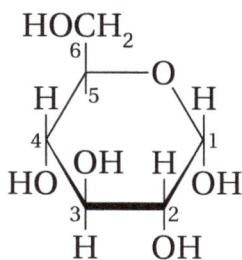

Ring form

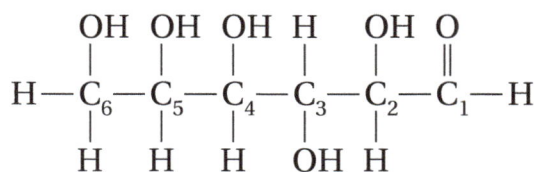

Chain form

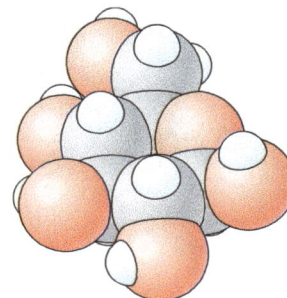

Figure 7.17 *Structural formulas for glucose. This carbohydrate, like most simple sugar and disaccharide molecules, is found primarily in ring form. How do ring and chain structures compare with one another?*

Condensation reactions were highlighted in Unit 3, pages 317–322, in the formation of esters and condensation polymers.

Sugar molecules composed of two monosaccharide units bonded together are called **disaccharides**. They are formed by a condensation reaction between two monosaccharides. Sucrose (table sugar, $C_{12}H_{22}O_{11}$) is a disaccharide composed of the ring forms of glucose and fructose, as illustrated in Figure 7.18.

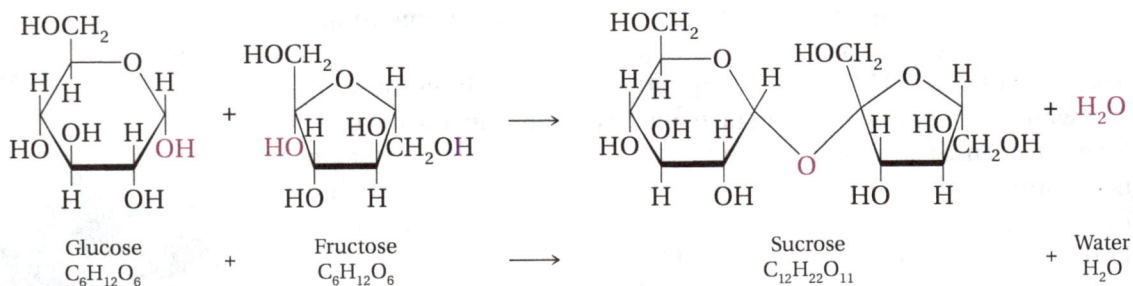

Glucose $C_6H_{12}O_6$ + Fructose $C_6H_{12}O_6$ → Sucrose $C_{12}H_{22}O_{11}$ + Water H_2O

Figure 7.18 *The formation of sucrose. Note that particular —OH groups react (shown in red), resulting in elimination of one H_2O molecule.*

All polysaccharides are polymers of monosaccharide molecules.

The reaction that forms disaccharides—a condensation reaction—can also cause monosaccharides to form polymers. Such polymers, not surprisingly, are called **polysaccharides** (see Figure 7.19). Starch, which is a major component of grains and many vegetables, is a polysaccharide composed of glucose units. Cellulose, which is the fibrous or woody material of plants and trees, is another polysaccharide formed from glucose.

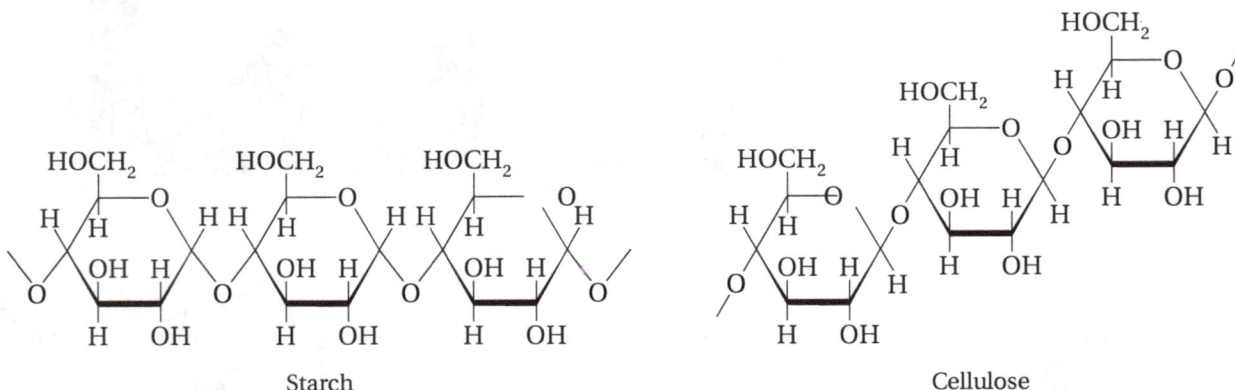

Starch

Cellulose

Figure 7.19 *Structural formulas for starch and cellulose, two polysaccharide molecules.*

The major types of carbohydrates are summarized in Table 7.2. Compare the bonds in starch with those in cellulose. Due to structural differences, starch is easily digested by the body (see Figure 7.20), whereas cellulose is indigestible by humans. Such indigestible types of carbohydrates are commonly called *fiber*.

Table 7.2

The Composition of Common Carbohydrates

Classification and Examples	Composition	Formula	Common Name or Source
Monosaccharides		$C_6H_{12}O_6$	
Glucose	–		Blood sugar
Fructose	–		Fruit sugar
Galactose	–		–
Disaccharides		$C_{12}H_{22}O_{11}$	
Sucrose	Fructose + glucose		Cane sugar
Lactose	Galactose + glucose		Milk sugar
Maltose	Glucose + glucose		Germinating seeds
Polysaccharides	Glucose polymers	$(C_6H_{10}O_5)_n$	
Starch			Plants
Glycogen			Animals
Cellulose			Plant fibers

Carbohydrates and fats are the primary high-energy substances in the human diet. One gram of a carbohydrate provides 4 Calories of food energy. Nutritionists recommend that about 45–65% of dietary Calories come from high-fiber carbohydrates. Worldwide, most people obtain carbohydrates by eating grains, such as rice, corn bread, wheat tortillas, bread, pasta, and beans. Other sources of carbohydrates include fruits, milk, and yogurt. Meats contain a small quantity of **glycogen**, the carbohydrate by which animals store glucose.

1 g carbohydrate = 4 Cal

Glycogen, similar in structure to plant starch, is used as a source of reserve energy by humans and animals.

Figure 7.20 *Breads and other foods contain starch, which is a digestible polysaccharide. Cellulose (found in wood, paper, and cotton) is a polysaccharide that humans cannot digest. What is a structural difference between digestible polysaccharides (such as starch) and cellulose? Which of these images show cellulose and which show starch?*

People in the United States tend to obtain more of their carbohydrates, on average, from wheat-based breads, potatoes, sugar-laden snacks, and desserts than is common in many other parts of the world. Currently, an average U.S. citizen consumes more than 82 kg (181 lb) of sugar annually, mainly from candy, desserts, and soft drinks.

B.2 FATS: ANOTHER WAY TO COMBINE C, H, AND O ATOMS

Unlike the terms *carbohydrate* and *protein*, *fat* has acquired its own general (and somewhat negative) meaning. From a chemical viewpoint, however, fats are just another major category of biomolecules with special characteristics and functions.

A significant part of a normal human diet, fats are present in meat, fish, poultry, oils, dairy products, nuts, and grains. When more food is consumed than is needed to satisfy energy requirements, much of the excess food energy is stored in the body as fat molecules. If food intake is not enough to meet the body's energy needs, then that stored fat is "burned" to release energy to make up the difference.

Fats are composed of carbon, hydrogen, and oxygen—the same three elements that compose carbohydrates. Fat molecules, however, contain fewer oxygen atoms and more carbon and hydrogen atoms. Thus, fats have a greater number of carbon–hydrogen bonds and fewer carbon–oxygen (and oxygen–hydrogen) bonds than do carbohydrate molecules. You can confirm this by comparing the structure of glyceryl tripalmitate, which is a typical fat molecule, shown in Figure 7.21, to that of glucose or starch molecules (Figure 7.17 on page 705 and Figure 7.19 on page 706, respectively).

> When an organic substance is burned, it reacts with oxygen gas to form carbon dioxide, water, and thermal energy. Similarly, when fats, carbohydrates, and protein are "burned" in the body, they react with oxygen gas and are converted into carbon dioxide, water, and energy.

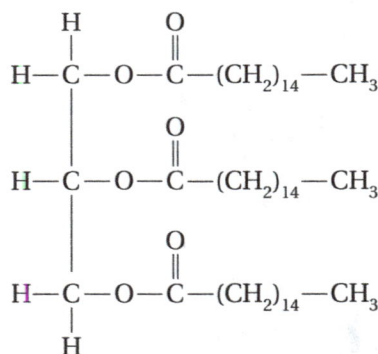

$$\begin{array}{c}
\text{H} \qquad\quad \text{O} \\
| \qquad\qquad \| \\
\text{H---C---O---C---(CH}_2)_{14}\text{---CH}_3 \\
| \qquad\qquad\quad \text{O} \\
\qquad\qquad\qquad \| \\
\text{H---C---O---C---(CH}_2)_{14}\text{---CH}_3 \\
| \qquad\qquad\quad \text{O} \\
\qquad\qquad\qquad \| \\
\text{H---C---O---C---(CH}_2)_{14}\text{---CH}_3 \\
| \\
\text{H}
\end{array}$$

Figure 7.21 *The structural formula of glyceryl tripalmitate, a typical fat.*

Generally, fat molecules are nonpolar and only sparingly soluble in water. As you can see from Figure 7.21, fats have long hydrocarbon portions that prevent them from dissolving in polar solvents such as water. Their low water solubility and high energy-storing capacity give fat molecules, unlike carbohydrates, chemical properties similar to those of hydrocarbons.

The chemical properties of a fat molecule are due, in part, to the *fatty acid* groups in the molecule. **Fatty acids** are a class of organic compounds composed of long hydrocarbon chains with a carboxylic acid group (—COOH) at one end. Two fatty acids are shown in Figure 7.22.

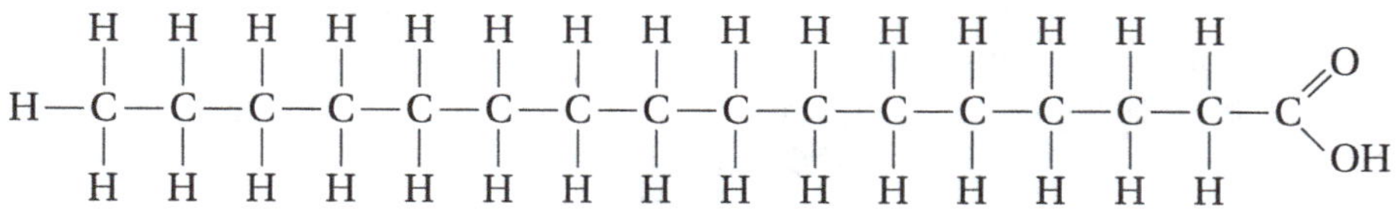

Palmitic acid

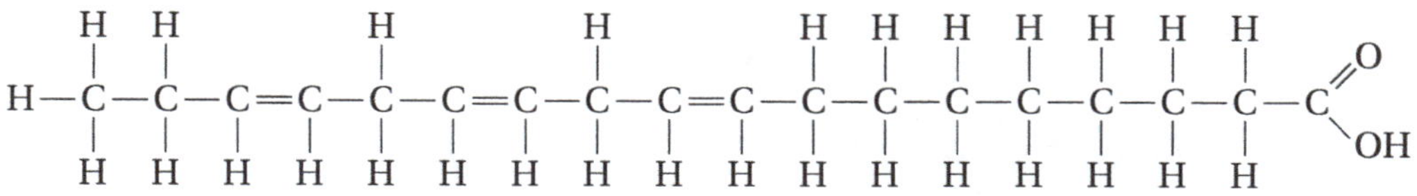

Linolenic acid

Figure 7.22 *Typical fatty acids.*

The typical fat molecule, a **triglyceride**, is a combination of a three-carbon molecule called *glycerol* and three fatty acid molecules.

1 Glycerol molecule	+	3 Fatty acid molecules	⟶	1 Triglyceride (fat) molecule	+	3 Water molecules

The formation of a triglyceride is shown in Figure 7.23. Each fatty acid molecule forms an ester linkage, as it reacts with an —OH group of glycerol. This produces one water molecule per ester linkage. The main product of this condensation reaction is a triglyceride composed of three ester groups. Although the three fatty acids depicted in Figure 7.23 are identical, fats can contain two or three different fatty acids.

In general, the terms "fat" and "triglyceride" are used interchangeably.

The reaction that produces a fat molecule is similar to the reaction you observed in Unit 3 that produced an ester, methyl salicylate (page 319).

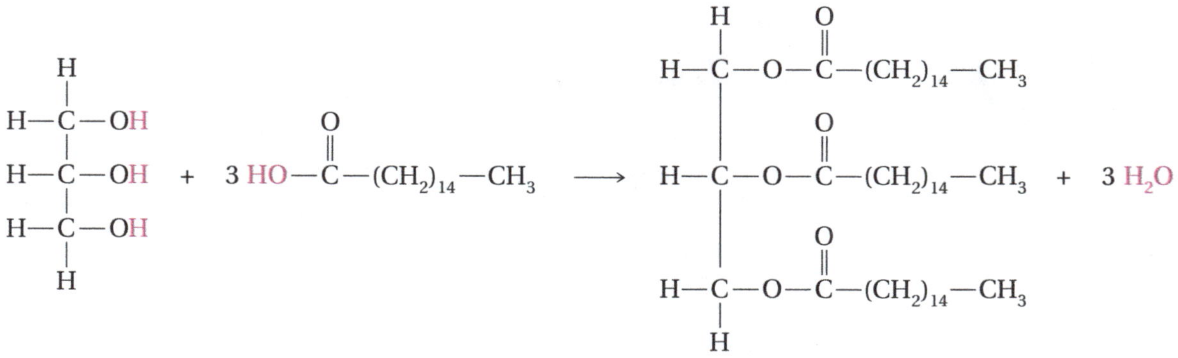

Glycerol   Palmitic acid   Glyceryl tripalmitate (a typical fat)   Water

Figure 7.23 *Formation of a typical fat molecule, a triglyceride. The colored atoms interact to form the fat molecule plus water.*

Because they have more carbon–hydrogen bonds than carbon–oxygen bonds, fats contain more stored energy per gram than do carbohydrates. In fact, one gram of fat contains over twice the energy stored in one gram of carbohydrate; 1 g fat is equivalent to 9 Cal, compared to 4 Cal for each gram of carbohydrate. Consequently, you must run more than twice as far or exercise twice as long to "work off" a given mass of fat as you must to "work off" the same mass of carbohydrate. For example, a glazed doughnut contains 11.6 g fat. Thus, the glazed doughnut has 11.6 g fat × 9 Cal/g fat = 104 Cal of energy from fat. It is not surprising that your body uses fat molecules to store excess food energy efficiently.

B.3 SATURATED AND UNSATURATED FATS

Recall from Unit 3 that hydrocarbons can be saturated (containing only single carbon–carbon bonds) or unsaturated (containing double or triple carbon–carbon bonds). Likewise, hydrocarbon chains in fatty acids are either saturated or unsaturated. Look again at the fatty acids in Figure 7.22 (page 709). Can you identify each as either saturated or unsaturated?

Fats containing saturated fatty acids are called **saturated fats**; those containing some unsaturated fatty acids are known as **unsaturated fats**. A **monounsaturated fat** contains just one carbon–carbon double bond in its fatty acid components. A **polyunsaturated fat** contains two or more C=C double bonds in the fatty acid portion of a triglyceride molecule. Based on these definitions, is the fat depicted in Figure 7.21 (page 708) saturated, monounsaturated, or polyunsaturated?

Triglycerides in animal fats are nearly all saturated and are solids at room temperature. However, fats from plant sources commonly are polyunsaturated or monounsaturated. In general, higher levels of unsaturation are associated with oils having lower melting points. At room temperature these polyunsaturated fats are liquids (see Figure 7.24).

Due to their C=C double bonds, unsaturated fats undergo addition reactions; saturated fats cannot undergo addition reactions. Thus, these two types of

> Oils are fats that are liquid at room temperature.

Dietary oil/fat

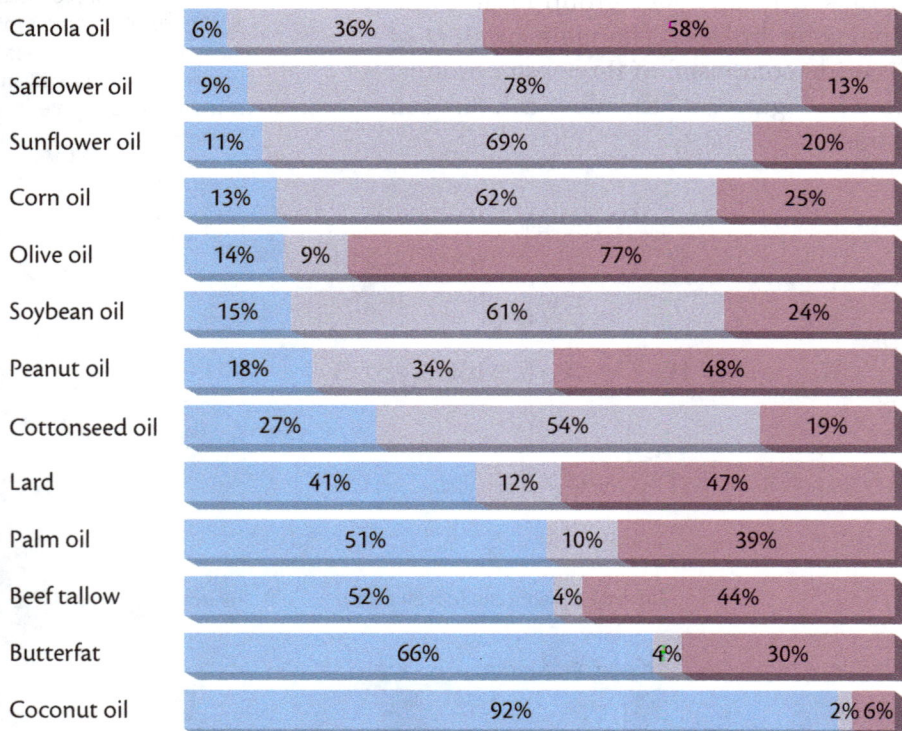

Oil/fat	Saturated fat	Polyunsaturated fat	Monounsaturated fat
Canola oil	6%	36%	58%
Safflower oil	9%	78%	13%
Sunflower oil	11%	69%	20%
Corn oil	13%	62%	25%
Olive oil	14%	9%	77%
Soybean oil	15%	61%	24%
Peanut oil	18%	34%	48%
Cottonseed oil	27%	54%	19%
Lard	41%	12%	47%
Palm oil	51%	10%	39%
Beef tallow	52%	4%	44%
Butterfat	66%	4%	30%
Coconut oil	92%	2%	6%

Key : ▪ Saturated fat   ▪ Polyunsaturated fat   ▪ Monounsaturated fat

Source : *Food Technology*, April 1989.

Figure 7.24 *The percent of polyunsaturated fat in fats from selected plant and animal sources.*

fats participate differently in body chemistry. Unsaturated fat molecules are much more chemically reactive. Increasing evidence suggests that saturated fats may contribute more to health problems than some unsaturated fats. Saturated fats are associated with formation of **arterial plaque**, which consists of fatty material that builds up in blood vessel walls (see Figure 7.25). The result is a condition commonly known as "hardening of the arteries," or **atherosclerosis**, which is a particular threat to coronary (heart) arteries and arteries leading to the brain.

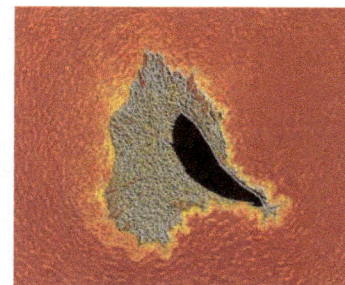

Figure 7.25 *An artery cross section, showing plaque (gray) on walls.*

DEVELOPING SKILLS

B.4 CALORIES FROM FAT

A sample of butter (Figure 7.26) provides the following nutritional values for one serving, which is defined as 1 tablespoon (tbsp) or 14 g:

Total fat	10.9 g	Polyunsaturated fat	0.4 g
Saturated fat	7.2 g	Calories	100

Figure 7.26 *What type of fat is primarily found in butter?*

Sample Problem: *Calculate the percent of polyunsaturated fats contained in butterfat.*

To find the answer, divide the mass of polyunsaturated fat by the mass of total fat and convert to a percent value:

$$\frac{0.4\ g}{10.9\ g} \times 100\% = 4\%$$

Thus, approximately 4% of butter's total fat is polyunsaturated.

1. Calculate the percent saturated fat in the total fat in butter.
2. The saturated and polyunsaturated fat percent values don't total 100%.
   a. What does the "missing" percent value represent?
   b. How many grams of fat does this represent?
3. Calculate the total percent fat in one serving of butter.
4. The *Dietary Guidelines for Americans 2010* suggest that from 20–35% of total Calories in human diets should come from fats. A way to evaluate this guideline for an individual food is to compare the total Calories from fat to the total Calories delivered in one serving of that food:
   a. Determine the total Calories from fat in 1 tbsp of butter.
   b. Calculate the percent of total Calories obtained from butter fat.
5. One serving (14 g) of margarine contains 10 g fat and 90 Cal.
   a. Find the percent of Calories obtained from margarine fat.
   b. Compare the percent of total Calories from margarine fat to the value from butter fat.

Most fats, according to the *Dietary Guidelines*, should come from polyunsaturated and monounsaturated fatty acids, supplied by fish, nuts, and legumes.

Recall that
1 g fat = 9 Calories.

B.5 HYDROGENATION

Partial **hydrogenation** adds hydrogen atoms to some C=C bonds in vegetable oil triglyceride molecules. The reaction with hydrogen converts a C=C double bond to a C—C single bond. Because this decreases the total unsaturated (C=C) sites, the partially hydrogenated product becomes more saturated, and the original liquid oil becomes semisolid. The ability to change oil from a liquid to a solid, depending on the extent of hydrogenation, allows food manufacturers to control the consistency and softness of their food products. Such partially hydrogenated fats are used in margarine, vegetable shortening, deep-fried foods, and snack foods.

The reaction between hydrogen and C=C double bonds in a polyunsaturated fat is typical of how C=C bonds in most alkenes react. Generally, C=C bonds in alkenes can be converted to C—C single bonds when they react with a variety of substances, including hydrogen.

You learned in Unit 3 (page 311) that a double bond in an alkene prevents adjacent carbon atoms in the double bond from rotating around the bond axis. The double-bonded carbon atoms align as shown in Figure 7.27. This inflexible arrangement between carbon atoms creates the possibility of *cis-trans isomerism*.

Ethene

Figure 7.27 *The structural formula for ethene.*

In **cis-trans isomerism**, two identical functional groups can be in one of two different molecular positions. Both groups can be on the same side of the double bond (the *cis* isomer), or they can be across the double bond from the other (the *trans* isomer), as illustrated in Figure 7.28.

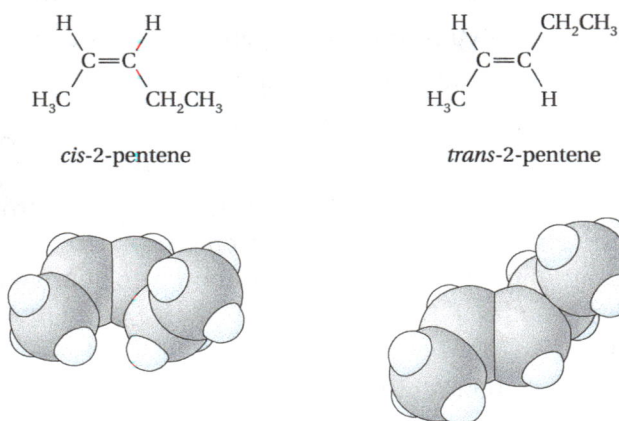

cis-2-pentene   *trans*-2-pentene

Figure 7.28 *A comparison of cis-2-pentene and trans-2-pentene. Note the positions of the functional groups.*

In the example of 2-pentene, each arrangement creates a different compound, even though the compounds have identical molecular formulas (C_5H_{10}). In *cis*-2-pentene, the molecule has the two hydrogen atoms on the same side (cis) of the double-bond plane (see Figure 7.28). In *trans*-2-pentene, the two hydrogen atoms are positioned on opposite sides (trans) of the double-bond plane.

Cis-trans isomerism, which was just described for an alkene, is also possible in unsaturated fats because fatty acid chains often contain C=C bonds. Unsaturated fatty acids in foods that have not been hydrogenated typically have their double bonds arranged in the cis arrangement, as shown in Figure 7.29. During hydrogenation, some cis double bonds break and reform in the trans arrangement. Although questions have arisen about nutritional safety of *trans*-fatty acids, studies addressing these concerns have not yet produced conclusive answers. However, the *Dietary Guidelines* advise that people should "keep *trans*-fatty acid consumption as low as possible."

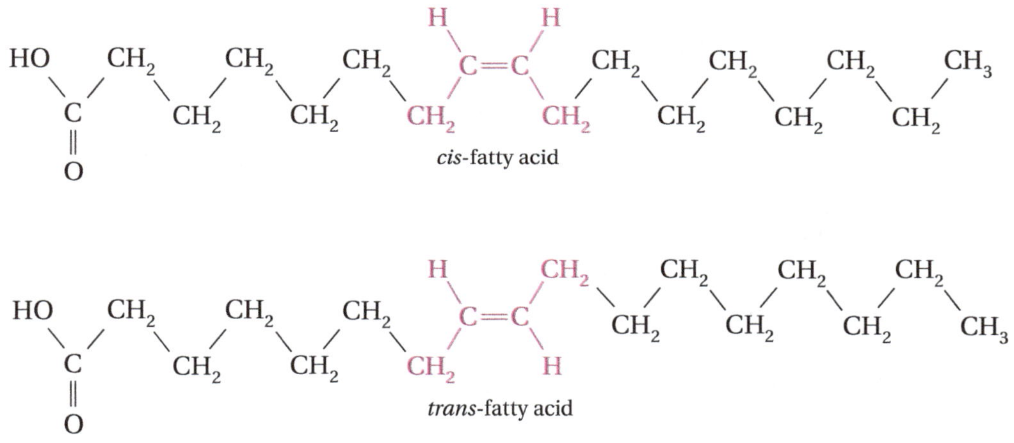

Figure 7.29 Cis- and trans-fatty acids.

Currently, most U.S. citizens obtain about 33% of their total food Calories from fats, within the upper limit of 35% recommended by *Dietary Guidelines*. Additionally, the *Guidelines* suggest that less than 7% of total food Calories should come from saturated fats.

High fat consumption is a factor in several modern health problems, including obesity and atherosclerosis. Most dietary fat consumed in the United States comes from processed meat, poultry, fish, and dairy products. Fast foods and deep-fried foods—such as hamburgers, french fries, fried chicken, and many snack items—add even more dietary fat. In addition, if your intake of food energy is higher than what you expend in physical activity, your body converts excess proteins and carbohydrates into fat for storage.

DEVELOPING SKILLS

B.6 FATS IN THE DIET

Your favorite ice-cream flavor just became available as frozen yogurt. The advertisement promotes frozen yogurt as a "reduced-fat alternative" to ice cream. Here are the nutritional data:

	Serving	Calories	Saturated Fat	Unsaturated Fat
Ice Cream	115 g	310	11 g	8 g
Frozen Yogurt	112 g	200	2.5 g	3.5 g

Sample Problem: *Calculate the percent saturated fat by mass in a serving of ice cream.*

To find the answer, divide the mass of saturated fat in a serving by the total mass of a serving and convert to a percent value:

$$\frac{11\ g}{115\ g} \times 100\% = 9.6\%$$

Thus, ~10% of a serving of ice cream is saturated fat.

Recall that 1 g of fat contains 9 Cal of food energy.

1. Calculate and compare the total percent fat (by mass) contained in each dessert.

2. Another way to compare these desserts is to examine the percent of Calories from fat in each.

   a. Does either ice cream or frozen yogurt meet the guideline of 35% or less Calories from fat?

   b. Does either meet the guideline of 10% or less Calories from saturated fat?

CHEMQUANDARY

FAT-FREE FOOD?

You just compared ice cream to frozen yogurt in terms of their fat content. You might consider *fat-free* ice cream as another frozen dessert. Consider this ice cream label carefully. Is fat-free ice cream actually fat free? How do you know? What will happen to the food energy provided by fat-free ice cream if it is not immediately used by the body? Do you think it is wise to base a diet entirely on fat-free foods? Why or why not?

Vanilla Chocolate Swirl
CONTAINS 100 CALORIES COMPARED TO 150 CALORIES IN OUR REGULAR ICE CREAM. A 33% REDUCTION.

Nutrition Facts
Serving Size: 1/2 Cup (71g)
Servings Per Container: 14

Amount Per Serving

Calories 100   Calories from Fat 0

	% Daily Value*
Total Fat 0g	1%
Saturated Fat 0g	0%
Cholesterol 0mg	0%
Sodium 50mg	2%
Total Carbohydrate 20g	7%
Dietary Fiber 6g	24%
Sugars 4g	
Sugar Alcohol 3g	
Protein 4g	

✓ concept check 4

1. Describe structural similarities and differences between carbohydrate and fat molecules.
2. Explain why there are no cis or trans saturated fats.
3. Do all chemical reactions proceed until there are no reactants left? Explain.

B.7 LIMITING REACTANTS

Biochemical reactions in your body convert the fats, carbohydrates, and—under extreme conditions—proteins in foods you eat into energy. These biochemical reactions, like all chemical reactions, require the presence of a complete set of reactants to produce the desired product. Furthermore, the amount of product produced by a chemical reaction depends upon the amounts of reactants present.

Think about baking a cake and consider the following recipe:

2 cups flour	1 tablespoon baking powder
2 eggs	1 cup water
1 cup sugar	1/3 cup oil

The proper combination of these quantities (and some thermal energy) will produce one cake. What if you have 14 cups flour, 4 eggs, 9 cups sugar, 10 tablespoons baking powder, 10 cups water, and 3 1/3 cups oil. How many complete cakes can you bake?

Well, 14 cups flour is enough for 7 cakes (2 cups flour per cake). And there is enough sugar for 9 cakes (1 cup sugar per cake). The supplies of baking powder, water, and oil are sufficient for 10 cakes (confirm this with the recipe). However, it is not possible to make 10, 9, or even 7 cakes with available ingredients.

Why? Because only 4 eggs are available, which is enough for 2 cakes. The supply of eggs limits the number of cakes you can make. Excess quantities of other ingredients (flour, sugar, baking powder, water, and oil) remain unused. If you want more than two cakes, you will need more eggs.

In chemical terminology, the eggs in this cake-making analogy represent the *limiting reactant* (also called the *limiting reagent*). The limiting reactant is the starting material (reactant) that is used up entirely when a particular chemical reaction occurs. This starting material limits how much product can be formed.

The idea of limiting reactants applies equally well to living systems. A shortage of a key nutrient or reactant can severely affect the growth or health of plants and animals. In many biochemical processes, a product from one reaction becomes a reactant for other reactions. If a reaction stops because one substance is completely consumed (the limiting reactant), all reactions that follow it will also stop.

> Recall that stoichiometry involves quantitative relationships among reactants and products in a reaction described by its chemical equation.

> Limiting reactants were first introduced in Unit 3, page 343.

Fortunately, in some cases, alternate reaction pathways are available. If the body's glucose supply is depleted, for example, glucose metabolism cannot occur. One backup system oxidizes stored body fat in place of glucose. More drastically, under starvation conditions, structural proteins are broken down and used for energy. Producing glucose from protein is much less energy efficient than producing glucose from carbohydrates; if dietary glucose becomes available later, glucose metabolism reactions start up again.

Alternate reaction pathways are not a permanent solution. If the intake of a vital nutrient is consistently inadequate, that nutrient may become a limiting reactant in biochemical processes and affect personal health.

MODELING MATTER

B.8 LIMITING-REACTANT ANALOGIES

In Unit 2, you considered a container filled with super-bounce balls as an analogy for the kinetic molecular behavior of gases (see page 184). Another analogy, a cake recipe, introduced you to the idea of limiting reactants. Think about this and other limiting-reactant analogies (See Figure 7.30) as you answer the following questions.

1. Consider the cake-making analogy (page 715). Assume that you have 26 eggs and the quantities of all other ingredients as previously specified

   a. Which ingredient now limits the total cakes that you can make?

   b. How many total cakes can you make under the conditions specified?

   c. When the limiting reactant is fully consumed, how much of each other ingredient will be left over?

2. A restaurant prepares carryout lunch boxes. Each box consists of 1 sandwich, 3 cookies, 2 paper napkins, 1 milk carton, and 1 container. The current inventory is 60 sandwiches, 102 cookies, 38 napkins, 41 milk cartons, and 66 cardboard containers.

   a. As carryout lunch boxes are prepared, which item will be used up first?

   b. Which item is the limiting reactant?

   c. How many complete carryout lunch boxes can be assembled from this inventory?

3. Now consider the cellular respiration of glucose:

   $$C_6H_{12}O_6 + 6\,O_2 \longrightarrow 6\,CO_2 + 6\,H_2O + \text{thermal energy}$$

   a. Suppose that 2 glucose molecules are available for this reaction. How many oxygen molecules would be needed to completely react with these 2 glucose molecules?

b. Assume that 15 oxygen molecules are available to react. How many carbon dioxide and water molecules can be made from that amount of oxygen and unlimited glucose? (*Hint:* Draw representations of all molecules available to react.)

4. Consider the cake-making and carryout lunch box analogies as they relate to chemical reactions.

   a. Using pictures or symbols, write a balanced equation to represent each of these analogies. (*Note:* Your pictures should convey information comparable to that conveyed by a chemical equation.)

   b. Describe one way in which these analogies help you to understand the notion of limiting reactants as applied to chemical reactions.

5. As you know, useful analogies also have limitations.

   a. Consider the lunch-box analogy. Even if you run out of one item, can you use the other items to prepare carryout lunches that people would still buy and eat? Explain.

   b. Consider the cake-making analogy. If you ran out of one ingredient, can you still use the other ingredients to make an edible cake? Explain.

   c. Consider the glucose reaction. If one reactant is used up, or if you have less than the reaction requires (for instance, one glucose molecule but only three oxygen molecules), could any product be produced? Explain.

   d. Considering your answers to Questions 5a through 5c, write a statement that summarizes a major limitation of both cake-making and lunch box analogies as they relate to actual chemical reactions.

6. Now it is your turn to devise a limiting-reactant analogy. Think about a limiting-reactant problem that you may encounter in everyday life.

   a. Identify what item or product you are trying to create.

   b. List all the reactants you will need.

   c. Identify the limiting reactant and how many other reactants are left over when the reaction is completed.

Figure 7.30 *What key ingredients are needed to make this triple-stacker s'more? How could the "limiting reactant" concept apply here?*

B.9 LIMITING REACTANTS IN CHEMICAL REACTIONS

In chemical reactions, just as in baking, materials react in fixed ratios. These ratios are referred to as the *stoichiometry* of the reaction. The ratios are indicated in balanced chemical equations by the coefficients of each substance. Consider the equation for cellular respiration ("burning") of glucose:

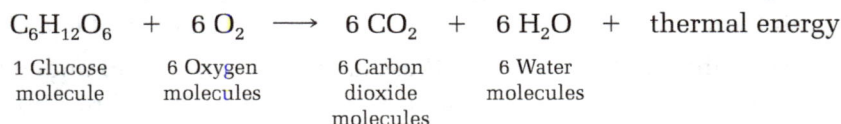

$$C_6H_{12}O_6 \ + \ 6\,O_2 \ \longrightarrow \ 6\,CO_2 \ + \ 6\,H_2O \ + \ \text{thermal energy}$$

| 1 Glucose molecule | 6 Oxygen molecules | 6 Carbon dioxide molecules | 6 Water molecules |

Suppose you have 5 glucose molecules available to react with 60 oxygen molecules. Which substance will become the limiting reactant in this reaction? From the equation, you can see that 1 glucose molecule reacts with 6 oxygen molecules. That means that 5 glucose molecules would require 30 oxygen molecules to react completely. On the other hand, to use up all 60 oxygen molecules, 10 glucose molecules are required. Which of those two scenarios is actually possible in this reaction? In other words, which reactant—oxygen or glucose—will be used up completely in this reaction?

Because 60 oxygen molecules are available, all the glucose can react, with some oxygen left unreacted. Alternatively, to react completely, the 60 oxygen molecules would require 10 glucose molecules; however, only 5 glucose molecules are available to the reaction. Thus, glucose is the limiting reactant. Because the reaction stops after the 5 glucose molecules react with 30 oxygen molecules, 30 oxygen molecules will remain unreacted ("excess") at the end of the reaction.

A chemical equation can be interpreted not only in terms of molecules but also in terms of *moles* and *grams*. For example, the oxidation reaction for glucose can be interpreted in terms of moles of reactants and products:

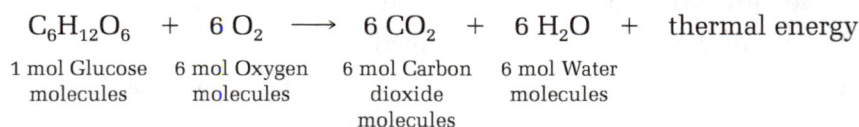

$$C_6H_{12}O_6 \ + \ 6\,O_2 \ \longrightarrow \ 6\,CO_2 \ + \ 6\,H_2O \ + \ \text{thermal energy}$$

| 1 mol Glucose molecules | 6 mol Oxygen molecules | 6 mol Carbon dioxide molecules | 6 mol Water molecules |

Molar masses can be used to convert those molar amounts to grams:

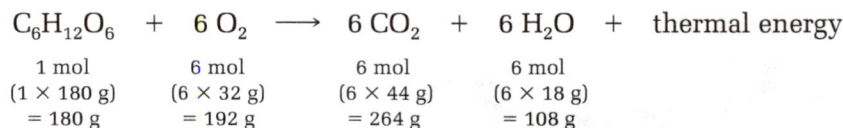

$$C_6H_{12}O_6 \ + \ 6\,O_2 \ \longrightarrow \ 6\,CO_2 \ + \ 6\,H_2O \ + \ \text{thermal energy}$$

1 mol	6 mol	6 mol	6 mol
$(1 \times 180\text{ g})$	$(6 \times 32\text{ g})$	$(6 \times 44\text{ g})$	$(6 \times 18\text{ g})$
$= 180\text{ g}$	$= 192\text{ g}$	$= 264\text{ g}$	$= 108\text{ g}$

> The sum of reactant masses (180 g + 192 g) equals the sum of product masses (264 g + 108 g). Why must that be so?

This equation provides information about the stoichiometric relationships among the molecules that remain the same regardless of the reacting amounts involved. For example: If only 90.0 g (0.500 mol) glucose is oxidized (half the mass above), only 96.0 g (3.00 mol) O_2 is needed.

DEVELOPING SKILLS

B.10 USING CHEMICAL EQUATIONS TO FIND LIMITING REACTANTS

As illustrated by the following sample problems, chemical equations can help you identify the limiting reactant.

Sample Problem 1: *How much CO_2 can be produced if 100.0 g glucose reacts as completely as possible with 100.0 g O_2? Which is the limiting reactant—oxygen or glucose?*

First write the chemical equation:

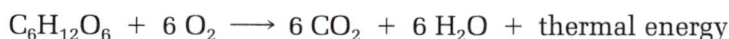

$$C_6H_{12}O_6 + 6\,O_2 \longrightarrow 6\,CO_2 + 6\,H_2O + \text{thermal energy}$$

Next, calculate how much CO_2 could be produced if *each* reactant were fully consumed. The total amount of product formed depends on the amount of reactant that is used up first (the limiting reactant). When the limiting reactant is totally consumed, no more product can be produced. Calculating the amount of product that theoretically could be produced from each individual reactant makes it easy to identify the limiting reactant. That is, by comparing the results of these calculations, the *least* amount of product formed identifies the limiting reactant.

Consider how much CO_2 could be produced if glucose fully reacts (assume that sufficient oxygen is available):

$$100.0 \text{ g glucose} \times \frac{1 \text{ mol glucose}}{180.0 \text{ g glucose}} \times \frac{6 \text{ mol } CO_2}{1 \text{ mol glucose}} = 3.33 \text{ mol } CO_2$$

Determine how much CO_2 could be produced if the O_2 fully reacts (assume that sufficient glucose is available):

$$100.0 \text{ g } O_2 \times \frac{1 \text{ mol } O_2}{32.0 \text{ g } O_2} \times \frac{6 \text{ mol } CO_2}{6 \text{ mol } O_2} = 3.12 \text{ mol } CO_2$$

We find that 100.0 g O_2 produces less CO_2 than 100.0 g glucose would produce. That suggests that when 100.0 g O_2 combines with 100 g glucose, O_2 is the reactant that is used up first. When all the O_2 has reacted, the reaction stops, with some excess glucose still present.

Sample Problem 2: *How much glucose remains unreacted when the reaction just described finally stops?*

Here is one way to determine this. First find out how much glucose reacts with O_2:

$$100.0 \text{ g } O_2 \times \frac{1 \text{ mol } O_2}{32.0 \text{ g } O_2} \times \frac{1 \text{ mol glucose}}{6 \text{ mol } O_2} \times \frac{180 \text{ g glucose}}{1 \text{ mol glucose}} = 93.7 \text{ g glucose}$$

Then, find how much glucose remains unreacted:

(100 g glucose initially) − (93.7 g glucose reacted) = 6.3 g glucose left over.

Now it is your turn to check your understanding of this concept.

1. Lactose, milk sugar, is a disaccharide (see Table 7.2, page 707). In the human body, the enzyme lactase (see Figure 7.31) helps hydrolyze a lactose molecule into one molecule each of two different monosaccharides, glucose and galactose:

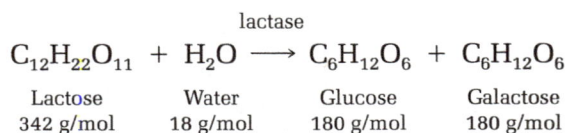

Figure 7.31 *Some dairy products contain added lactase, an enzyme that promotes digestion of lactose, for "lactose-intolerant" individuals.*

$$\overset{\text{lactase}}{C_{12}H_{22}O_{11} + H_2O \longrightarrow C_6H_{12}O_6 + C_6H_{12}O_6}$$

Lactose	Water	Glucose	Galactose
342 g/mol	18 g/mol	180 g/mol	180 g/mol

Suppose that you are investigating this reaction. During one trial, you combine 1.5 g lactose and 10.0 g water with a lactase enzyme.

   a. Identify the limiting reactant in this trial.

   b. How many moles of glucose will this reaction produce?

   c. How many grams of glucose will be produced?

   d. After the reaction stops, which reactant will be left over; that is, which reactant was in excess?

   e. How many moles of that substance will be left unreacted?

2. Imagine you are also investigating how the body digests fat. The process starts with hydrolyzing the triglyceride. Each molecule reacts with three molecules of water to form three molecules of fatty acid and one molecule of glycerol:

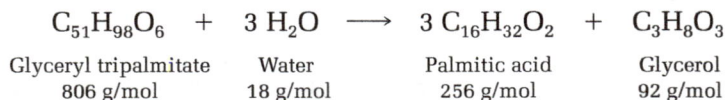

$$C_{51}H_{98}O_6 + 3 H_2O \longrightarrow 3 C_{16}H_{32}O_2 + C_3H_8O_3$$

Glyceryl tripalmitate	Water	Palmitic acid	Glycerol
806 g/mol	18 g/mol	256 g/mol	92 g/mol

During one trial, you combine 4.0 g glyceryl tripalmitate with 1.0 g water.

   a. Identify the limiting reactant in this trial.

   b. How many moles of palmitic acid will this reaction produce?

c. What mass (in grams) of palmitic acid will be produced?

d. After the reaction stops, which reactant will be left over; that is, which reactant was initially in excess?

e. How many moles of that substance will remain unreacted?

▋MAKING DECISIONS

B.11 ANALYZING FATS AND CARBOHYDRATES

In Section A, you analyzed the energy content of a three-day diet. Using information in provided in this section, you can now identify the main sources of this energy.

1. Using appropriate resources, determine the mass of carbohydrates (in grams) contained in each item on the three-day food inventory. Then, calculate the percent of total Calories provided by carbohydrates. (*Note:* You calculated average Calories per day on page 700.) Report both sets of data in a suitable summary table.

2. If possible, identify the mass (in grams) of fiber in each item, as well. Record these data.

3. Using appropriate resources, determine the mass of fat (in grams) in each item in the inventory. If possible, identify the total mass of saturated and unsaturated fat in each item. Record these data.

4. Calculate the average mass of fat supplied daily by the three-day diet described in the food inventory.

5. a. Calculate the average daily energy (in Calories) supplied by fat.

   b. Based on the value calculated in Question 5a, is food energy from fat less than 30% of the total food energy supplied?

6. Identify possible ways to reduce the quantity of fat in the food inventory you are analyzing.

7. Now think of two possible Riverwood High School vending machine items that are *not* on your inventory.

   a. Find the total mass of carbohydrates (in grams) in each item. If possible, also report the total mass (in grams) of fiber in each item, as well. Record these data.

   b. Determine the mass of fat (in grams) in each item. If possible, report the total mass of saturated and unsaturated fat in each item. Record these data.

   c. Is the food energy from fat less than 30% of the total energy supplied by each item?

SECTION B SUMMARY

Reviewing the Concepts

> Carbohydrates, composed of carbon, hydrogen, and oxygen, include sugars, starches, and cellulose.

1. What is the origin of the term *carbohydrate*?

2. Using a shaded oval to represent a single monosaccharide molecule, sketch a model of a
   a. monosaccharide.
   b. disaccharide.
   c. polysaccharide.

3. Name two common monosaccharides and two common disaccharides.

4. All sugars are carbohydrates, but not all carbohydrates are sugars. Explain.

5. What kind of chemical reaction causes monosaccharides to link and produce more complex carbohydrates?

6. Why is starch classified as a polysaccharide?

7. How many Calories would be provided by 25 g carbohydrate?

> Fats and carbohydrates are both composed of carbon, hydrogen, and oxygen atoms, but they differ in structure and function.

8. What are fatty acids?

9. Consider the reaction that forms a typical fat molecule.
   a. List the substances that are combined in this reaction.
   b. How would you classify this reaction?
   c. Name the new functional group formed in this reaction.

10. Fats are examples of triglycerides. Why does that name provide an appropriate description of a fat molecule?

11. How many Calories would be provided by 10.6 g of olive oil? Olive oil is 100% fat.

12. What aspect of their structure makes fats sparingly soluble in water?

> Fats may be saturated or unsaturated. Unsaturated fats can be made more highly saturated through hydrogenation.

13. Define, in terms of their carbon—carbon bonds, the terms *saturated* fat and *unsaturated* fat.

14. Distinguish between monounsaturated and polyunsaturated fats.

15. With what is a saturated fat actually saturated?

16. How does the degree of saturation affect a fat molecule's properties?

17. The ingredient labels on some brands of margarine state that the oil is partially hydrogenated.
   a. What property does the product possess as a result of partial hydrogenation?
   b. Why do you think a manufacturer might decide against completely hydrogenating margarine?

18. What is the difference between cis and trans isomers?

19. Why do cis and trans isomers occur with a C=C double bond and not with a C—C single bond?

20. What are some natural sources of
   a. saturated fats?
   b. unsaturated fats?

21. Draw structural formulas for cis and trans isomers of 2-butene, CH_3—CH=CH—CH_3.

22. Draw a structural formula of a fat molecule containing both *cis-* and *trans-*fatty acids.

> Differences in structure between fat and carbohydrate molecules account for their different properties and energy content.

23. How are functional groups in fat molecules different from functional groups in carbohydrates? Use structural formulas to clarify your explanations.

24. From a chemical standpoint, explain why fat molecules contain more food energy per gram than do carbohydrate molecules.

25. List two chemical characteristics shared by all
    a. fats.   b. carbohydrates.

26. What property of fats makes them good energy-storage molecules?

> A chemical reaction involves substances interacting in specific, stoichiometric relationships. The limiting reactant determines how much product can be produced.

27. Is it ever possible to use up *all* reactants available for a chemical reaction? Explain.

28. How can a *limiting reactant* be identified?

29. Suppose you want to make a batch of s'mores. The "equation" for producing one s'more is as follows:

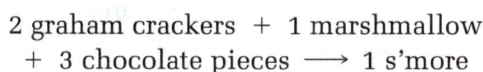

    2 graham crackers + 1 marshmallow + 3 chocolate pieces ⟶ 1 s'more

    If you have 12 graham crackers, 25 marshmallows, and 12 chocolate pieces:
    a. How many s'mores can you make?
    b. Which is the limiting "reactant"?
    c. How much of each excess reactant will be left over?

30. Consider the reaction of hydrogen gas and oxygen gas to produce water:

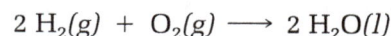

    $$2 H_2(g) + O_2(g) \longrightarrow 2 H_2O(l)$$

    a. If 4.0 mol hydrogen gas are combined with 4.0 mol oxygen gas and the reaction described occurs, how many moles of water could be produced?
    b. What is the limiting reactant in this reaction?
    c. How much of the excess reactant will be left over?

31. One step in an early soap-making method was to allow fireplace wood ashes (containing potash, K_2O) to react with water, producing a highly basic potassium hydroxide (KOH) solution. The equation is as follows:

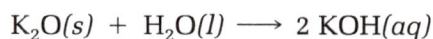

    $$K_2O(s) + H_2O(l) \longrightarrow 2 KOH(aq)$$

    Assume that a 5.4-g sample of K_2O reacts as completely as possible with 9.0 g water.
    a. What will be the limiting reactant?
    b. What mass (in grams) of potassium hydroxide will be produced?

What role does molecular structure play in metabolism of carbohydrates and fats?

In this section, you have learned about the chemistry of nutrition and how your body uses the food you consume to sustain life. You have explored the chemical structures and properties of carbohydrate and fat molecules and the difference in energy available from the metabolism of these molecules. Think about what you have learned, then answer the question in your own words in organized paragraphs. Your answer should demonstrate your understanding of the key ideas in this section.

Be sure to consider the following in your response: structures of simple and complex carbohydrates; chemical composition and structures of fat molecules; the nutritional value of carbohydrates and fats; nutritional guidelines; and limiting reactants.

Connecting the Concepts

32. Why would it be unhelpful to label a container "sugar" in a chemical storeroom?

33. Explain how a relatively small number of subunits (fatty acids and monosaccharides) can create so many different kinds of fats and carbohydrates.

34. How do cis-trans isomers compare to the structural isomers you have learned about? (Recall that isomers were discussed in Unit 3, page 290.)

35. Polysaccharides are examples of naturally occurring polymers.

    a. How are they similar to synthetic polymers?

    b. How are they different from synthetic polymers?

36. Compare the chemical properties of fats to the chemical properties of alkanes.

37. Why is the term "burning fat" sometimes used to describe exercising, even though no actual burning is involved?

38. How are chemical equations similar to recipes used for cooking? How are they different?

39. In thinking about the reaction $HCl + NaOH \longrightarrow NaCl + H_2O$, one student concluded that 1.0 g HCl should react completely with 1.0 g NaOH. Explain why this reasoning is incorrect.

40. Consider the reaction described in Question 30. Specify a pair of starting masses (in grams) for hydrogen gas and oxygen gas so that *both* reactants would completely react in the chemical change.

Extending the Concepts

41. Research three different triglycerides and list the fatty acids in each.

42. Examine the ingredient listing on several food packages in your home. Identify the names of any specific carbohydrates or fats that you recognize.

43. Human consumption of *trans*-fatty acids may be associated with health risks. Investigate and summarize current research regarding this nutritional concern.

44. Fat-like products such as Olestra were designed and marketed as fat replacements in some foods. Investigate the chemistry of these fat substitutes and relate that information to health risks and benefits.

45. Assume that you currently consume 3000 Cal of food energy daily and want to lose 30 lb of body fat within two months (60 days).

a. If you decide to lose that weight only by reducing your intake of food energy (with no extra exercise), how many food Calories, on average, would you need to remove from your daily diet?

b. How many food Calories could you still consume daily?

c. Would the scheme described in Questions 45a and 45b, if followed, be a sensible way to lose weight? Why?

46. Some well-publicized diet programs use terms such as *net carbs*, *counting carbs*, and *glycemic index*. Research the meanings of these terms and analyze their dietary usefulness based on the chemistry that you have learned.

CHEMISTRY *AT WORK*
Q&A

Mark Dewis, Flavor Chemist at International Flavors and Fragrances in Union Beach, New Jersey

Imagine a juicy steak or a ripe peach. These foods are much more than vehicles for calories and nutrients to get into our bodies. They're also tasty foods that make eating enjoyable. It's the job of flavor chemists to make sure that the foods we eat—even those that are changed from their original form, such as beef soup or peach yogurt—are delicious and taste like what we'd expect. Part of flavor chemistry is developing the flavors that we perceive in foods and making these flavors blend with the other ingredients in a food, whether it's yogurt, a soda, or chewing gum. Read on to see how one chemist makes what we eat a tasty way to get what we need.

Q. What is flavor chemistry?

A. At a basic and practical level, flavor chemistry is about understanding why food in its natural state tastes so good. If you take a tomato or grilled steak or wine or butter—or any natural food that people have been consuming for hundreds of years—the flavor chemistry industry tries to replicate this natural food experience. We try to figure out how nature makes food taste and how we can mimic the natural taste in a product.

Q. How did you become a flavor chemist?

A. When I was a high school student, I learned what I was good at and interested in so I wouldn't have to struggle in school. I'm lucky that chemistry happened to be one of those things. I majored in chemistry in college and went on to get a doctorate in synthetic organic chemistry. By the time I graduated, I had several options for careers, including pharmaceuticals, agricultural chemistry, and food and flavor chemistry. The last one was particularly fascinating—I found it compelling that you can do science to re-create food or improve taste.

Umami comes from a Japanese word for flavorful. The taste of mushrooms is umami—not quite salty, sour, sweet, or bitter.

Q. What are flavors, and where do they come from?

A. What the flavor industry and consumers call flavor is a combination of smell and taste. One aspect is aroma, which is kind of like an edible fragrance. We usually think of taste as one of the five basic tastes: bitter, salty, sour, sweet, and umami (or savory). We re-create a lot of flavors from natural sources, such as essential oils, extracts, and fermentation or enzymatic products from real food. We don't need all the flavor compounds in a food to re-create its flavor. For example, coffee has more than 500 flavor compounds, but we can re-create good coffee flavor from just a few of these. Some flavor compounds are made completely in the lab—the artificial flavors you might have seen on food packaging. Most ingredients in a flavor do come from natural sources, although if one ingredient is synthetic, the whole flavor is called artificial.

Q. What considerations go into making a new flavor?

A. First, you have to know what your target is because the same flavor may not work in different kinds of food and drinks. For example, the lemon-lime flavor that works in candy might not work in soda. A lemon-lime candy might be flavored with lemon and lime oils, which you isolate by pricking the fruits' skins. These are 90 percent limonene. Consumers are expecting a lemon-lime soda to be clear, so you can't use limonene because it's not soluble. To make a flavor realistic, you have to mix and match different flavor chemicals to give the product the flavor you want. For a strawberry flavor, is it a fresh strawberry or a candy strawberry? Different chemical combinations will give a different result.

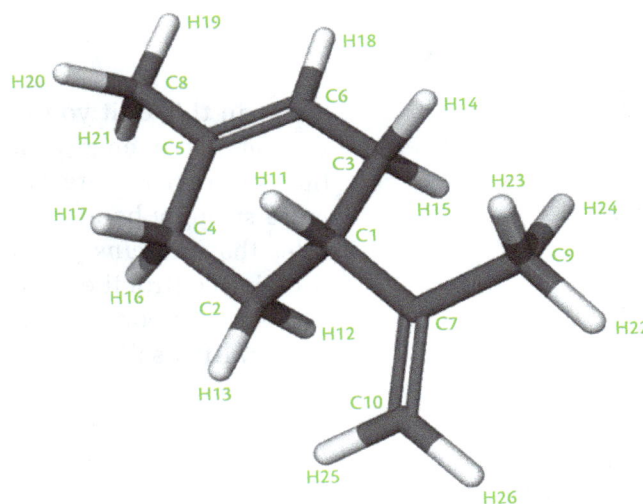

The limonene molecule gives oranges, lemons, and limes their fragrance. Limonene is a kind of hydrocarbon called a terpene.

Q. What advice do you have for students interested in flavor chemistry?

A. Take lots of chemistry classes, especially organic chemistry, because most flavor compounds tend to be organic compounds. It's important to take biology classes, too, including microbiology, biochemistry, and enzymology. You can get some good hands-on experience by getting internships in product-development labs at big food companies that make many different products. It gives you some insight into how flavors work in real life—for example, the different compounds that go into making apple-cinnamon flavors for cereal versus apple pie. You can also get some valuable experience right now. Flavor chemists taste each component that goes into a flavor to learn the set of ingredients they work with, and you can learn quite a bit by tasting different products that are supposed to have the same flavor. For example, chicken soup and cooked chicken, or chocolate and a chocolate milkshake. Once you start paying attention, you'll be surprised how different the "same" flavors taste.

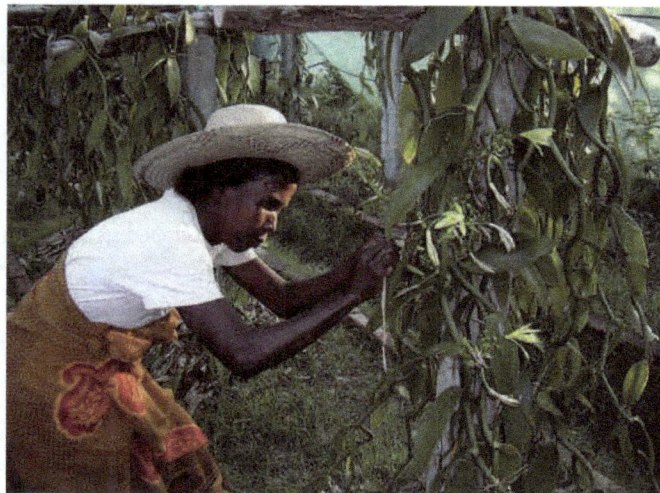

A woman in Madagascar pollinates vanilla flowers by hand. The pods will be harvested and turned into a common flavor essence—vanilla extract.

PROTEINS

Why are protein molecules essential to living organisms?

So far in this unit you have explored how carbohydrates and fats that you eat deliver energy that helps to keep you alive. The chemical bonds that store energy are broken, and energy is released in forming new, more strongly bonded molecules. In this section, you will investigate roles that *proteins* play in maintaining your well-being. For instance, proteins control the release of energy stored in foods and provide structure and support to maintain your cells and organs; in short, helping sustain life itself.

GOALS

- Describe the general structure of a protein molecule.
- Describe how amino acids in food are linked by peptide bonds to form proteins within the body.
- Describe how some protein molecules (enzymes) serve as catalysts.
- Explain why complete or complementary proteins are an essential part of the human diet.

concept check 5

1. Fats and carbohydrates are both made of carbon, hydrogen, and oxygen atoms. Why are the properties of fats and carbohydrates so different?
2. According to the collision theory for chemical reaction rates, what must happen for a chemical reaction to occur?
3. Could proteins be considered polymers? Explain.

C.1 PROTEINS—FUNCTION AND STRUCTURE

Whenever you look at a living creature, most of what you see is protein—skin, hair, feathers (Figure 7.32), eyeballs, fingernails, and claws. **Protein** also is a major structural component of human tissue. Inside your body, muscles, cartilage, tendons, and ligaments are composed of protein.

The word *protein* is based on the Greek *proteios*, which means "of prime importance."

Figure 7.32 *This hummingbird feather is composed of many different proteins.*

At the cellular level, proteins help make it possible to transport materials into and out of cells. Your immune system depends on the ability of protein molecules to identify foreign substances. The rates of many chemical reactions that your cells require would be too slow were it not for special proteins called **enzymes**. Enzymes are specialized molecules that act as catalysts for biochemical reactions. Many proteins act as enzymes in the body. Your body contains tens of thousands of different proteins. Table 7.3 lists some major roles that proteins play in the human body.

Table 7.3

Major Groups of Proteins in the Human Body		
Type	**Function**	**Examples**
Structural proteins		
Muscle	Contraction, movement	Myosin
Connective tissue	Support, protection	Collagen, keratin
Chromosomal proteins	Part of chromosome structure	Histones
Membranes	Control influx and outflow, communication	Pore proteins, receptors
Transport proteins	Carrying of needed substances (such as O_2)	Hemoglobin
Regulatory proteins		
Fluid balance	Maintain pH, water, and salt content of body fluid	Serum albumin
Enzymes	Control metabolism	Proteases
Hormones	Regulate body functions	Insulin
Protective proteins	Attack foreign bodies	Antibodies (gamma globulin)

Proteins are polymers built from smaller molecules called **amino acids**. As Figure 7.33 shows, amino acid molecules contain carbon, hydrogen, oxygen, and nitrogen; a few, like cysteine, also contain sulfur. Just as monosaccharide molecules serve as building blocks for more complex carbohydrates, 20 different amino acids serve as the structural units of all proteins.

All amino acids have several structural features in common, but each amino acid is a unique molecule; four amino acids are shown in Figure 7.33. Every amino acid has two functional groups, the amino group ($-NH_2$) and the carboxylic acid group ($-COOH$).

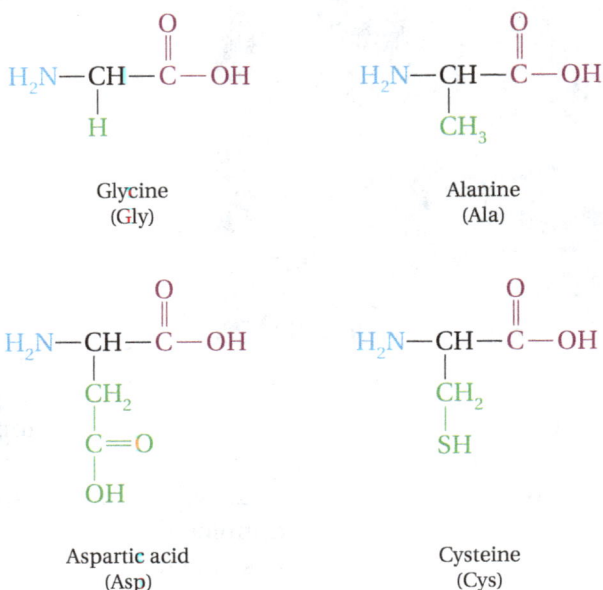

Glycine
(Gly)

Alanine
(Ala)

Aspartic acid
(Asp)

Cysteine
(Cys)

Figure 7.33 *Structural formulas of glycine, alanine, aspartic acid, and cysteine. The amino groups are highlighted in blue; carboxylic acid groups are in red; and green highlights unique side groups that distinguish amino acids from each other.*

Even though hundreds of amino acids have been identified in nature, the human body uses only 20 amino acids.

Like starch, nylon, and polyester, proteins are condensation polymers (see pages 321–322). The combination of two amino acid molecules with loss of one water molecule, as illustrated in Figure 7.34, is a typical condensation reaction. The bonds linking amino acid units together are **peptide bonds**. Proteins are chains of amino acids that vary in length, from ten to several thousand amino acids. Just as the 26 letters of the alphabet combine in different ways to form hundreds of thousands of words, the 20 amino acids can combine in a nearly infinite number of ways to form different proteins that meet many needs of living organisms.

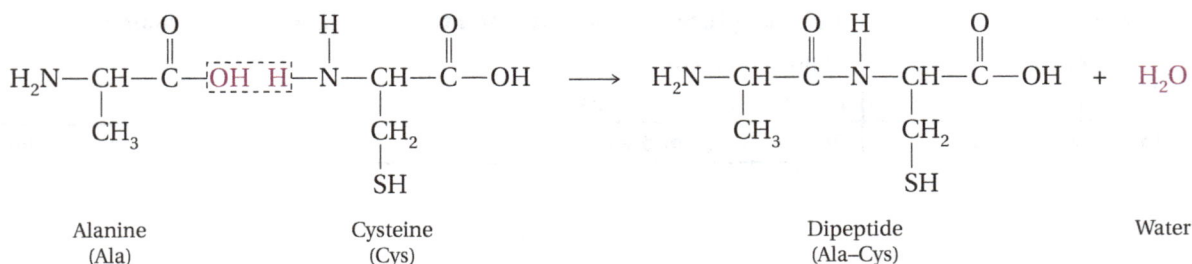

Alanine
(Ala)

Cysteine
(Cys)

Dipeptide
(Ala–Cys)

Water

Figure 7.34 *Formation of a dipeptide from two amino acids (Ala and Cys).*

MODELING MATTER

C.2 MOLECULAR STRUCTURE OF PROTEINS

Protein molecules differ from one another in the number and types of amino acids they contain and in the sequence in which the amino acids are bonded (see Figure 7.35). However, the way in which amino acids bond to one another to form these long polymers is the same for every protein molecule. You will now investigate how amino acids bond and how the sequence of amino acids affects a protein's structure.

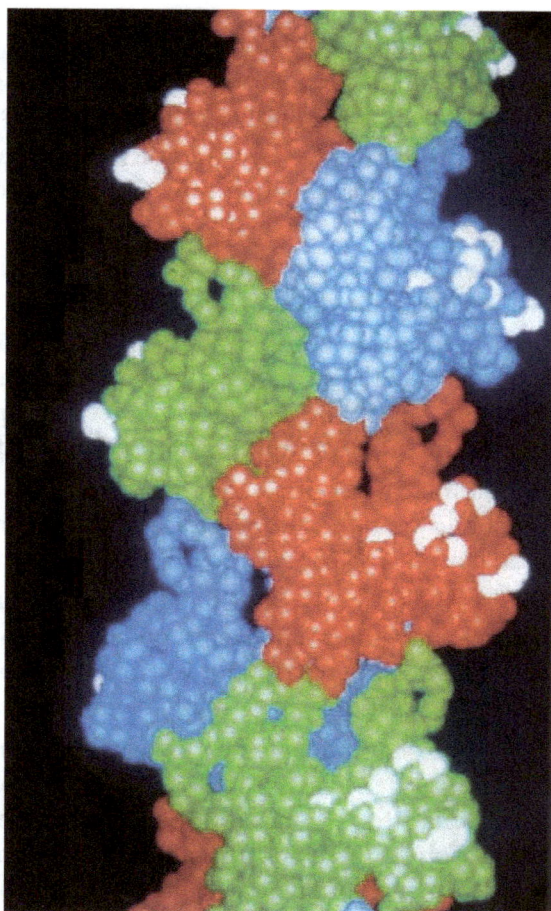

Figure 7.35 *Computer-generated space-filling model depicting a polymer of the protein actin. For clarity, each monomer (the repeated polymer unit) is shown in a different color.*

1. Draw structural formulas for glycine and alanine on a sheet of paper. Refer to Figure 7.33.

   a. Circle and identify the functional groups in each molecule.

   b. How are the two molecules alike?

   c. How do the two molecules differ?

2. Proteins are polymers of amino acids. Examine the equations in Figures 7.34 (page 730) and Figure 7.36 to see how two amino acids join. Notice that the amino group on one amino acid is bonded to the carboxylic acid group on another. This linkage, a peptide bond, is shown here:

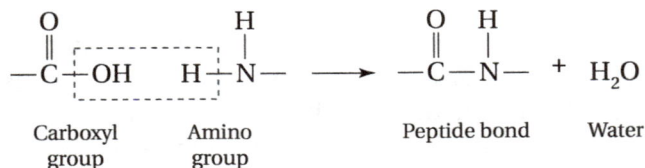

$$-\underset{\substack{\|\\ \text{O}}}{\text{C}}\!\!-\!\!\text{OH} \quad \text{H}\!-\!\text{N}\!-\!\underset{\substack{|\\ \text{H}}}{} \quad \longrightarrow \quad -\underset{\substack{\|\\ \text{O}}}{\text{C}}\!-\!\underset{\substack{|\\ \text{H}}}{\text{N}}\!- \;+\; H_2O$$

Carboxyl   Amino   Peptide bond   Water
group   group

Figure 7.36 *A carboxyl group combing with an amino group.*

When two amino acid units are linked by a peptide bond, the product is called a **dipeptide**. Because an amino acid contains at least one amino group and one carboxylic acid group, amino acids can form a peptide bond at either end.

3. Using structural formulas, write the equation for the reaction between two glycine molecules to form a dipeptide. Circle the peptide bond in the dipeptide.

4. Using structural formulas, write equations for possible reactions between a glycine molecule and an alanine molecule.

5. Examine structural formulas of the dipeptide products identified in Question 4. Note that each dipeptide still contains a reactive amino group and a reactive carboxylic acid group. These dipeptides can then react with other amino acids, forming even more peptide linkages.

6. Remembering that the carboxylic acid group of the first amino acid forms a peptide bond with the amino group of the second amino acid, explain why the amino acid sequence A-B-C is different from C-B-A.

7. Assuming that you have supplies of three different amino acid molecules—A, B, and C—and that each type of amino acid can be used only once, how many different tripeptides (three amino acids linked together) can be formed? Write all possible combinations.

8. Writing the sequence of amino acids by using their chemical names is a time-consuming and tedious process for most protein molecules. Consequently, chemists have devised three-letter (and even one-letter) abbreviations for each amino acid in proteins. For example, a tripeptide with a sequence of glycine–aspartic acid–cysteine is abbreviated Gly-Asp-Cys. Using three-letter abbreviations, write down all other possible sequences for these three amino acids.

9. How many tetrapeptides (four amino acids bonded together) could be formed from the four amino acids shown in Figure 7.33 (page 730)? Write the possible combinations by using their three-letter abbreviations.

10. The cells in your body build proteins from 20 different amino acids. If each protein can be as long as 10 000 amino acid units, would you estimate that the theoretical total of the different proteins that your cells could produce to be in the hundreds, thousands, or millions? Explain your answer.

11. How would a living organism benefit from its ability to synthesize so many different proteins? What potential problems do you envision?

C.3 PROTEIN IN YOUR DIET

What foods are the main protein sources in your diet? To which food groups shown in Figure 7.2 (page 687) do the protein-containing foods in Figure 7.37 belong?

When foods containing proteins reach your stomach and small intestine, peptide bonds break. The separated amino acid units then travel through the intestinal walls to the bloodstream, the liver, and then to the rest of the body. Individual cells can use these amino acids as building blocks for new proteins.

If you eat more protein than your body requires, or if your body needs to use protein because carbohydrates and fats are in short supply, amino acids are metabolized in the liver. Nitrogen atoms are removed and converted in the liver to urea (Figure 7.38), which is excreted through the kidneys in urine. The remaining portions of the amino acid molecules are either converted to glucose and oxidized, releasing 4 Cal of energy per gram, or stored as fat.

The human body can normally synthesize adequate supplies of 11 of the 20 required amino acids. The other nine, called **essential amino acids**, must be obtained from the diet. See Table 7.4 on pages 734–735.

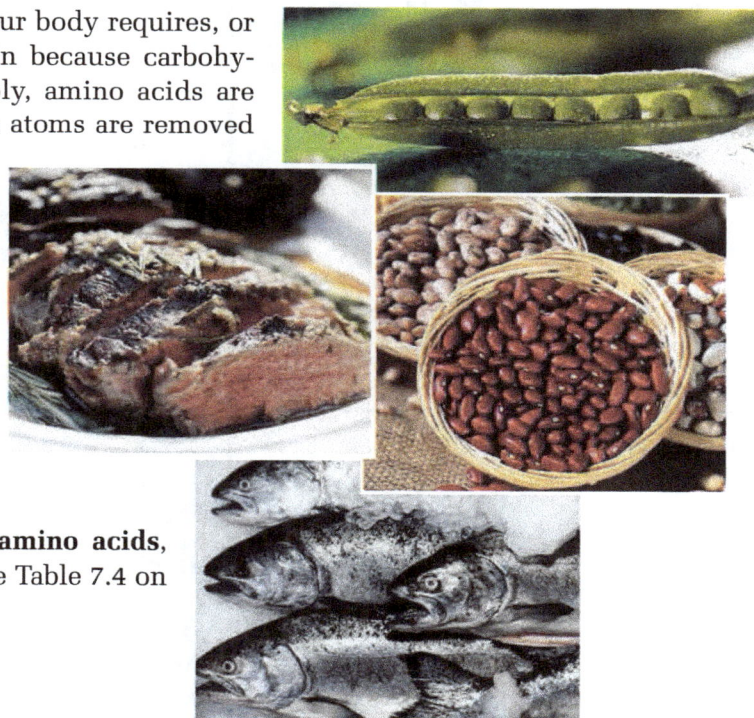

Figure 7.37 *Foods, including plant sources, rich in protein.*

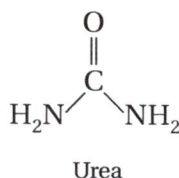

Urea

Figure 7.38 *The structural formula for urea.*

Table 7.4

Amino Acids

Amino Acid	3-Letter Code	Structure	Amino Acid	3-Letter Code	Structure
Alanine	Ala		Glutamic acid	Glu	
Arginine	Arg		Glutamine	Gin	
Asparagine	Asn		Glycine	Gly	
Aspartic acid	Asp		*Histidine*	*His*	
Cysteine	Cys		*Isoleucine*	*Ile*	

NOTE: Amino acids in italics are essential.

Amino Acids

Amino Acid	3-Letter Code	Structure	Amino Acid	3-Letter Code	Structure
Leucine	*Leu*		Serine	Ser	
Lysine	*Lys*		*Threonine*	*Thr*	
Methionine	*Met*		*Tryptophan*	*Trp*	
Phenylalanine	*Phe*		Tyrosine	Tyr	
Proline	Pro		*Valine*	*Val*	

NOTE: Amino acids in italics are essential.

If an essential amino acid is in short supply in the diet, that amino acid can become a limiting reactant in building any protein containing that amino acid. When this happens, the only way the body can make that protein on its own is by breaking down one of its proteins containing that essential amino acid.

> Of course, the diet can be altered so that it includes that missing essential amino acid.

Any dietary source of protein that contains adequate amounts of all essential amino acids represents a source of **complete protein**. Most sources of animal protein contain all nine essential amino acids in quantities sufficient to meet dietary requirements. Plant protein and some sources of animal protein are incomplete, that is, they do not contain adequate amounts of all nine essential amino acids.

No single plant can provide adequate amounts of all essential amino acids, but certain combinations of plants can. These combinations of foods, which are said to contain **complementary proteins**, are part of many diets around the world (see Figure 7.39).

> Complementary proteins do not need to be consumed at every meal. However, sufficient complementary proteins must be consumed so that the daily protein intake is adequate.

Because the body cannot store amino acids, you must consume some every day. The recommended level of protein intake according to the *Dietary Guideline for Americans 2010* is 10–35% of total daily Calories consumed. The consumption of too much protein carries potential health risks. Excess protein causes stress on the two organs that metabolize amino acids: the liver and the kidneys. Excess protein also increases excretion of calcium ions (Ca^{2+}) that function in nerve transmission and help build bones and teeth. A diet with too much protein can even cause dehydration because more fluids are needed for the urinary excretion of urea. This problem is particularly relevant to athletes. Finally, consuming excess protein may lead to inadequate intake of other nutrients.

Figure 7.39 *Complementary proteins provide adequate amounts of all essential amino acids when consumed together in a single day.*

In 1997, the Food and Nutrition Board of the National Academy of Sciences created Dietary Reference Intakes (DRIs). The DRIs are quantitative estimates of nutrient intakes for planning and assessing the diets for healthy people. The DRIs for protein shown in Table 7.5, as well as for other nutrients not shown in the table, depend on a person's age and gender. DRIs are based on median heights and weights for a U.S. population of designated age and gender.

Food labels list nutritional information, as shown in Figure 7.40. Daily values are based on a daily intake of 2000 Calories and must be adjusted for an individual's body size, age, level of physical activity, and energy demands related to such exercise.

Figure 7.40 *A nutrition label specifying protein content.*

Table 7.5

Dietary Reference Intakes (DRIs) for Protein

Life Stage Group	RDA/AI* g/d
Infants	
0–6 months	9.1*
7–12 months	11.0
Children	
1–3 years	13
4–8 years	19
Males	
9–13 years	34
14–18 years	52
19–70 years	56
> 70 years	56
Females	
9–13 years	34
14–70 years	46
> 70 years	46
Pregnant or Nursing	71

NOTE: Recommended Dietary Allowances (RDAs) are in **bold** type, Adequate Intakes (AIs) in ordinary type followed by an asterisk (*). RDAs and AIs may both be used as goals for individual intake. RDAs are set to meet the needs of almost all (97 to 98 percent) individuals in a group.

* Based on 1.5 g/kg/day for infants, 1.1 g/kg/day for 1–3 y, 0.95 g/kg/day for 4–13 y, 0.85 g/kg/day for 14–18 y, 0.8 g /kg/day for adults, and 1.1 g/kg/day for pregnant (using pre-pregnancy weight) and lactating women.

SOURCE: *Dietary Reference Intakes for Energy, Carbohydrate. Fiber, Fat, Fatty Acids, Cholesterol, Protein, and Amino Acids (2002/2005).* This report may be accessed via www.nap.edu.

DEVELOPING SKILLS

C.4 DAILY PROTEIN REQUIREMENTS

Sample Problem: *What mass of protein should a woman over 70 consume each day, on average?*

From Table 7.5, females > 70 years should consume an average of 46 g/d of protein.

Use information from Table 7.5 to answer the following questions:

1. What mass of protein should a person of your age and gender consume each day, on average?

2. For each kilogram of body mass, infants require more protein than adults do. Why should protein values per kilogram of body mass be highest for infants and become lower as a person ages (Figure 7.41)?

3. a. What food do infants consume that meets most of their relatively high protein needs?

   b. What evidence in Table 7.5 supports your answer? Explain.

Figure 7.41 *Why do an individual's age and developmental stage influence his or her protein needs?*

concept check 6

1. Amino acids have both an amino group and a carboxylic acid group. Why are both necessary for the formation of polypeptides?
2. Why does your body need a steady supply of amino acids from the protein in your diet?
3. Do catalysts exist within biological systems? What are they called and what is their function?

INVESTIGATING MATTER

C.5 ENZYMES

Preparing to Investigate

Think of a time you had to respond quickly, such as catching an object that fell unexpectedly or getting ready for school when you were late. How do your cells get the energy they need to respond quickly? You have already learned that food is "burned" to meet the body's continuous energy needs. The rate of "burning" can be adjusted from low to high, literally within a heartbeat.

What is the secret behind this impressive performance? It lies with biological catalysts called *enzymes*. Enzymes are able to speed up specific reactions without undergoing any lasting change themselves. In this investigation, you will monitor the rate of an enzyme-catalyzed reaction. In particular, you will investigate how the enzyme catalase affects the rate of decomposition of hydrogen peroxide (H_2O_2):

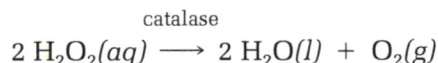

$$2\ H_2O_2(aq) \xrightarrow{\text{catalase}} 2\ H_2O(l)\ +\ O_2(g)$$

You will be assigned a particular food—apple, potato, or liver—to test for the presence of catalase. You will test a fresh piece of the food and one that has been boiled to investigate whether either material catalyzes the decomposition of hydrogen peroxide.

Making Predictions

1. What evidence will indicate that a gas is produced in the reactions?
2. Predict whether one or both pieces of the food, fresh or boiled, will contain catalase.

Gathering Evidence

1. Before you begin, put on your goggles and wear them properly throughout the investigation.

2. Obtain two pieces of your assigned food sample: one fresh and one boiled. (***Caution:*** *Never taste or eat anything in the laboratory.*)

3. Label two 16 × 125-mm test tubes: one "fresh," the other "boiled."

4. Add 5 mL 3% hydrogen peroxide (H_2O_2) solution to each test tube. Do you see evidence that a reaction has taken place? Record your observations.

5. Add a portion of the fresh food sample to the appropriate test tube. Insert a stopper containing a segment of glass tubing into the mouth of the test tube, and arrange tubing as shown in Figure 7.42. Be sure the end of the glass tubing is submerged in the beaker of water.

6. Record the estimated number of bubbles formed per minute in the test tube for a total of 2 min.

7. Repeat Steps 5 and 6 with a sample of boiled food and the second test tube of hydrogen peroxide solution.

8. Discard and dispose of the samples and solutions as directed by your teacher.

9. Wash your hands thoroughly before you leave the laboratory.

Figure 7.42 *Setup for enzyme catalysis investigation.*

Interpreting Evidence

1. Compare your experimental data with those of other class members who used

   a. the same material.   b. a different material.

2. What evidence for the presence of catalase do you see in your results?

Making Claims

3. What can you conclude about the enzyme's activity when comparing observations of the test with fresh material and boiled material? Cite evidence for your conclusion.

Reflecting on the Investigation

4. Why does commercial hydrogen peroxide contain preservatives?

5. Someone tells you that when hydrogen peroxide is put on a skin cut, the observed foaming shows that the cut is infected. Explain the actual chemical reason for this foaming.

If the same amount of original hydrogen peroxide solution were left undisturbed without adding any catalase-containing material, it would take days for the same extent of hydrogen peroxide decomposition to occur. How do enzymes work so well at catalyzing reactions? Read on.

C.6 HOW ENZYMES WORK

The names of most enzymes include the suffix -ase. As with any catalyst, the enzyme name is included above the arrow in the equation representing the reaction. This shows that although the enzyme is involved in the reaction, it is not consumed by it.

The speed of an enzyme-catalyzed reaction is hard to comprehend. In a single second, one molecule of the enzyme amylase helps to release 18 000 energy-rich glucose molecules from starch.

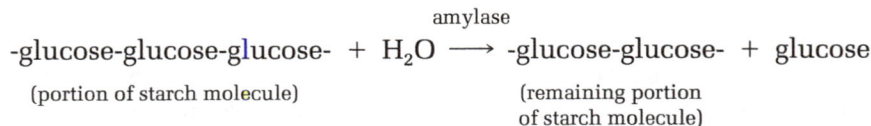

$$\text{-glucose-glucose-glucose-} + H_2O \xrightarrow{\text{amylase}} \text{-glucose-glucose-} + \text{glucose}$$

(portion of starch molecule)   (remaining portion of starch molecule)

How do enzymes (or catalysts in general) work? According to the collision theory of reaction rates, chemical reactions can occur only when two reactant atoms or molecules collide with proper energy and orientation. When reactant molecules are large, the appropriate groups on the molecules that react must come together in the correct orientation. If molecules collide randomly, this is not likely to happen; thus the reaction rate will be very slow.

Catalysts cause reactions to occur more quickly by properly orienting reactant molecules. Reactions in living cells often involve very large molecules reacting at only one site, so catalysts are essential. Without enzymes or catalysts, many reactions would not occur at all (or at least not at reasonable rates at normal temperatures). Living systems manufacture a large variety of enzymes, each tailored to assist a particular cellular reaction (Figure 7.43).

In general, enzymes function as follows:

The discussion of automobile catalytic converters on pages 211–213 describes how catalysts function; the reaction is speeded up because the catalyst lowers the activation-energy barrier.

- A reactant molecule—known as the **substrate**—and the enzyme come together. The substrate molecule fits into the enzyme at an **active site**, where the substrate molecule's key functional groups are properly positioned. (Figure 7.44).

- The three-dimensional structure of the enzyme—how the amino acids in the protein polymer arrange themselves and interact—forms an active site. The shape of this site determines which substrate molecules can bind to the active site and what kind of reaction the enzyme is able to catalyze. The active site is not completely rigid and can change shape to accommodate different substrate molecules. Some enzymes have active sites that are fairly flexible and can accommodate a wide range of substrates. Other enzymes have rigid active sites, so only a few substrates can fit. This description of enzyme activity is known as the "induced-fit" model.

- The enzyme interacts with the substrate, weakening critical bonds and making the reaction more energetically favorable.

- The substrate is changed into a product or products as weakened bonds break and stronger ones form. One or more products then depart from the enzyme surface, freeing the enzyme to interact with other substrate molecules. Each enzyme molecule can participate in numerous reactions without any permanent change to its structure. Because of this regenerating (recycling), enzymes are needed in much smaller amounts than reactants.

Figure 7.43 *Without enzymes, this germinating avocado seed would be unable to develop into a mature plant. As the plant grows, it secretes enzymes that catalyze digestions of nutrients stored in the seed. The resulting smaller molecules are absorbed, thus promoting plant growth.*

Scientific American Conceptual Illustration

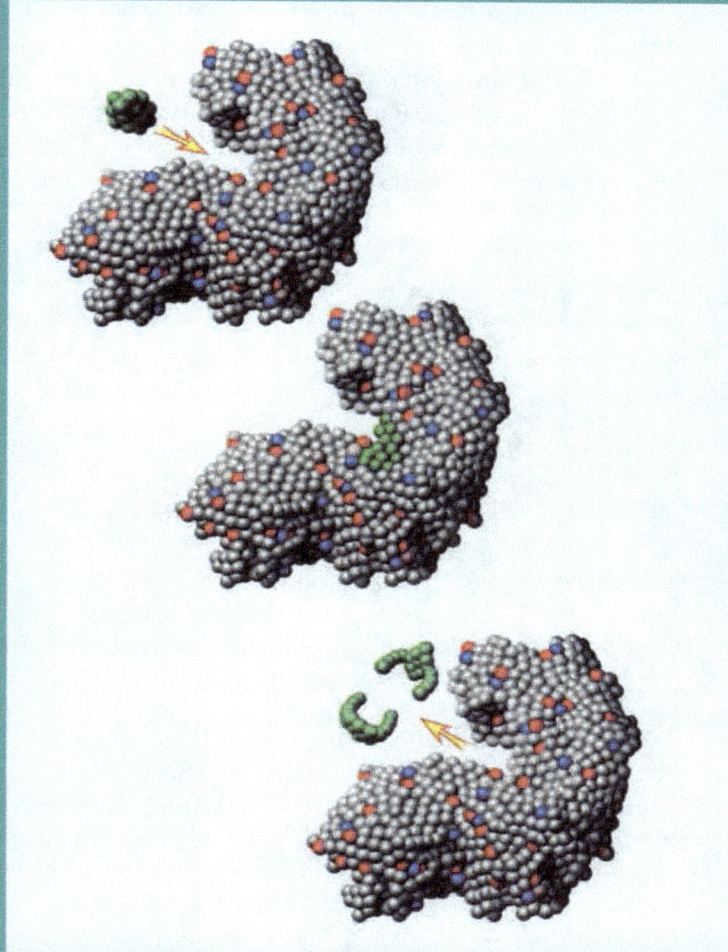

Figure 7.44 *An enzyme is a molecule produced by a living organism that catalyzes a specific chemical reaction. An active site within each enzyme accommodates (fits) only certain molecules, called substrates, depicted here in green. As the substrate binds to the enzyme's active site (upper left), the enzyme lowers the energy needed to break chemical bonds in the substrate and facilitates new bond formation. After a substrate is chemically altered, it detaches from the enzyme (lower right) and the enzyme is available to catalyze the transformation of more substrate molecules. Some enzymes can transform over 100 000 substrate molecules per second. The enzyme's entire three-dimensional structure maintains the shape and specificity of its active site. Thus, an enzyme's effectiveness depends upon careful control of surrounding factors that can alter its shape, such as pH and temperature. The shapes of most enzymes are significantly altered (denatured) under extreme conditions. In some cases, the enzyme becomes permanently denatured and is consequently unable to catalyze its specific reaction even if surrounding conditions later return to normal. While enzymes are not altered as they catalyze reactions, they are subject to degradation by certain reactions. Thus, they must be continually synthesized by cells that depend upon their role as biochemical catalysts.*

Each enzyme is as selective as it is fast. It catalyzes only certain reactions, even though substrates for many other reactions are also available. How does an enzyme "know" what to do? For example, why does amylase help to break starch molecules into thousands of glucose molecules instead of decomposing hydrogen peroxide molecules into water and oxygen gas?

The answers to these questions are the same: The active site of a given enzyme has a three-dimensional shape that only allows certain properly shaped molecules or functional groups to occupy it. Some enzymes are much more selective than others. Each enzyme catalyzes only one particular class or type of reaction. Because of the interaction of an enzyme's active site with its substrate or substrates, cells can precisely control which reactions take place.

CHEM**QUANDARY**

A PROBLEM TO CHEW ON

Chew an *unsalted* cracker for a minute or two before swallowing. How would you describe the taste when you first begin to chew? How would you describe the taste a minute or two later? If you noted a change, what causes the cracker flavor to change over time? What's happening at the particulate level?

Your body's enzymes help convert large molecules into many smaller molecules, releasing energy in the process. Each smaller molecule stores chemical energy that is released by your cells through other enzyme-catalyzed reactions. In the next investigation, you will explore the performance of an enzyme involved in this process.

■ INVESTIGATING MATTER

C.7 AMYLASE TESTS

Preparing to Investigate

In this investigation, you will explore how temperature and pH affect the performance of the enzyme amylase. Amylase contained in saliva helps to break down starch molecules into individual glucose molecules.

Glucose reacts with Benedict's reagent to produce a yellow-to-orange precipitate. The color and amount of precipitate formed is a direct indication of the glucose concentration generated by the enzyme-catalyzed reaction.

Your laboratory team will explore the performance of amylase either at room temperature or at a lower temperature. Before you begin, read *Gathering Evidence* and prepare a data table with appropriately labeled columns specifying data to be collected by each group member.

Making Predictions

1. Predict how temperature will affect the rate of the enzyme-catalyzed reaction.

2. Predict how pH will affect the rate of the enzyme-catalyzed reaction.

Gathering Evidence

Day 1. Preparing the Samples

1. Before you begin, put on your goggles, and wear them properly throughout the investigation.

2. Label five test tubes near the top with the temperature your group is assigned to investigate and with pH values of 2, 4, 7, 8, and 10. Also mark each label so that your test tubes can be distinguished from those of other groups (see Figure 7.45).

3. Using the pH solutions provided by your teacher, add 5 mL of pH 2, 4, 7, 8, and 10 solutions to the appropriate test tubes.

4. Add 2.5 mL of starch suspension to each tube.

5. Add 2.5 mL of 0.5% amylase solution to each tube.

6. Insert a stopper into each tube. Hold the stopper in place with your thumb or finger and shake each tube well for several seconds.

7. Leave the room-temperature test tubes in the laboratory overnight as directed by your teacher.

8. Give your teacher the test tubes that are to be refrigerated.

9. Wash your hands thoroughly before leaving the laboratory.

Figure 7.45 *Remember to label all test tubes, so that you can identify them later in the investigation.*

Day 2: Evaluating the Results

10. Before you begin, put on your goggles, and wear them properly throughout the investigation.

11. Prepare a hot-water bath by adding about 100 mL of tap water to a 250-mL beaker. Add a boiling chip. Warm the beaker on a hotplate. (Heat just to below boiling—hot, but not boiling.)

12. Add 5 mL of Benedict's reagent to each tube. Replace each stopper, being careful not to mix the stoppers. Hold the stopper in place with a thumb or finger and shake each tube well for several seconds.

13. Ensure that all test tubes are still clearly labeled. Remove the stoppers and place the test tubes into the hot-water bath.

14. Heat the test tubes in the hot-water bath until the solution in at least one tube has turned yellow or orange. Then continue heating for two to three more minutes.

15. Use tongs to remove the test tubes from the hot-water bath. Arrange them in a test-tube rack in order of increasing pH.

16. Observe and record the color of the contents of each tube.

17. Share your data with your classmates as directed by your teacher.

18. Wash your hands thoroughly before leaving the laboratory.

Benedict's reagent also provides a test for the presence of glucose in urine, a sympton of diabetes.

Interpreting Evidence

1. For each temperature, record the color of the solution in each test tube.

2. You know that Benedict's solution will react with glucose to form a yellow-orange precipitate. Based on your observations, rank the test tubes in order of least glucose present to most glucose present.

3. Using your ranking from Question 2, decide at which pH the amylase was the most effective. At which pH values (if any) was the amylase ineffective? Share your data with the class.

Making Claims

4. Make a scientific claim about how pH and temperature affect the ability of amylase to catalyze the decomposition of starch.

Reflecting on the Investigation

5. Many cells in the body only function properly within a certain pH range. For example, your blood is buffered to maintain a constant pH of about 7.4. Certain medical conditions can cause the pH of the blood to increase or decrease. Why do you think this is potentially dangerous?

6. What would be the effect on an individual if body temperature were to
   a. decrease significantly (as in hypothermia)?
   b. increase drastically (as with a severe fever)?

7. Do you think an enzyme that functions in the bloodstream would also function in the stomach? Explain.

Your cells synthesize many enzymes and other kinds of proteins to keep you alive. The amino acids used to synthesize these proteins are best obtained through a diet that provides appropriate amounts of protein. You will now decide whether a particular diet meets those protein needs.

■ MAKING DECISIONS

C.8 PROTEIN CONTENT

You have analyzed your three-day food inventory in terms of energy the diet provides and fat and carbohydrate molecules it delivers. Now consider whether the food provided supplies the recommended amounts of a key building block of living material—protein.

Use the food inventory to answer the following questions. Refer to Table 7.5 (page 737), if necessary.

1. What is the average total mass of protein (in grams) consumed daily?

2. What other information would you need to know regarding the person who follows this food-intake pattern to evaluate the appropriateness of protein supplied by these foods?

3. a. Does the person consume a good balance of essential amino acids? Explain your answer.

   b. What types of snack foods are high in protein (especially types that might be included in the vending machines at Riverwood High School)?

SECTION C SUMMARY

Reviewing the Concepts

> Proteins, major structural components of living creatures, fulfill many cellular roles.

1. Name three types of tissue in your body for which protein is the main structural component.

2. The name *protein* comes from a Greek word that means "of prime importance." Why is this name appropriate?

3. List five cellular functions where proteins are particularly important.

4. Name three food items composed primarily of protein.

5. Why are proteins considered polymers?

6. What chemical elements do proteins contain?

7. How many Calories would be provided by the metabolism of 3.6 g protein? (Recall that 1 g protein = 4 Cal.)

8. What is the chemical composition of an enzyme?

> Amino acids are the chemical subunits that make up proteins.

9. How does the relatively small number of different amino acids account for the vast variety of proteins found in nature?

10. What is a peptide bond? Use structural formulas to illustrate your answer.

11. How does the total number of amino acids vary within protein molecules?

12. What is the protein DRI value for
    a. a 4-month-old infant?
    b. a 36-year-old male of median height and weight?

13. Write structural formulas for the following molecules:
    a. A dipeptide of glycine and cysteine.
    b. A tripeptide abbreviated Asp-Ala-Cys.

14. Explain the meaning of the following terms:
    a. complete protein
    b. essential amino acid
    c. complementary proteins

15. On which two functional groups is the name "amino acid" based?

> Some protein molecules function as enzymes, that is, biological catalysts that speed up cellular reactions.

16. a. How are enzymes like other catalysts?
    b. How are enzymes different from other catalysts?

17. What would be the effect if all enzyme activity in the human body suddenly stopped? Explain.

18. Describe how high temperatures affect the ability of most enzymes to function.

19. Explain why enzymes can speed up only certain chemical reactions.

20. a. Explain the interaction of an active site and a substrate in terms of an induced-fit model.
    b. Describe at least one limitation to the induced-fit model for enzyme activity.

Why are protein molecules essential to living organisms?

In this section you have learned about the structure and function of proteins and the important role that proteins play in biochemical processes. Think about what you have learned and how these ideas are related, then answer the question in your own words in organized paragraphs. Your answer should demonstrate your understanding of the key ideas in this section.

Be sure to consider the following in your response: the structure of amino acids, peptide bonds, the role of proteins in the diet, and enzyme function.

Connecting the Concepts

21. a. In what ways are proteins similar to carbohydrates and fats?

    b. In what ways are proteins different?

22. If a steak is left on a barbecue grill too long, it turns black. What does that observation suggest about the chemical composition of protein in the meat?

23. Although many plants contain high levels of protein, vegetarians must be more concerned than non-vegetarians about including adequate protein in their diets.

    a. How can vegetarians ensure that they obtain all chemical building blocks needed to build required proteins?

    b. Explain your answer.

24. Using simple diagrams, including structural formulas, sketch how an enzyme might help form a peptide bond between two amino acids.

25. If a person followed a daily diet of 55 g fat, 75 g protein, and 85 g carbohydrate, how much energy would be provided through the metabolism of each? Show your calculations.

26. In Unit 4, you learned that fish species need a water pH range within which they can live. Given what you know about enzymes from the amylase tests in this section, explain why this is so.

27. Explain how the protein you actually eat—whether from beef, turkey, beans, nuts, or tofu—is transformed into human-body proteins.

Extending the Concepts

28. The genetic code in DNA carries blueprints for making proteins in the body. Explain how DNA helps determine the body's physical development and functioning.

29. The phrase "form follows function" is particularly applicable to enzymes. Explain how form and function are closely related for enzyme molecules.

30. Obtain information on the condition known as *ketosis*. What are its causes and effects? How can a high-protein diet lead to ketosis?

31. Explain why hydrogen peroxide is an effective antiseptic. How does your answer relate to enzyme action?

32. Research and report on chemical and physical properties of one or more amino acids used by the human body in building protein molecules.

33. What is a *zwitterion*? Under what conditions do amino acids become zwitterions?

VITAMINS, MINERALS, AND ADDITIVES

What roles do vitamins, minerals, and additives play in foods we eat?

The focus of this unit so far has been on macronutrients in foods: proteins, carbohydrates, and fats. However, there are other substances in food, often in trace amounts: *vitamins, minerals,* and sometimes *additives.* Dietary guidelines imply that vitamins and minerals play vital roles within your body. These micronutrients occur naturally in many foods, and may also be added to foods, such as cereal or bread, to improve their quality as sources of various micronutrients. Food manufacturers use food additives for different reasons, as you will soon learn. What do vitamins and minerals do in your body? What purposes do food additives serve?

GOALS

- List key vitamins and describe their importance to the body.
- Explain why daily intake requirements of vitamins depend on the function and solubility properties of the vitamins as well as characteristics (such as age and gender) of individuals.
- List key minerals and explain their importance to health.
- List and describe main uses of food additives.
- Describe how titration and chromatography can be used to analyze food products.

concept check 7

1. What are the primary roles of each of the following nutrients in a person's diet?
   a. carbohydrates
   b. fats
   c. proteins
2. a. Describe in your own words what an enzyme is and what it does.
   b. Then, speculate what a "co-enzyme" might be.
3. How does the mass of vitamins and minerals you need to consume each day compare to the mass of proteins, fats, and carbohydrates in your daily diet?

D.1 VITAMINS

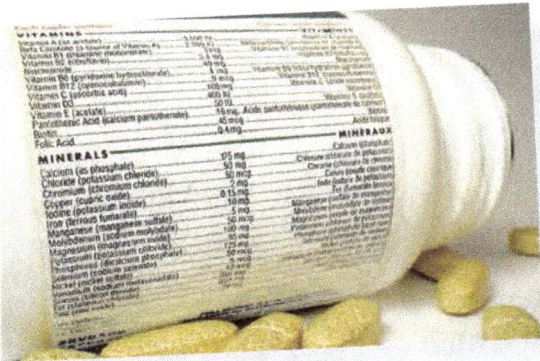

Figure 7.46 *Vitamin supplements are useful whenever sufficient vitamins are not provided in one's diet.*

Vitamins are biomolecules necessary for growth, reproduction, health, and life. Each vitamin is required in only a tiny amount. The total quantity of *all* vitamins required daily by an adult is only about 0.2 g; "a little goes a long way" with vitamins (see Figures 7.46 and 7.47).

How much is "enough"? That depends on age and gender, as is suggested by Table 7.6.

Figure 7.47 *Many foods such as leafy greens, liver, milk, eggs, and whole grains are rich in vitamins.*

Table 7.6

Age or Condition	Vit A (μg/d)	Vit D (μg/d)	Vit E (mg/d)	Vit K (μg/d)	Vit C (mg/d)	Vit B_1 (mg/d)	Vit B_2 (mg/d)	Vit B_3 (mg/d)	Vit B_6 (mg/d)	Vit B_9 (μg/d)	Vit B_{12} (μg/d)
Males											
9–13 yrs	600	5*	11	60*	45	0.9	0.9	12	1.0	300	1.8
14–18 yrs	900	5*	15	75*	75	1.2	1.3	16	1.3	400	2.4
19–30 yrs	900	5*	15	120*	90	1.2	1.3	16	1.3	400	2.4
31–50 yrs	900	5*	15	120*	90	1.2	1.3	16	1.3	400	2.4
51–70 yrs	900	10*	15	120*	90	1.2	1.3	16	1.7	400	2.4
> 70 yrs	900	15*	15	120*	90	1.2	1.3	16	1.7	400	2.4
Females											
9–13 yrs	600	5*	11	60*	45	0.9	0.9	12	1.0	300	1.8
14–18 yrs	700	5*	15	75*	65	1.0	1.0	14	1.2	400	2.4
19–30 yrs	700	5*	15	90*	75	1.1	1.1	14	1.3	400	2.4
31–50 yrs	700	5*	15	90*	75	1.1	1.1	14	1.3	400	2.4
51–70 yrs	700	10*	15	90*	75	1.1	1.1	14	1.5	400	2.4
> 70 yrs	700	15*	15	90*	75	1.1	1.1	14	1.5	400	2.4
Pregnant	770	5*	15	90*	85	1.4	1.4	18	1.9	600	2.6
Nursing	1300	5*	19	90*	120	1.6	1.6	17	2.0	500	2.8

Note: Recommended Dietary Allowances (RDAs) are in **bold** type. RDAs are established to meet the needs of almost all (97 to 98%) individuals in a group. Adequate Intakes (AIs) are followed by an asterisk (*). AIs are believed to cover the needs of all individuals in the group, but a lack of data prevent being able to specify with confidence the percent of individuals covered by this intake.

Source: Food and Nutrition Board, National Academy of Sciences: National Research Council, *Dietary Reference Intakes 2004.*

Vitamins perform very specialized tasks. Vitamin D, for example, helps move calcium ions from your intestines into the bloodstream. Without vitamin D, your body would not use much of the calcium you ingest. Some vitamins function as **coenzymes**, which are organic molecules that interact with enzymes and enhance their activity. For example, the B-vitamins act as coenzymes in releasing energy from food molecules. Figure 7.48 illustrates how a coenzyme functions.

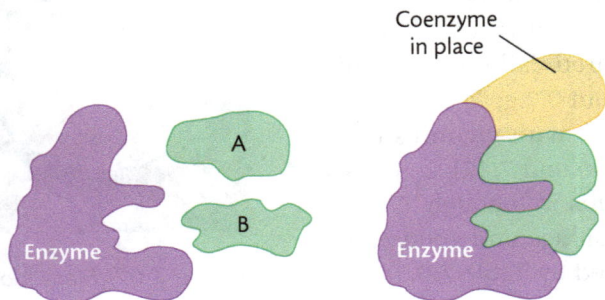

Figure 7.48 *A vitamin serving as a coenzyme.*

Long before the term *vitamin* was introduced early in the last century, people had discovered that small quantities of certain substances were necessary to maintain health. One example of vitamin deficiency is scurvy, once common among sailors; this condition is characterized by swollen joints, bleeding gums, and tender skin. Although early seafarers (in the 1700s and 1800s) did not know what caused scurvy, they commonly loaded citrus fruit on board, which they ate during long voyages to prevent scurvy. Scurvy is now known to be caused by vitamin C deficiency. Vitamin C is supplied by citrus fruit. In addition to vitamin C, about a dozen different vitamins have been identified over the past century, each critical to reactions occurring within the human body. Table 7.7 documents how some of those vitamins support human life.

Vitamins are classified as fat-soluble or water-soluble (see Table 7.7 and Figure 7.49). Water-soluble vitamins with polar functional groups pass directly into the bloodstream. They are not stored in the body; they must be ingested daily. Some water-soluble vitamins, including the B vitamins and vitamin C, are also destroyed by heat in cooking.

Recall that "like dissolves like." See page 424.

Water-soluble vitamins: **Fat-soluble vitamins:**

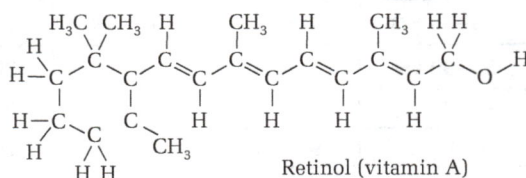

B$_2$ (riboflavin)   B$_3$ (niacin)   Retinol (vitamin A)   Vitamin D

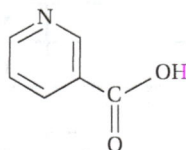

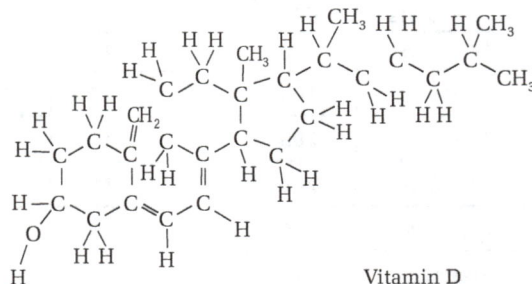

Figure 7.49 *Structures of two water-soluble and two fat-soluble vitamins.*

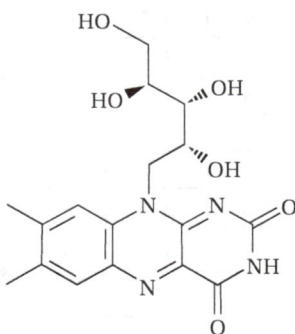

Your body absorbs fat-soluble vitamins into the blood from the intestine with assistance from fats in the food you eat. Because the nonpolar structures of fat-soluble vitamins allow them to be stored in body fat, it is not necessary to consume fat-soluble vitamins daily. In fact, because fat-soluble vitamins accumulate within the body, they can build up to toxic levels if taken in excessively large quantities (megadoses).

Table 7.7

Vitamins by Category, Showing Sources and Deficiency Conditions

Vitamin (Name)	Main Sources	Deficiency Condition
Water-soluble		
B₁ (thiamine)	Liver, milk, pasta, bread, wheat germ, lima beans, nuts	Beriberi: nausea, severe exhaustion, paralysis
B₂ (riboflavin)	Red meat, milk, eggs, pasta, bread, beans, dark green vegetables, peas, mushrooms	Severe skin problems
B₃ (niacin)	Red meat, poultry, enriched or whole grains, beans, peas	Pellagra: weak muscles, no appetite, diarrhea, skin blotches
B₅ (pantothenic acid)	Liver, kidneys, yeast, egg yolk, broccoli, whole grains, yogurt, legumes, avocados, sweet potatoes	Anemia
B₆ (pyridoxine)	Muscle meats, liver, poultry, fish, whole grains	Depression, nausea, vomiting
B₇ (biotin)	Kidneys, liver, egg yolk, yeast, nuts	Dermatitis
B₉ (folic acid)	Kidneys, liver, leafy green vegetables, wheat germ, peas, beans	Anemia
B₁₂ (cobalamin)	Red meat, liver, kidneys, fish, eggs, milk	Pernicious anemia, exhaustion
C (ascorbic acid)	Citrus fruits, melon, tomatoes, green peppers, strawberries	Scurvy: tender skin; weak, bleeding gums; swollen joints
Fat-soluble		
A (retinol)	Liver, eggs, butter, cheese, dark green and deep orange vegetables	Inflamed eye membranes, night blindness, scaling of skin, faulty teeth and bones
D (calciferol)	Fish-liver oils, fortified milk	Rickets: soft bones
E (tocopherol)	Liver, wheat germ, whole-grain cereals, margarine, vegetable oil, leafy green vegetables	Breakage of red blood cells in premature infants, oxidation of membranes
K (menaquinone)	Liver, cabbage, potatoes, peas, leafy green vegetables	Hemorrhage in newborns; anemia

DEVELOPING SKILLS

D.2 VITAMINS IN THE DIET

Sample Problem: Use Table 7.7 (page 751) to identify three water-soluble vitamins and one fat-soluble vitamin that are helpful in preventing anemia.

Referring to Table 7.7, water-soluble vitamins B_5, B_9, and B_{12}, and fat-soluble vitamin K have deficiency conditions that include forms of anemia.

1. Carefully planned vegetarian diets are nutritionally balanced. Individuals who follow a vegan diet do not consume any animal products, including eggs and milk. Because of their dietary limitations, vegans must ensure that they obtain the recommended daily allowances of two particular vitamins.

   a. Use Table 7.7 (page 751) to identify these two vitamins, and briefly describe the effect of their absence in the diet.

   b. How might individuals following a vegan diet avoid this problem?

2. Complete the following table about yourself, using data from Tables 7.6 (page 749) and 7.8.

Vegetable (one-cup serving)	Your RDA		Total Serving to Supply Your RDA	
	B_1	C	B_1	C
Green peas				
Broccoli				

   a. Would any of your entries change if you were of the opposite gender? If so, which entry or entries?

   b. Based on your completed table, why do you think variety is essential in a person's diet?

   c. Why might vitamin deficiencies pose problems even if people receive adequate supplies of food Calories?

3. Nutritionists recommend eating fresh fruit rather than canned fruit, and raw or steamed vegetables instead of canned or boiled vegetables (see Figures 7.50 and 7.51).

   a. What does food freshness have to do with vitamins?

   b. To what types of vitamins might nutritionists be referring?

   c. Is such advice sound? Explain.

Figure 7.50 *How does the vitamin content of raw broccoli compare to that of steamed broccoli?*

Figure 7.51 *Why do dietary guidelines usually favor raw fruit over canned fruit?*

Table 7.8

Vitamin B₁ and C Content of Some Vegetables		
	Vitamin (in mg)	
Vegetable (one-cup serving)	B₁ (thiamine)	C (ascorbic acid)
Green peas	0.387	58.4
Lima beans	0.238	17
Broccoli	0.058	82
Potatoes	0.15	30

4. Examine the structural formulas for the two water-soluble vitamins and two fat-soluble vitamins in Figure 7.49 (page 750).

   a. You investigated solubility characteristics in Unit 4 in an effort to make sense of the saying, "Like dissolves like." Identify particular features in the structures of the water-soluble vitamins that make them "like" water, and thus able to dissolve in water.

   b. What structural features of the fat-soluble vitamins make them "like" fats, and thus able to dissolve in fats? (Another way to think about this is to ask what structural features make them "unlike" water, and thus not very soluble in water.)

   c. How does the concept of intermolecular forces help to explain how a water-soluble vitamin dissolves in water but does not dissolve well in fat?

5. Below is the structural formula for one of the vitamins listed in Table 7.7 (page 751). This vitamin can be found in liver, peas, and green leafy vegetables, among other sources. Deficiency of this vitamin in a person's diet can result in anemia.

   a. Use Table 7.7 (page 751) to identify vitamins that match the description of this vitamin's sources and deficiency conditions.

   b. Based on its structure, do you think this vitamin is water-soluble or fat-soluble? Explain.

   c. Based on your answers to 5a and 5b, identify the vitamin.

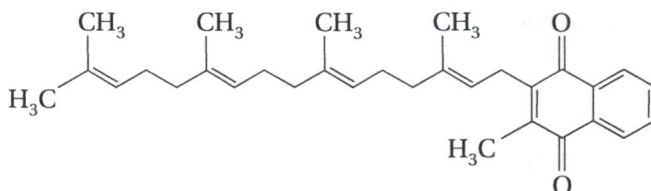

■ INVESTIGATING MATTER

D.3 VITAMIN C

Preparing to Investigate

Vitamin C, also called ascorbic acid, is a water-soluble vitamin. It is also among the least stable vitamins because it reacts readily with oxygen gas, and exposure to light or heat can decompose it. In this investigation, you will find out how much vitamin C is contained in some popular beverages, including fruit juices, milk, and soft drinks.

This investigation is based on a chemical reaction of ascorbic acid (vitamin C) with iodine (I_2). A colored solution of iodine (I_2) oxidizes ascorbic acid, forming the colorless products dehydroascorbic acid, hydrogen ions, and iodide ions:

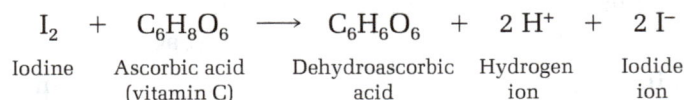

$$I_2 \ + \ C_6H_8O_6 \ \longrightarrow \ C_6H_6O_6 \ + \ 2\ H^+ \ + \ 2\ I^-$$

| Iodine | Ascorbic acid (vitamin C) | Dehydroascorbic acid | Hydrogen ion | Iodide ion |

Figure 7.52 shows the structures of ascorbic acid and dehydroascorbic acid.

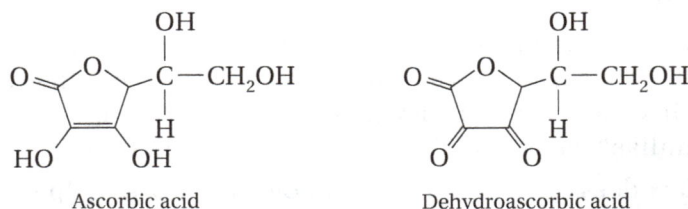

> You may have heard vitamin C described as an "antioxidant." Based on the reaction with I_2 shown here, do you think that means that vitamin C oxidizes or reduces other substances?

Figure 7.52 *Molecular structures of ascorbic acid and dehydroascorbic acid.*

You will conduct a titration (Figure 7.53), a common laboratory procedure to determine concentrations of substances in solution. This investigation involves adding a known amount of one reactant (iodine solution) slowly from a Beral pipet to a second reactant (ascorbic acid in the beverage) in a 24-well plate until just enough has been added for a complete reaction. The completion of the reaction, the **endpoint**, is signaled by a color change. Knowing the chemical equation for this reaction, you can then calculate the unknown amount of the second reactant (ascorbic acid) from the measured volume and concentration of the iodine-solution **titrant**.

> You also conducted titrations in Unit 4, page 464.

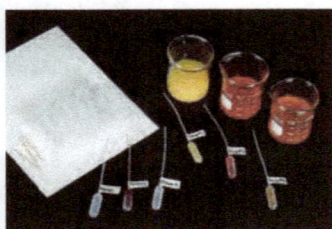

The titration endpoint in the beverage-containing well of the well plate is the point where a dark blue-black color appears and does not disappear with additional stirring. This color is due to the reaction of excess iodine with starch. First, you add a starch indicator suspension to the beverage sample to be tested. Next, an iodine solution of known concentration is added drop by drop from a Beral pipet.

The chemical equation shows that as long as ascorbic acid is present, the iodine is quickly converted to colorless iodide ions; you will observe no

Figure 7.53 *Materials needed for fruit-juice titration.*

bluish black iodine-starch product. When all the available ascorbic acid has been oxidized to colorless dehydroascorbic acid, the next drop of iodine solution added reacts with starch, producing the blue-black color, signaling the endpoint.

You will begin by completing a titration involving a solution of known vitamin C concentration. Once you have data from this titration, you can determine the mass of ascorbic acid that reacts with one drop of iodine solution. You can then calculate the mass (in milligrams) of vitamin C present in a 25-drop sample of each beverage. This information will allow you to rank the tested beverages in terms of the mass of vitamin C that each beverage contains.

Before starting this investigation, prepare a suitable data table. Leave room in the data table to record the total drops of vitamin C solution in 1.0 mL (Step 1) and the total drops of iodine needed to reach the endpoint (Step 5). Provide a horizontal row for each beverage that you will investigate.

Gathering Evidence

Part I. Standardizing the Iodine Solution

1. Before you begin, put on your goggles, and wear them properly throughout the investigation.

2. Fill a Beral pipet with vitamin C solution. Then determine how many drops of vitamin C solution delivered by that pipet represents a volume of 1.0-mL. (*Note:* The volume per drop may depend on the angle at which you hold the pipet. Keep the angle at which you hold your pipet consistent throughout the investigation.)

3. Fill a second Beral pipet with iodine solution.

4. Add 25 drops vitamin C solution into a well of a clean 24-well plate. The vitamin C solution has a known concentration of 1.0 mg vitamin C per milliliter of solution.

5. Add 1 drop of starch suspension to the same well.

6. Place a sheet of white paper underneath the well plate; it will help you detect the appearance of color.

7. Add iodine solution, one drop at a time, carefully counting drops, to the well containing starch and vitamin C mixture (see Figure 7.54). After adding each drop of iodine solution, use a toothpick to gently stir the resulting mixture.

8. Add and count the iodine solution drop by drop, with stirring, until the solution in the well remains bluish black for 20 seconds. If the color fades before 20 seconds have elapsed, add another drop of iodine solution.

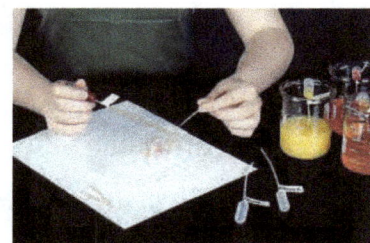

Figure 7.54 *Proper technique for a well plate titration.*

9. Record the total drops of iodine solution needed to reach the endpoint (the appearance of the first "permanent" bluish black color).

10. Repeat Steps 3–9 two times and record your data.

Part II. Determining the Quantity of Vitamin C in Beverages

You will now use titration to determine the unknown quantity of vitamin C in beverages assigned by your teacher. Adapt the procedure used in Part I, considering these points:

- Colored beverages may not produce a true bluish black endpoint color. For example, red beverages may make the endpoint appear purple.
- You will need to conduct three trials and record appropriate data for each beverage analyzed. Construct an appropriate data table prior to beginning Part II.

Wash your hands thoroughly with soap and water before leaving the laboratory.

Analyzing Evidence

1. Determine the mass (in mg) of vitamin C that reacts with 1 drop of iodine solution. Use your data from Part I to perform the following calculations. The Sample Problem below illustrates how to set up the final calculation.

   a. What volume of standardized vitamin C solution did you use for each trial in Part I?

   b. How many milligrams of vitamin C did that volume of solution contain? The concentration of vitamin C solution used in Part I is 1.0 mg/mL.

   c. Determine the average number of drops of iodine used to react with the vitamin C in your three trials.

   d. Calculate the mass (in mg) of vitamin C that reacts with 1 drop of iodine solution. See the Sample Problem below.

Sample Problem: *Suppose you found that there were 30 drops of vitamin C solution in 1.0 mL and that it took 22 drops of iodine solution to reach the endpoint.*

These data lead to the calculated result that 0.038 mg vitamin C reacts with 1 drop iodine solution.

$$25 \text{ drops vitamin C} \times \frac{1 \text{ mL vitamin C}}{30 \text{ drops vitamin C}} \times \frac{1 \text{ mg vitamin C}}{1 \text{ mL vitamin C}} \times \frac{1}{22 \text{ drops I}_2}$$

$$\text{(step 1)} \qquad \text{(step 5)}$$

$$= 0.038 \text{ mg vitamin C per drop I}_2 \text{ solution}$$

2. Determine the mass of vitamin C (in mg) contained in 25 drops of each beverage you analyzed in Part II.

Interpreting Evidence

1. Rank the tested beverages in terms of how much vitamin C each contains, from the highest quantity to the lowest.

2. Among the beverages tested, in your opinion, were any vitamin C levels

   a. unexpectedly low? If so, explain.

   b. unexpectedly high? If so, explain.

3. Imagine that you added too many drops of iodine solution during the titration and missed the true endpoint. Will this procedural error increase or decrease your calculated milligrams of vitamin C in the sample? Explain.

Reflecting on the Investigation

4. Describe how the concept of limiting reactants applies to titration as an analytical technique. In the titrations you performed, what was the limiting reactant?

5. Supposed you heated the beverages, let them cool, and then performed the titrations in Part II.

   a. Would you expect the number of drops of I_2 solution required to titrate each beverage to be higher or lower than the data you recorded?

   b. Explain.

D.4 MINERALS: ESSENTIAL WITHIN ALL DIETS

Of the more than 100 known elements, only 32 are believed essential to support human life. In nutrional terms, **minerals** refer to these elements (including calcium, sodium, chlorine, selenium, and zinc) in their ionic forms. For convenience, essential minerals are divided into **major minerals** (also called **macrominerals**), and **trace minerals** (also called **microminerals**). Some are quite common; others are likely to be found in large quantities only on research-laboratory shelves.

Minerals have several functions in the body. Some minerals become part of the body's structural material, such as bones and teeth. Others help enzymes do their jobs. Still others help maintain the health of the heart and other organs. The thyroid gland, for example, uses only a miniscule quantity of iodine (only millionths of a gram daily) to produce the vital hormone thyroxine (Figure 7.55). The field of bioinorganic chemistry explores how minerals function within living systems.

Figure 7.55 *The thyroid gland does not function properly without proper levels of iodine in one's body. Iodine deficiency causes goiter (depicted here).*

Table 7.9

Dietary Minerals

Mineral	Typical Food Sources	Deficiency Condition
Macrominerals		
Calcium (Ca)	Milk, dairy products, canned fish	Rickets in children; osteomalacia and osteoporosis in adults
Chlorine (Cl)	Table salt, meat, salt-processed foods	—
Magnesium (Mg)	Seafood, cereal grains, nuts, dark green vegetables, cocoa	Heart spasms, anxiety, disorientation
Phosphorus (P)	Meat, dairy products, nuts, seeds, beans	Blood cell disorders, gastrointestinal tract and renal dysfunction
Potassium (K)	Orange juice, bananas, dried fruits, potatoes	Poor nerve function, irregular heartbeat, sudden death during fasting
Sodium (Na)	Table salt, meat, salt-processed food	Headache, weakness, thirst, poor memory, appetite loss
Sulfur (S)	Protein (e.g., meat, eggs, legumes)	Conditions related to deficiencies in sulfur-containing essential amino acids
Trace minerals		
Chromium (Cr)	Animal and plant tissue, liver	Loss of insulin efficiency with age
Cobalt (Co)	Animal protein, liver	Conditions related to deficiencies in cobalt-containing vitamin B_{12}
Copper (Cu)	Egg yolk, whole grains, liver, kidney	Anemia in malnourished children
Fluorine (F)	Seafood, fluoridated water	Dental decay
Iodine (I)	Seafood, iodized salt	Goiter
Iron (Fe)	Meat, green leafy vegetables, whole grains, liver	Anemia; tiredness and apathy
Manganese (Mn)	Whole grains, legumes, nuts, tea, leafy vegetables, liver	—
Molybdenum (Mo)	Whole grains, legumes, leafy vegetables, liver, kidney	Weight loss, dermatitis, headache, nausea, disorientation
Selenium (Se)	Meat, liver, organ meats, grains, vegetables	Muscle weakness, Keshan disease (heart-muscle disease)
Zinc (Zn)	Shellfish, meat, wheat germ, legumes, liver	Anemia, growth retardation

Your body contains rather large quantities, at least 100 mg per kilogram of body mass, of each of the seven macrominerals. Each trace mineral is present in relatively small quantities, less than 100 mg per kilogram of body mass in an average adult. However, trace minerals are just as essential in a human diet as macrominerals. Any essential mineral, whether macromineral or trace mineral, can become a limiting reactant if it is not present in sufficient quantity.

The essential minerals and their dietary sources and deficiency conditions are listed in Table 7.9. Several other minerals, including arsenic (As), cadmium (Cd), and tin (Sn), are known to be needed by laboratory test animals. These minerals and perhaps other trace minerals may be essential to human life. You may be surprised to learn that the widely known poison arsenic might be an essential mineral. In fact, many substances beneficial in low doses become toxic in higher doses. Table 7.10 summarizes recommended daily doses for several macrominerals and trace minerals.

Table 7.10

Dietary Reference Intakes (DRIs) for Selected Minerals

Age or Condition	Calcium (mg/d)	Phosphorus (mg/d)	Magnesium (mg/d)	Iron (mg/d)	Zinc (mg/d)	Iodine (µg/d)
Males						
9–13 yrs	1300*	1250	240	8	8	120
14–18 yrs	1300*	1250	410	11	11	150
19–30 yrs	1000*	700	400	8	11	150
31–50 yrs	1000*	700	420	8	11	150
51–70 yrs	1200*	700	420	8	11	150
> 70 yrs	1200*	700	420	8	11	150
Females						
9–13 yrs	1300*	1250	240	8	8	120
14–18 yrs	1300*	1250	360	15	9	150
19–30 yrs	1000*	700	310	18	8	150
31–50 yrs	1000*	700	320	18	8	150
51–70 yrs	1200*	700	320	8	8	150
> 70 yrs	1200*	700	320	8	8	150
Pregnant	1000*	700	350	27	11	220
Nursing	1000*	700	310	9	12	290

Note: Recommended Dietary Allowances (RDAs) are in **bold** type. RDAs are established to meet the needs of almost all (97 to 98%) individuals in a group. Adequate Intakes (AIs) are followed by an asterisk (*). AIs are believed to cover the needs of all individuals in the group, but a lack of data prevent being able to specify with confidence the percent of individuals covered by this intake.

Source: Food and Nutrition Board, National Academy of Sciences: National Research Council, *Dietary Reference Intakes 2004.*

DEVELOPING SKILLS

D.5 MINERALS IN THE DIET

Use the values in Table 7.10 (page 759) to answer these questions.

Sample Problem: *How many cups of broccoli would a 16-year-old female need to eat each day to reach her daily iron allowance? (1 cup broccoli = 1.1 mg iron) Assume that broccoli is her only dietary source of iron.*

The iron DRI for a 16-year-old female is 15 mg:

$$15 \text{ mg iron} \times \frac{1 \text{ cup broccoli}}{1.1 \text{ mg iron}} = 14 \text{ cups broccoli}$$

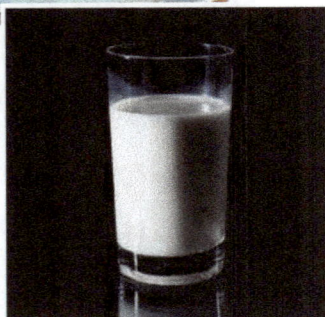

1. One slice of whole wheat bread contains 0.8 mg iron.
   a. How many slices of whole-wheat bread would supply your daily iron allowance? (Assume that this is your only dietary source of iron.)
   b. Predict health consequences of consuming an inadequate quantity of iron.

2. One cup of whole milk contains 288 mg calcium. How much milk would you need to drink daily to meet your daily allowance for that mineral, if milk were your only dietary source of calcium?

3. One medium pancake (Figure 7.56) contains about 27 mg calcium and 0.4 mg iron.
   a. Does a pancake provide a greater percent of your DRI for calcium or for iron?
   b. Explain your answer.

4. The following questions focus on calcium and phosphorous.
   a. What total mass of each of these minerals do you need to consume daily?
      i. calcium.   ii. phosphorus.
   b. Why are the values in the answer to Question 4a higher than DRI values for other listed essential minerals? (*Hint:* Consider how calcium and phosphorus are used in the body.)
   c. List several good dietary sources of
      i. calcium.   ii. phosphorus.
   d. Predict the health consequences of a deficiency of
      i. calcium.   ii. phosphorus.
   e. Would a particular age group or gender be especially affected by the consequences you listed in 4d? If so, which and why?

5. Most table salt, sodium chloride (NaCl), includes a small amount of added potassium iodide (KI).
   a. Why do you think KI is added to table salt?
   b. If you decide not to use iodized salt, what other kinds of food could you use as sources of iodine?

Figure 7.56 *Which foods are good sources of minerals such as calcium?*

"Iodized salt" refers to such products.

concept check 8

1. Consider minerals and vitamins as nutrients.
   a. Which of these nutrients is usually found in ionic form?
   b. Which of these nutrients is usually found in covalent form?
   c. How else would you describe similarities or differences between vitamins and minerals to a friend?
2. Vitamins and minerals occur naturally in foods. Why then are some foods labeled as "enriched" with particular vitamins or minerals?
3. Beyond minerals and vitamins, identify one food additive with which you are familiar. Why might that additive be included in a food product?

D.6 FOOD ADDITIVES

Small amounts of vitamins and minerals occur naturally in food. Some foods, especially processed foods such as packaged snacks or frozen entrees, also contain small amounts of **food additives**. Manufacturers add those substances during processing to increase the nutritive value of foods or to enhance their storage life, visual appeal (see Figure 7.57), flavor, or ease of production. A food label might provide this ingredient information:

Sugar, bleached flour (enriched with niacin, iron, thiamine, and riboflavin), semisweet chocolate, animal and/or vegetable shortening, dextrose, wheat starch, monocalcium phosphate, baking soda, egg white, modified corn starch, salt, nonfat milk, cellulose gum, soy lecithin, xanthan gum, mono- and diglycerides, BHA, BHT.

Figure 7.57 *Food-coloring additives are often used to make food items more attractive and appealing.*

This list shows quite a collection of ingredients! You probably recognize the major ingredients, such as sugar, flour, shortening, and baking soda, and some additives such as vitamins (niacin, thiamine, and riboflavin) and minerals (iron and monocalcium phosphate). However, you probably do not recognize the food additives xanthan gum (an emulsifier that helps produce uniform, non-separating water–oil mixtures (see Figure 7.58, page 762) and BHA and BHT (butylated hydroxyanisole and butylated hydroxytoluene, which are antioxidants that act as preservatives).

Scientific American Working Knowledge Illustration

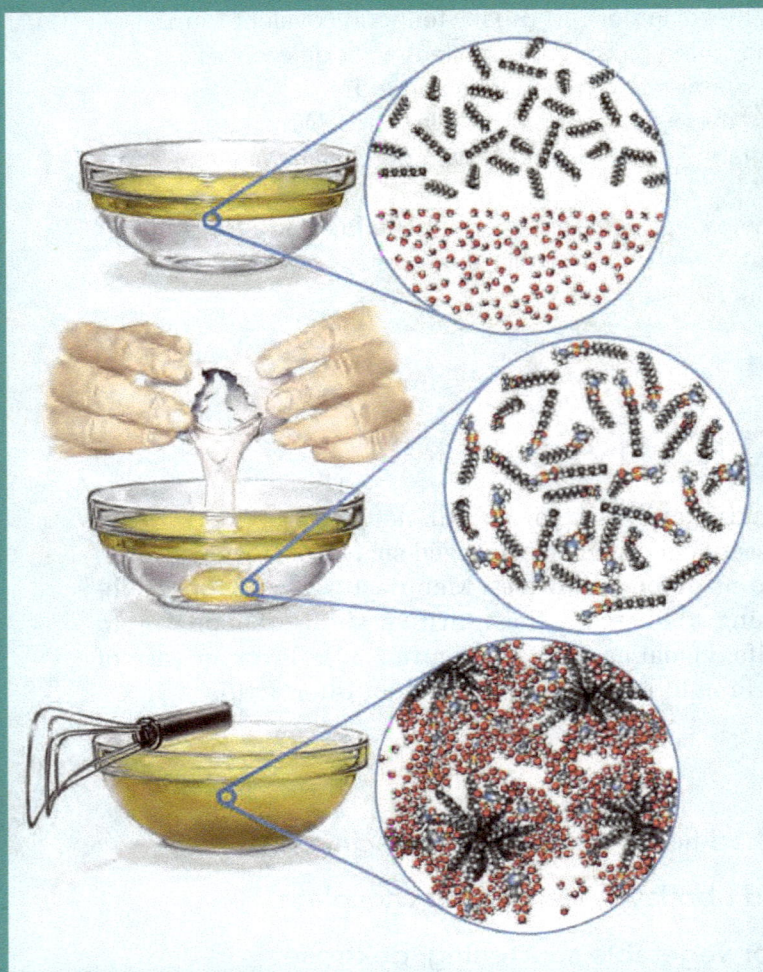

Figure 7.58 *Emulsifier action.* *These illustrations depict how lecithin, a molecule contained in egg yolk, emulsifies (mixes together) an oil–water system. In the top view, oil is layered over water. The particulate-level enlargement reveals that oil molecules are long, nonpolar carbon chains quite different in structure and polarity from the small, polar, water molecules, also shown. Water–water intermolecular forces are much stronger than are oil–water intermolecular forces. Hence these two liquids do not mix. Each lecithin molecule, the emulsifying agent (middle view), has a long, nonpolar carbon chain and a smaller, polar region involving phosphorous, oxygen, nitrogen, and hydrogen atoms. Lecithin molecules orient themselves so that they simultaneously interact with both oil and water molecules. Once an egg is cracked open and added and the entire mixture is whisked, micelles form (bottom view). Each micelle is a spherical arrangement of oil molecules surrounded by and attracted to the nonpolar region of lecithin molecules. The polar regions of lecithin molecules align at the outside of a micelle, causing the entire sphere to intermingle among polar water molecules. This stable arrangement permits oil molecules to remain evenly dispersed through water. Emulsification is used to make numerous food products, including mayonnaise and ice cream.*

Food additives have been used since ancient times. For example, salt has been used for centuries to preserve foods, and spices helped disguise the flavor of food that was no longer fresh. To make foods easier and less expensive to distribute and store, most manufacturers, especially those of processed foods, rely on food-preservation additives. Table 7.11 summarizes the major categories of food additives. The structural formulas of two common additives are shown in Figure 7.59.

Color and taste additives often enhance the commercial appeal of food products. In the following investigation, you will analyze several commonly used food-coloring agents (dyes).

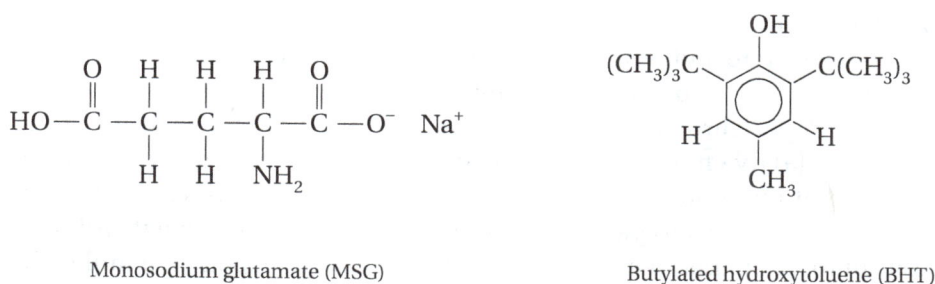

Monosodium glutamate (MSG)   Butylated hydroxytoluene (BHT)

Figure 7.59 *The molecular structures of MSG and BHT.*

Table 7.11

Food Additive Types, Purposes, and Examples

Additive Type	Purpose	Examples
Anticaking agents	Keep foods free-flowing	Sodium ferrocyanide
Antioxidants	Prevent fat rancidity	BHA and BHT
Bleaches	Whiten foods (flour, cheese); hasten cheese maturing	Sulfur dioxide, SO_2
Coloring agents	Increase visual appeal	Carotene (natural yellow color); synthetic dyes
Emulsifiers	Improve texture, smoothness; stabilize oil-water mixtures	Cellulose gums, dextrins
Flavoring agents	Add or enhance flavor	Salt, monosodium glutamate (MSG), spices
Humectants	Retain moisture	Glycerin
Leavening agents	Give foods light texture	Baking powder, baking soda
Nutrients	Improve nutritive value	Vitamins, minerals
Preservatives and antimycotic agents (growth inhibitors)	Prevent spoilage, microbial growth	Propionic acid, sorbic acid, benzoic acid, salt
Sweeteners	Impart sweet taste	Sugar (sucrose), dextrin, fructose, aspartame, sorbitol, mannitol

■ INVESTIGATING MATTER

D.7 ANALYZING FOOD-COLORING ADDITIVES

Preparing to Investigate

Many candies contain artificial coloring agents to increase their visual appeal. Colorless candies would be quite dull! In this investigation, you will analyze the food dyes in two commercial candies and compare them with dyes in food-coloring products.

You will separate and identify the food dyes using **paper chromatography**. This technique uses a solvent (called the *mobile phase*) and paper (the *stationary phase*). Paper chromatography is based on relative differences in attraction between (a) dye molecules and solvent and (b) dye molecules and paper. As the solvent and dye mixture moves up the paper, dye molecules that are more strongly attracted to paper will more readily leave the solvent and separate out onto the paper, leaving a colored spot on the paper. Dye molecules less attracted to paper (and more attracted to solvent) will separate out onto the paper later. Thus, characteristic areas representing different dye molecules will appear on the paper.

To analyze the results of this chromatography investigation, you will calculate the R_f value for each spot. The R_f value, as illustrated in Figure 7.60, is a ratio involving the distance traveled by each dye compared to the distance the solvent has moved up the paper. The actual distance that the dye and solvent travel might vary from trial to trial, but the R_f value will remain constant for a particular dye. Thus chemists use the R_f value, not the actual distance traveled, to identify the molecules.

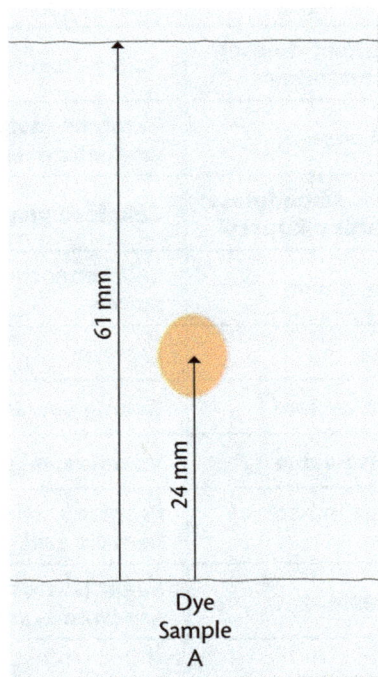

Calculating R_f for Dye Sample A

$$R_f = \frac{24\ mm}{61\ mm} = 0.39$$

Figure 7.60 *Paper chromatograph and calculation of R_f values.*

Read through *Gathering Evidence* and prepare an appropriate data table for this investigation.

Gathering Evidence

1. Before you begin, put on your goggles, and wear them properly throughout the investigation.

2. Obtain one piece each of two different commercial candies from your teacher, as well as a food-coloring sample. The food coloring and candies should all be the same color.

3. Put each candy into a separate well of a well plate. Note which candy is in each well. Add 5–10 drops of water to each well. Stir the mixture in each well with a separate toothpick until the colored coating completely dissolves from the candy. Add 3 to 4 drops of the food coloring sample to a third well. Observe and record the initial colors of each sample.

4. Obtain a strip of chromatography paper, handling it only by its edges. With a pencil (do not use a pen), draw a horizontal line 2 cm from the bottom of the paper strip and another horizontal line 3 cm from the top.

5. Next, place a spot of dye solution on the bottom line; the spot should not be large. Place a drop of the first candy's colored solution with a toothpick, as shown in Figure 7.61.

6. Allow the drop of solution to sit until the spot it makes stops spreading out on the paper. Then apply a second drop of the same sample on top of the spot.

7. Obtain a graduated cylinder to be used as a chromatography chamber. Fill the chamber/cylinder with the solvent (water) to a depth of 1 cm.

8. Lower the spotted chromatography paper's bottom edge into the chromatography chamber until the edge rests evenly in the solvent (water). Be sure the colored spots remain above the solvent surface. Cover the chromatography chamber. See Figure 7.62.

9. Repeat this procedure (Steps 4–8) for the second candy's colored solution and also for the food-coloring sample.

10. Allow the solvent in each cylinder to move up the paper until solvent reaches the top penciled line (3 cm from the top). Then remove the paper from the chambers and, using a pencil, mark the farthest point of solvent travel. Allow the paper to air-dry overnight.

11. After the paper has dried, record the colors observed for the dye sample and candy solutions.

12. Measure the distance (in cm) from the initial pencil line where you placed the spots to the center of each dye spot. Record these distances in your data table.

13. Measure and record the distance (in cm) that the solvent moved.

Figure 7.61 *Spotting a sample on chromatography paper.*

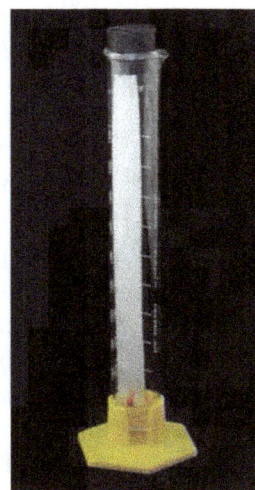

Figure 7.62 *Prepared chromatography paper in a graduated cylinder chromatography chamber. Note that the sample spot is above the solvent level.*

Analyzing Evidence

1. Calculate the R_f value of each dye spot you investigated.
2. Which sample solution, if any, created a single spot rather than several spots?

Making Claims

1. Which dye in the sample solutions had the greatest attraction for the paper? Use evidence to support your claim.
2. Which dye in the sample solutions had the greatest attraction for the solvent? Use evidence to support your claim.
3. Based on your data, do any of the three samples contain the same dyes? Use evidence to support your claims.

Reflecting on the Investigation

4. The candy and food-coloring packages list each dye that they contain. Compare this information with your experimental results.
   a. What similarities did you find?
   b. What differences did you find?
   c. If you found differences, what are some possible reasons for them?
5. Why was it important to use a pencil rather than a pen to mark lines on the chromatography paper in Step 3?

D.8 REGULATING ADDITIVES

Both processed and unprocessed foods may contain contaminants that were not deliberately added, such as pesticides, mold, antibiotics used to treat animals, insect fragments, food-packaging materials, or dirt. We presume that food purchased in grocery stores and restaurants is safe to eat. In most cases, that is true. Nonetheless, in the past, some food additives and contaminants were identified as or suspected of posing hazards to human health.

Safe food in the United States is required by law—the Federal Food, Drug and Cosmetic Act of 1938. This act authorized the Food and Drug Administration (FDA) to monitor food's safety, purity, and wholesomeness. This act has been amended to address concerns about pesticide residues; artificial colors and food dyes; potential cancer-causing agents (carcinogens); and **mutagens**, which are agents that cause mutations, or changes in DNA. Food manufacturers must complete a battery of tests and provide extensive evidence about the safety of any proposed food product or additive. Any new food product must earn FDA approval before it is marketed.

According to the amended Federal Food, Drug and Cosmetic Act, ingredients that are known not to be hazardous and were in use for a long time prior to the act were exempted from testing. These substances, rather than legally defined as additives, constitute the "generally recognized as safe" (GRAS) list. The GRAS list, periodically reviewed in light of new findings, includes items such as salt, sugar, vinegar, vitamins (for example, vitamin C and riboflavin), and some minerals.

In accord with the Delaney Clause, which was added to the Act in the 1950s, every proposed new food additive must be tested on laboratory animals (usually mice). The Delaney Clause specifies that "no additive shall be deemed to be safe if it is found to induce cancer when it is ingested by man or animal." Thus, approval of a proposed additive is denied if it causes cancer in the test animals.

Since the 1950s, great advances have occurred in science and technology. Improvements in chemical-analysis techniques permit scientists to detect even smaller amounts of potentially harmful substances. Consequently, scientists can now detect and study food contaminants that have always been present but previously were undetected.

This new information has resulted in greater understanding of potential risks associated with human exposure to particular food additives. People now recognize that amounts of additives comparable to those causing cancer in test animals are often vastly greater than would ever be encountered in a human diet. In light of this, Congress passed the Food Quality and Protection Act in 1996. This legislation, which replaced the Delaney Clause, states that manufacturers may use food additives that present a "negligible risk."

Many concerns about food additives remain. Sodium nitrite ($NaNO_2$), for instance, is a color stabilizer and spoilage inhibitor used in many cured meats, such as hot dogs and lunch meats. Nitrites are particularly effective in inhibiting growth of the bacterium *Clostridium botulinum*, which produces botulin toxin. This toxin is the cause of botulism, an often fatal disease. Sodium nitrite, however, may be a carcinogen. In the stomach, nitrites are converted to nitrous acid:

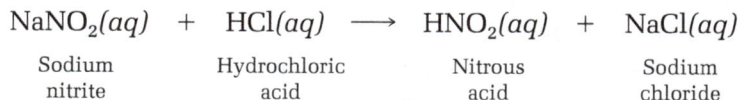

$$NaNO_2(aq) \;+\; HCl(aq) \longrightarrow HNO_2(aq) \;+\; NaCl(aq)$$

| Sodium nitrite | Hydrochloric acid | Nitrous acid | Sodium chloride |

Nitrous acid can then react with compounds formed during protein digestion, producing nitrosoamines, known as potent carcinogens. An example of this reaction is shown below. The concentration of carcinogenic compounds produced, however, is generally low, below the toxic threshold.

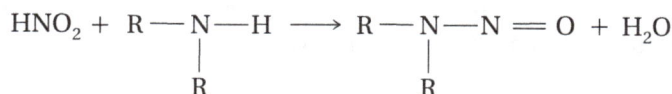

$$HNO_2 + R\!-\!\overset{\displaystyle |}{\underset{\displaystyle R}{N}}\!-\!H \longrightarrow R\!-\!\overset{\displaystyle |}{\underset{\displaystyle R}{N}}\!-\!N\!=\!O + H_2O$$

Testing a new food additive usually requires millions of dollars and years of research.

Figure 7.63 *Nitrites are used to prevent spoilage and preserve flavor in some meat products.*

CHEM**QUANDARY**

NITRITE ADDITIVES

Consider these benefits and risks of using nitrites to preserve meats:

	Using Nitrites	**Eliminating Nitrites**
Benefit	Decreases risk of formation of botulin toxin	Decreases risk of formation of possible carcinogens
Risk	Increases risk of formation of possible carcinogens	Increases risk of formation of botulin toxin

Based on this information, do you think people should keep using nitrites to preserve meats?

D.9 ARTIFICIAL SWEETENERS

Because many people believe they should reduce the amount of sugar in their diets, many use low-Calorie sweeteners. Saccharin (found in Sweet 'N' Low™) was the first sugar substitute used extensively in the United States. There has been some controversy over saccharin use because early investigations conducted with rats suggested that massive amounts of saccharin may cause cancer. Further investigation indicated that a link between saccharin and cancer was very weak. Currently, it is generally believed that saccharin is safe for human consumption. Today, saccharin is used in many candies, baked goods, jellies, and jams.

The sugar-substitute aspartame (NutraSweet™ and Equal™) is an ingredient in many diet beverages and thousands of other food products. Aspartame (Figure 7.64) is a chemical combination of two natural amino acids, aspartic acid and phenylalanine, neither of which, by itself, tastes sweet. One gram of aspartame contains roughly the same food energy as one gram of table sugar (4 Cal), but aspartame tastes 200 times sweeter than sugar. Because we need smaller quantities of aspartame to sweeten a product, it as a "low-Calorie" alternative to sugar—one that is also safe for people with diabetes. Annually, thousands of tons of aspartame are used in the United States to sweeten diet drinks and foods. Aspartame decomposes at cooking temperatures. However, Splenda, an artificial sweetener based on sucralose (a chlorinated carbohydrate; see Figure 7.64), can be used directly in cooking.

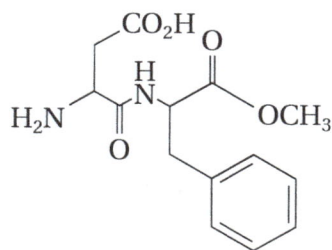
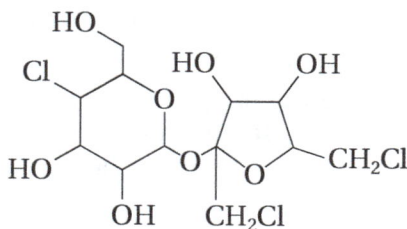

Aspartame   Sucralose

Figure 7.64 *Molecular structures of aspartame and sucralose.*

Although no serious warning regarding aspartame has been issued for the general population, aspartame does pose a health hazard to phenylketonurics—individuals who cannot properly metabolize phenylalanine. An FDA-required warning, *Phenylketonurics: Contains Phenylalanine,* on the label of foods with aspartame highlights the potential risk (see Figure 7.65).

Individuals with specific medical conditions may need to avoid certain foods and food additives. For instance, people with diabetes (see Figure 7.66) must regulate their intake of carbohydrates, including sugars. People with high blood pressure (hypertension) must avoid excess sodium. Some people have food allergies. If such restrictions apply to you, you must always read food labels. Sometimes a new ingredient will be used in a food product that you have used safely in the past; this new ingredient may put you at risk.

Figure 7.65 *The warning to phenylketonurics found on a can of diet drink containing aspartame.*

Figure 7.66 *A diabetic individual checks her blood glucose level.*

DEVELOPING SKILLS

D.10 FOOD ADDITIVE SURVEY

Being aware of what you eat is a wise habit to develop. In this activity, you will investigate food labels and consider additives.

1. Collect labels from five or six packaged foods in your home. Select no more than two samples of the same type of food. For example, no more than two breakfast cereals or two canned soups. Bring your set of labels to class.

2. From the ingredient listings on the labels, select three additives.

3. Complete a summary table with this format:
   a. List the three additives in a vertical column along the left side.
   b. Make four vertical columns to the right of the additives with these headings: "Food Product Where Found," "Purpose of Additive (if known)," "Chemical Formula," and "Other Information."

4. Use Table 7.11 (page 763) to review the purposes of particular food additives. Then answer the following question for each food additive in your summary table: Why do you think this additive is included in this food?

5. What alternatives to food additives can you propose to prevent food spoilage?

MAKING DECISIONS

D.11 ANALYZING VITAMINS AND MINERALS

So far, you have analyzed the three-day food inventory in terms of fats, carbohydrates, and proteins. Now you will decide whether the inventoried foods provide adequate quantities of vitamins and minerals.

1. Your teacher will specify two vitamins and two minerals for you to analyze.

2. Using references provided by your teacher, analyze each food item in your three-day inventory for the presence of those two vitamins and two minerals. Record these values.

3. Determine the average intake of those two vitamins and two minerals over the three days covered in the three-day inventory. Record these data.

4. a. Does your analyzed food inventory provide enough of each specified vitamin and mineral?

   b. If not, what types of foods can be added to increase intake of these vitamins and minerals? Identify the food groups to which these food types belong.

5. Some people believe that eating a well-balanced diet provides all the vitamins a person needs to remain healthy. Others suggest that vitamin supplements can enhance personal health. Proponents of both opinions have engaged in considerable advertising to promote their viewpoints.

   a. After considering your data, which viewpoint do your results support?

   b. What could you do to help people make more informed decisions about this issue?

SECTION D SUMMARY

Reviewing the Concepts

> Vitamins are organic molecules necessary for basic life functions.

1. Vitamins are not considered foods, yet they are vital to healthful diets. Explain.
2. What is a coenzyme?
3. Give an example of a function served by a vitamin.
4. List some typical symptoms of dietary deficiency of
   a. vitamin B_{12}.
   b. vitamin A.
   c. vitamin D.

> The required daily intake of a vitamin depends on properties of the vitamin as well as characteristics of individuals.

5. Vitamins are micronutrients. What does this mean about the quantity of vitamins required daily?
6. What molecular properties determine whether a vitamin is fat-soluble or water-soluble?
7. Why should people take water-soluble vitamins more regularly than fat-soluble ones?
8. Give two examples of how DRIs for vitamins and minerals vary for individuals, based on age and gender.

> Minerals are elements that are essential for human life.

9. Name three essential minerals.
10. Of the more than 100 known elements, how many are believed essential for human life?
11. What is the difference between a *macromineral* and a *trace mineral*?
12. What roles do minerals play in the body?
13. List a macromineral and a trace mineral and describe the deficiency condition for each.
14. What are some dietary sources of
    a. magnesium?   c. iron?
    b. potassium?   d. molybdenum?
15. Regarding minerals in the diet, some people believe that "if a little is good, more must be better." Evaluate this idea, particularly considering such minerals as arsenic (As) and cadmium (Cd).

> Food additives are used to increase the nutritive value, storage life, visual appeal, taste, or ease of producing foods.

16. What determines whether a substance is considered an additive or a basic component of a food?

17. List two typical food additives and their functions.

18. Give an example of a food additive used to
    a. increase nutritive value.
    b. improve storage life.
    c. enhance visual appeal.
    d. ease production.

19. a. What is a carcinogen?
    b. What is a mutagen?

20. How does the Food Quality and Protection Act differ from the Delaney Clause?

21. Aspartame's energy value is the same as that of sugar, 4 Cal/g. Why, then, is aspartame useful as a low-Calorie artificial sweetener?

22. Explain why aspartame can't be used in baking.

> Titration is a common laboratory procedure used to determine the amount of solute in a particular solution. Paper chromatography is a technique for separating and identifying the components of a solution.

23. Describe the key steps in a vitamin C titration, and explain how this process allows you to determine the concentration of vitamin C in a beverage.

24. What is meant by a titration's *endpoint*?

25. What reactions allowed you to detect the presence of vitamin C (usually colorless) in solution in Investigating Matter D.3?

26. Suppose you complete a vitamin-C titration and determine that 0.035 mg vitamin C reacts with 1 drop iodine solution. How much vitamin C is in a particular sample that turns bluish-black after adding
    a. 1 drop of iodine?
    b. 12 drops of iodine?

27. For what juices would the vitamin-C titration endpoint be difficult to observe? Explain.

28. a. How is an R_f value determined?
    b. What does its R_f value indicate about a particular solute?

29. In a paper-chromatography investigation, the solvent moves 5.8 cm, and a food-dye component moves 3.9 cm. What is the R_f value for the food-dye? Show calculations to support your answer.

30. What properties allow chromatography to separate components of a solution?

31. A student tests two samples of dye solution with paper chromatography. In A, the solvent moves 6.2 cm, while the dye moves 4.1 cm. In B, the solvent moves 5.3 cm, and the dye moves 4.1 cm. Do these data support the idea that A and B are the same dye? Explain.

What roles do vitamins, minerals, and additives play in the foods we eat?

In this section you have studied the structure, sources, and functions of vitamins and minerals, considered reasons for adding substances to foods, and used quantitative techniques to investigate the presence of vitamins and additives in food products. Think about what you have learned, then answer the question in your own words in organized paragraphs. Your answer should demonstrate (and perhaps clarify!) your understanding of the key ideas in this section.

Be sure to consider the following in your response: vitamins, minerals, food additives, and methods for detecting and quantifying components of food products.

Connecting the Concepts

32. Although high temperatures can destroy some vitamins in foods, they rarely affect the quality of minerals in foods. Explain.

33. Why do food standards use the term "generally recognized as safe" (GRAS), rather than "always safe" or "100% guaranteed safe" with certain food additives?

34. Would paper chromatography provide any useful information if one or more sample components were colorless? Explain.

35. How could you modify Investigating Matter D.3 to find out how much vitamin C is lost during cooking a food item?

36. One cup of spinach provides 51 mg calcium and 1.7 mg iron. Does spinach provide a greater percent of your DRI for calcium or for iron? Show calculations to support your answer.

37. Suppose a government banned all food additives. Overall, do you think the ban would produce positive or negative results? Explain.

38. Suppose one component in a chromatography sample was quite volatile, tending to evaporate when placed on paper. What effect might this property have on the experimental results?

39. All substances described in this unit consist of atoms, molecules, or ions. With that in mind, why do you think some people describe food additives as "chemicals," but do not describe carbohydrates and fats also as chemicals?

40. In a paper chromatography investigation, a sample failed to move up the paper when water is used as a solvent. How could the investigation be modified to address this problem?

Extending the Concepts

41. Investigate some additional types of chromatography, such as gas chromatography and column chromatography. Explain how each works.

42. Modern food additives are generally safe and healthful. That was not always the case in the past. Research the history of food additives and the unexpected impact of some early additives on human health.

43. When administered in large doses, a prospective food additive is found to cause cancer in laboratory rats. Should that evidence be used to withhold the approval for its use in human foods? Explain your answer.

44. Research some food-preparation methods that tend to preserve vitamin content. Discuss the chemical concepts that can account for the effectiveness of such methods.

45. Liver is a rich source of many trace minerals. Explain this based on liver physiology.

46. Research *goiter*. When and where has it been most prevalent? Why? What specific product makes it much less common today?

47. Some historians claim that the most significant contribution to the success of the British Navy in the 1700s was adding sauerkraut to shipboard food supplies. Investigate and explain the possible connection.

PUTTING IT ALL TOGETHER

GUIDING THE PTSA'S DECISIONS

Now that you have completed a series of analyses of the three-day food and activity inventory list, it is time to meet with other members of the Vending Machine Policy Planning Committee to consider next year's Riverwood High School food vending machine policies and to convey recommendations to the PTSA. To guide your writing, review the results of your analyses and think of points you would like to raise about students' eating habits. Use guidelines for a healthful diet provided by the PTSA at the start of this unit (page 685) to decide on the content and presentation of your report.

As a group, the team (your class) should make recommendations regarding the types or classes of foods that should be supplied in Riverwood High School food vending machines. Focus on generalized recommendations, such as: "The vending machines should offer food options high in calcium." In your discussion, be sure to support proposed recommendations with evidence and/or data from your three-day food inventory analyses.

Once your team has prepared its list of recommendations, each team member should write a report to submit to the PTSA. Here are guidelines for writing your report:

Thinking About Your Audience

As you studied this unit, you learned much about food chemistry, so you may decide to include some technical terminology in your report. If you do, make sure you explain these terms, so that PTSA members can understand your findings and recommendations. Many of them have studied chemistry (they know about bonding and other fundamental chemistry concepts), but are unfamiliar with food chemistry.

Writing

Your written report should include these sections:

1. *Introduction*—a brief summary highlighting features of the three-day food inventory. In one or two paragraphs, provide readers with an overview of what you found in your food-inventory analysis. In addition, list your top three vending machine policy recommendations. Hold the details of your study and detailed explanations of your recommendations until later.

2. *Background Information*—information that PTSA members need to understand your report. Here are some questions you may need to answer for readers:

- What are fats, carbohydrates, and proteins? You may include information about each substance's molecular structure, energy content, and function in the human body.

- Why do people need fats, carbohydrates, and proteins in their diets?

- What roles do vitamins and minerals play in the diet?

- You may include structural formulas and other visual aids, but remember that you need to guide your readers in what to look for in any visuals.

3. *Data Analysis*—the body of your report. In this section, you bring together detailed data you gathered concerning energy, fats, carbohydrates, protein, and vitamins and minerals. Organize and present the information so that it will make sense to readers. Include graphs and tables as appropriate. You may wish to address questions such as these:

- What is the average daily Calorie intake? How reasonable is this value? Does the average accurately represent each day in the inventory, or does the Calorie intake vary widely from day to day?

- How does daily Calorie intake compare with daily Calorie expenditure?

- What percent of total Calories are supplied by carbohydrates, fats, and protein? Are these percent values in line with USDA recommendations?

- What percent of fats are saturated and unsaturated? Do these levels meet USDA recommendations?

- Which vitamins and minerals are present at adequate levels? Which are lacking?

4. *Conclusions*—here you will provide further details regarding recommendations you highlighted at the start of your report. Discuss each recommendation in light of your findings and the guidelines for a healthful diet summarized at the start of this unit. List representative vending-machine food items that (a) meet one or more of your proposed recommendations, and (b) do not meet your recommendations, and explain how you decided on the particular items. Keep in mind that nutrition is not the only consideration—taste, appearance, and whether students would actually eat the items are also considerations.

LOOKING BACK

This unit focused on food, a mixture of chemical substances that you necessarily must encounter daily. You can now explain how food that you eat provides energy for daily living and structural components for growth. You can attach deeper chemical meaning to terms such as *carbohydrate*, and can better evaluate consequences associated with deciding to consume or not consume certain foods. The next time you hear someone remark, "You are what you eat," smile and remark that you know some of the chemistry behind the true meaning of their words!

The Scientific Method versus Scientific Methods

Scientists deepen their knowledge and understanding of the natural world by observing and manipulating their environment. The inquiry approach used by scientists to solve problems and seek knowledge has led to vast increases in understanding how nature works. Many efforts have been made to formalize and list the steps that scientists use to generate and test new knowledge. You may have been asked to learn the "steps" of the Scientific Method, such as Make Observations, Define the Problem, and so on. But no one comes to an understanding of how scientific inquiry works by simply learning a list of steps or definitions of words.

Although you may not go on to become a research scientist, it is important that all students acquire the ability to conduct scientific inquiry. Why? Everyone is confronted daily with endless streams of facts and claims. What should be accepted as true? What should be discarded? Having well-developed ways to evaluate and test claims is essential in deciding between valid and deceptive information.

What abilities are needed to conduct scientific inquiry? According to the National Science Education Standards (NSES), they include the abilities to

- identify questions and concepts that guide scientific investigations.
- design and conduct scientific investigations.
- use technology and mathematics to improve investigations and communications.
- formulate and revise scientific explanations and models using logic and evidence.
- recognize and analyze alternative explanations and models.
- communicate and defend a scientific argument.

These skills are necessary for both doing and learning science. They are also important skills to evaluate information in daily living. You can only acquire these skills through practice—by doing exercises and investigations such as those contained in this textbook.

Even with all the abilities listed above, doing science is a complex behavior. The NSES outlines the ideas that all students should know and understand about the practices of science:

- Scientists usually inquire about how physical, living, or designed systems function.
- Scientists conduct investigations for a wide variety of reasons.
- Scientists rely on technology to enhance the gathering and manipulation of data.
- Mathematics is essential in scientific inquiry.
- Scientific explanations must adhere to criteria such as: a proposed explanation must be logically consistent; it must abide by the rules of evidence; it must be open to questions and possible modification; and it must be based on historical and current scientific knowledge.
- Results of scientific inquiry—new knowledge and methods—emerge from different types of investigations and public communication among scientists.

The last statement above acknowledges that—despite generalizations that can be listed—there are many paths to gaining new scientific knowledge. That is what makes studying scientific processes so important. Learning how to acquire the abilities to do and understand scientific inquiry may be the most important and useful thing you learn in this course.

Numbers in Chemistry

Chemistry is a quantitative science. Most chemistry investigations involve not only measuring but also a search for the meaning among the measurements. Chemists learn how to interpret as well as perform calculations using these measurements.

Scientific Notation

Chemists often deal with very small and very large numbers. Instead of using many zeros to express very large or very small numbers, they often use scientific notation. In scientific notation, a number can be rewritten as the product of a number between 1 and 10 and an exponential term—10^n, where n is a whole number. The exponential term is the number of times 10 would have to be multiplied or divided by itself to yield the appropriate number of digits in the number. For instance, 10^3 is $10 \times 10 \times 10$, or 1000; $3.5 \times 10^3 = 3500$.

> **Sample Problem 1:** *Express the distance between New York City and San Francisco, 4 741 000 meters, using scientific notation.*
>
> $$4\ 741\ 000 \text{ m} = (4.741 \times 1\ 000\ 000) \text{ m or } \mathbf{4.741 \times 10^6} \text{ m}$$

> **Sample Problem 2:** *Express the amount of ranitidine hydrochloride in a Zantac tablet, 0.000 479 mol, using scientific notation.*
>
> $$0.000\ 479 \text{ mol} = 4.79 \times 0.0001 \text{ mol or } \mathbf{4.79 \times 10^{-4}} \text{ mol}$$

It is easier to assess magnitude and to perform operations with numbers written in scientific notation than with numbers fully written out. As you will see, it is also easier to communicate the precision of the measurements involved.

Rules for *adding and subtracting* using scientific notation:

Step 1. Convert the numbers to the same power of 10.

Step 2. Add (subtract) the non-exponential portion of the numbers. *The power of ten remains the same.*

> **Sample Problem 3:** *Add $(1.00 \times 10^4) + (2.30 \times 10^5)$.*
>
> **Step 1.** A good rule to follow is to express all numbers in the problem to the highest power of ten. Convert 1.00×10^4 to 0.100×10^5.
>
> **Step 2.** $(0.100 \times 10^5) + (2.30 \times 10^5) = 2.40 \times 10^5$

Rules for *multiplying* using scientific notation:

Step 1. Multiply the nonexponential numbers.

Step 2. Add the exponents.

Step 3. Convert the answer to scientific notation.

> *Sample Problem 4:* *Multiply (4.24×10^2) by (5.78×10^4).*
>
> **Steps 1 and 2.** $(4.24 \times 5.78) \times (10^{2+4}) = 24.5 \times 10^6$
>
> **Step 3.** Convert to scientific notation $= 2.45 \times 10^7$

Rules for *dividing* using scientific notation:

Step 1. Divide the nonexponential numbers.

Step 2. Subtract the denominator exponent from the numerator exponent.

Step 3. Express the answer in scientific notation.

> *Sample Problem 5:* *Divide (3.78×10^5) by (6.2×10^8).*
>
> **Steps 1 and 2.** $\dfrac{3.78}{6.2} \times (10^{5-8}) = 0.61 \times 10^{-3}$
>
> **Step 3.** Convert to scientific notation $= 6.1 \times 10^{-4}$

Practice Problems*

1. Convert the following numbers to scientific notation.

   a. 0.000 036 9

   b. 0.0452

   c. 4 520 000

   d. 365 000

2. Carry out the following operations:

   a. $(1.62 \times 10^3) + (3.4 \times 10^2)$

   b. $(1.75 \times 10^{-1}) - (4.6 \times 10^{-2})$

   c. $\dfrac{6.02 \times 10^{23}}{12.0}$

   d. $\dfrac{(6.63 \times 10^{-34}\ \text{J·s}) \times (3.00 \times 10^8\ \text{m s}^{-1})}{4.6 \times 10^{-9}\ \text{m}}$

*Answers to odd-numbered problems can be found on page ANS-1.

Dimensional Analysis

Dimensional analysis, also called the *factor-label method*, is used by scientists to keep track of units in calculations and to help guide their work in solving problems. The method is helpful in setting up problems and also in checking work.

Dimensional analysis consists of three basic steps:

Step 1. Identify equivalence relationships in order to create suitable conversion factors.

Step 2. Identify the given unit(s) and the new unit(s) desired.

Step 3. Arrange each conversion factor so that each unit to be converted can be divided by itself (and thus cancelled).

Sample Problem 1: *In an exercise to determine the volume of a rectangular object, your laboratory partner measured the object's length as 12.2 in (inches). However, measurements of the object's width and height were recorded in centimeters. In order to calculate the object's volume, convert the object's measured length to centimeters.*

Step 1. Find the equivalence relating centimeters and inches.
$$2.54 \text{ cm} = 1 \text{ in}$$

Step 2. Identify the given unit and the new unit.
Given unit = in   new unit = cm

Step 3. Create a fraction so that the "given" unit (in) can be divided and thus cancelled.
$$12.2 \text{ in} \times \frac{2.54 \text{ cm}}{1 \text{ in}} = 31.0 \text{ cm}$$

Sample Problem 2: *How many seconds are in 24 hours?*

Step 1. Identify the equivalence:
1 hr = 60 min   1 min = 60 s

Step 2. Given unit: hr   new unit: s

Step 3. Arrange for the given unit to cancel and progress to the desired unit:
$$24 \text{ hr} \times \frac{60 \text{ min}}{1 \text{ hr}} \times \frac{60 \text{ s}}{1 \text{ min}} = 86\ 400 \text{ s}$$

Practice Problems*

3. The distance between two European cities is 4 741 000 m. That may sound impressive, but to put all those digits on a car odometer is slightly inconvenient. Kilometers are a better choice for measuring distance in this case. Convert the distance to kilometers.

4. The density of aluminum is 2.70 g/cm^3. What is the mass of 234 cm^3 of aluminum?

Significant Figures

Significant figures are all the digits in a measured or calculated value that are known with certainty plus the first uncertain digit. Numerical measurements have some inherent uncertainty. This uncertainty comes from the measurement device as well as from the human making the measurement. No measurement is exact. When you use a measuring device in the laboratory, read and record each measurement to one digit beyond the smallest marking interval on the scale.

Guidelines for Determining Significant Figures

Step 1. All digits recorded from a laboratory measurement are called significant figures.

The measurement of 4.75 cm has three significant figures.

Note: If you use a measuring device that has a digital readout, such as a balance, you should record the measurement just as it appears on the display.

Measurement	Number of Significant Figures
123 g	3
46.54 mL	4
0.33 cm	2
3 300 000 nm	2
0.033 g	2

Step 2. All non-zero digits are considered significant.

---

*Answers to odd-numbered problems can be found on page ANS-1.

Step 3. There are special rules for zeros. Zeros in a measurement or calculation fall into three types: middle zeros, leading zeros, and trailing zeros.

Middle zeros are always significant.
303 mm   a middle zero—always significant. This measurement has three significant figures.

A leading zero is never significant. It is only a placeholder, not a part of the actual measurement.
0.0123 kg   two leading zeros—never significant. This measurement has three significant figures.

A trailing zero is significant when it is to the right of a decimal point. This is not a placeholder. It is a part of the actual measurement.
23.20 mL   a trailing zero—significant to the right of a decimal point. This measurement has four significant figures.

The most common errors concerning significant figures are (1) reporting all digits found on a calculator readout, (2) failing to include significant trailing zeros (14.150 g), and (3) considering leading zeros to be significant—0.002 g has only one significant figure, not three.

Practice Problem*

5. How many significant figures are in each of the following?
   a. 451 000 m
   b. 4056 V
   c. 6.626×10^{-34} J·s
   d. 0.0065 g
   e. 0.0540 mL

*Answers to odd-numbered problems can be found on page ANS-1.

Using Significant Figures in Calculations

Addition and subtraction: The number of *decimal places* in the answer should be the same as in the measured quantity with the smallest number of *decimal places*.

> *Sample Problem 1:* Add the following measured values and express the answer to the correct number of significant figures.
>
> $$\begin{array}{r} 1259.1 \ \ \text{g} \\ 2.365 \ \ \text{g} \\ +\ 15.34 \ \ \text{g} \\ \hline 1276.805 \ \ \text{g} \end{array} \quad = \textbf{1276.8 g}$$

Multiplication and division: The number of *significant figures* in the answer should be the same as in the measured quantity with the smallest number of *significant figures*.

> *Sample Problem 2:* Divide the following measured values and express the answer to the correct number of significant figures.
>
> $$\frac{13.356 \text{ g}}{10.42 \text{ mL}} = 1.2817658 \text{ g/mL} = \textbf{1.282 g/mL}$$

Practice Problem

6. Report the answer to each of these using the correct number of significant figures.

   a. 16.27 g + 0.463 g + 32.1 g

   b. 42.04 mL − 3.5 mL

   c. 15.1 km × 0.032 km

   d. $\dfrac{13.36 \text{ cm}^3}{0.0468 \text{ cm}^3}$

Note: Only measurements resulting from scale readings or digital readouts carry a limited number of significant figures. However, values arising from direct counting (such as 25 students in a classroom) or from definitions (such as 100 cm = 1 m, or 1 dozen = 12 things) carry unlimited significant figures. Such "counted" or "defined" values are regarded as exact.

Interpreting Graphs

Graphs are of four basic types: pie charts, bar graphs, line graphs, and *x-y* plots. The type chosen depends on the characteristics of the data displayed.

Pie charts show the relationship of the parts to the whole. This presentation helps a reader visualize the magnitude of difference among various parts. They are made by taking a 360° circle and dividing it into wedges according to the percent of the whole represented by each part.

Bar graphs and **line graphs** compare values within a category or among categories. The horizontal axis (*x*-axis) is used for the quantity that can be controlled or adjusted. This quantity is the **independent variable.** The vertical axis (*y*-axis) is used for the quantity that is influenced by the changes in the quantity on the *x*-axis. This quantity is the **dependent variable.** For example, a bar graph could present a visual comparison of the fat content (dependent variable on the *y*-axis) of types of cheese (independent variable on the *x*-axis). Such a graph would make it easy to choose a cheese snack with a low-fat content. Bar graphs can also be useful in studying trends over time.

Graphs involving ***x-y* plots** are commonly used in scientific work. Sometimes it is difficult to decide if a graph is a line graph or an *x-y* plot. In an *x-y* plot, it is possible to determine a mathematical relationship between the variables. Sometimes the relationship is the equation for a straight line ($y = mx + b$), but other times it is more complex and may require transformation of the data to produce a simpler graphical relationship. The first example below refers to a straight-line or direct relationship.

Sample Problem 1: *A group of entrepreneurs was considering investing in a mine that was said to contain gold. To verify this claim, they gave several small, irregular mined particles to a chemist, who was told to use nondestructive methods to analyze the samples.*

The chemist decided to determine the density of the small samples. The chemist found the volume of each particle and determined its mass. The data collected are shown in the table below.

Particle	Volume (mL)	Mass (g)
1	0.006	0.116
2	0.012	0.251
3	0.015	0.290
4	0.018	0.347
5	0.021	0.386

Use the x-y plot of these data to evaluate whether the particles are gold.

Gold Particle Data

Mass/Volume

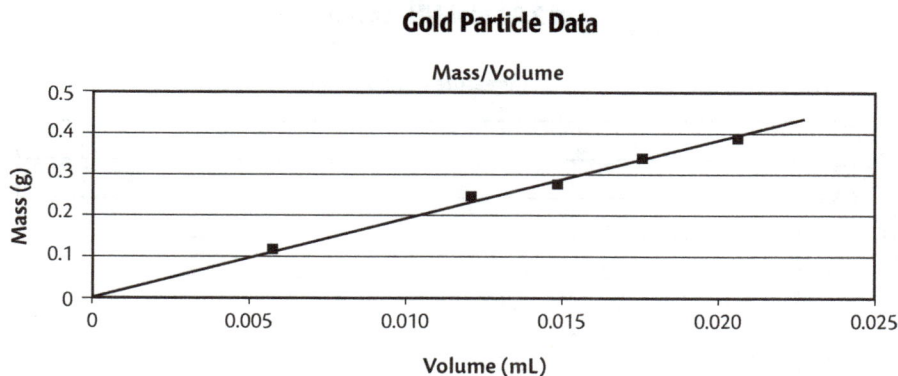

Since the *x-y* plot is linear, the sample materials are likely the same. If point (0, 0) is included (a sample of zero volume has zero mass), the slope is about 19 g/mL, close to gold's density (19.3 g/mL), so these are likely gold particles.

Sample Problem 2: *The graph below is a plot of data gathered at constant temperature involving the volume of 2.00 mol of ammonia (NH₃) gas measured at various pressures. Determine whether the measurements are related by a simple mathematical relationship.*

Volume of 2.00 mol NH₃ at Different Pressures

Pressure/Volume Relationship

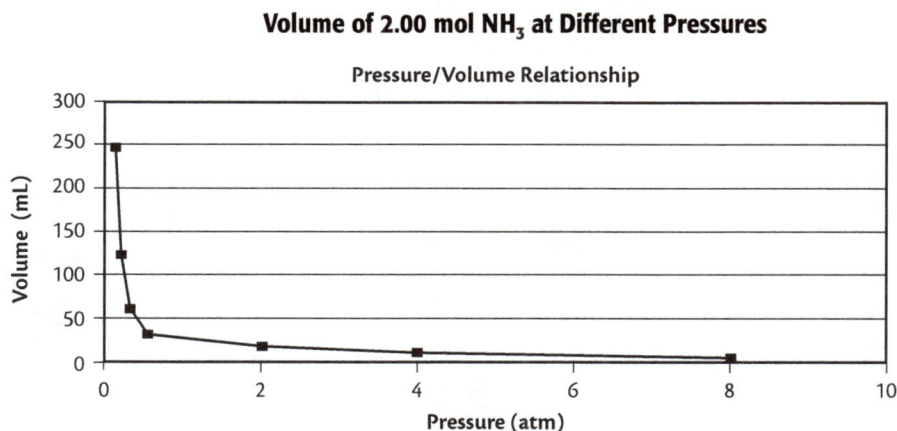

When the best smooth graph line is not a straight line (or direct relationship), the data can be manipulated to see if any other simple mathematical relationship is possible. Several of the most common types of mathematical relationships are inverse, exponential, and logarithmic. Each relationship has unique characteristics that can often be identified from the graphical presentation. Knowing the mathematical relationship allows scientists to interpret the data. In this case, it appears that as the pressure increases the volume decreases, which is a characteristic of an inverse relationship. In testing this theory, the value of 1/V (the inverse of the volume) can be calculated, recorded in another column of the table, and plotted versus the pressure.

$$\frac{1}{Volume} \text{ of 2.00 mol NH}_3 \text{ at Different Pressures}$$

Pressure vs. 1/Volume

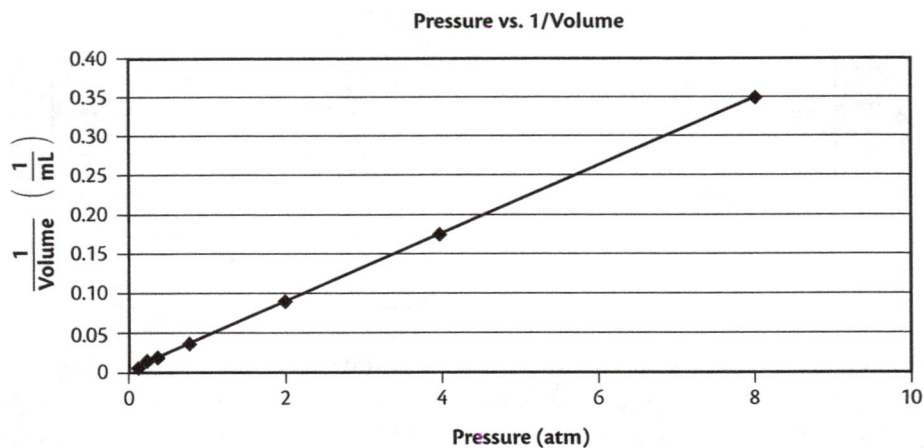

This graph exhibits a straight line showing that pressure is directly related to the inverse of the volume. This leads to the mathematical result that pressure and volume are inversely related. If this mathematical manipulation had not resulted in a straight line, some other reasonable relationships might have been considered and tested.

Equations, Moles, and Stoichiometry

Moles and Molar Mass

The **mole** is regarded as a counting number–a number used to specify a certain number of objects. **Pair** and **dozen** are other examples of counting numbers. A mole equals 6.02×10^{23} objects. Most often, things that are counted in units of moles are very small—atoms, molecules, or electrons.

The modern definition of a mole specifies that one mole is equal to the number of atoms contained in exactly 12 g of the carbon-12 isotope. This number is named after Amedeo Avogadro, who proposed the idea, but never determined the number. At least four different types of experiments have accurately determined the value of Avogadro's number—the number of units in one mole. Avogadro's number is known to eight significant figures, but three will be enough for most of your calculations—6.02×10^{23}.

The modern atomic weight scale is also based on C-12. Compared to C-12 atoms with a defined atomic mass of exactly 12, hydrogen atoms have a relative mass of 1.008. Therefore, one mole of hydrogen atoms has a mass of 1.008 g. One mole of oxygen atoms equals 15.9994 g—also called the molar mass of oxygen. The total mass of one mole of a compound is found by adding the atomic weights of all of the atoms in the formula and expressing the sum in units of grams.

Sample Problem: Calculate the mass of one mole of water.

$$2 \text{ mol H} \times \frac{1.008 \text{ g H}}{1 \text{ mol H}} = 2.016 \text{ g H}$$

$$1 \text{ mol O} \times \frac{16.00 \text{ g H}}{1 \text{ mol O}} = 16.00 \text{ g O}$$

molar mass of water = 2.016 g + 16.00 g = 18.02 g

Practice Problems*

Find the molar mass (in units of g/mol) for each of the following:
1. Acetic acid, CH_3COOH
2. Formaldehyde, $HCHO$
3. Glucose, $C_6H_{12}O_6$
4. 2-Dodecanol, $CH_3(CH_2)_9CH(OH)CH_3$

*Answers to odd-numbered problems can be found on page ANS-1.

Gram–Mole Conversions

Conversions between grams and moles can be readily accomplished by using the technique of dimensional analysis (see Appendix B).

Sample Problem 1: *What mass (in grams) of water contains 0.25 mol H_2O?*

The mass of one mole of water (18.02 g) is found as illustrated earlier. Two factors can be written, based on that relationship:

$$\frac{1 \text{ mol } H_2O}{18.02 \text{ g } H_2O} \text{ and } \frac{18.02 \text{ g } H_2O}{1 \text{ mol } H_2O}$$

The second conversion factor is chosen so that each unit to be converted is divided by itself and thus cancelled. Units for the answer are g H_2O, as expected.

$$0.25 \text{ mol } H_2O \times \frac{18.02 \text{ g } H_2O}{1 \text{ mol } H_2O} = 4.5 \text{ g } H_2O$$

Sample Problem 2: *How many moles of water molecules are present in a 1.00-kg sample of water?*

$$1.00 \text{ kg } H_2O \times \frac{1000 \text{ g } H_2O}{1 \text{ kg } H_2O} \times \frac{1 \text{ mol } H_2O}{18.02 \text{ g } H_2O} = 55.5 \text{ mol } H_2O$$

Practice Problems*

5. Acetic acid, CH_3COOH, and salicylic acid, $C_7H_6O_3$, can be chemically combined to form aspirin. If a chemist uses 5.00 g salicylic acid and 10.53 g acetic acid, how many moles of each compound are involved?

6. Calcium chloride hexahydrate, $CaCl_2 \cdot 6 \text{ } H_2O$, can be sprinkled on sidewalks to melt ice and snow. How many moles of that compound are in a 5.0-kg sack of that substance?

A Quantitative Understanding of Chemical Formulas

Calculating percent composition from a formula

The percent by mass of each component found in a sample of material is called its **percent composition.** To find the percent of an element in a particular compound, first calculate the molar mass of the compound. Then find the total mass of the element contained in one mole of the compound. Then divide the mass of the element by the molar mass of the compound and multiply the result by 100%.

Sample Problem 1: *Calculate the molar mass of sucrose, $C_{12}H_{22}O_{11}$*

12 mol C (12.01 g/mol)	=	144.1 g
22 mol H (1.008 g/mol)	=	22.18 g
11 mol O (16.00 g/mol)	=	176.0 g
Molar mass of $C_{12}H_{22}O_{11}$	=	342.3 g

*Answers to odd-numbered problems can be found on page ANS-1.

Sample Problem 2: Find the mass percent of each element in sucrose (rounded to significant figures).

$$\% \ C = \frac{\text{mass C}}{\text{mass C}_{12}\text{H}_{22}\text{O}_{11}} \times 100\% = \frac{141.1 \text{ g C}}{342.3 \text{ g C}_{12}\text{H}_{22}\text{O}_{11}} \times 100\% = 42.10\% \ C$$

$$\% \ H = \frac{\text{mass H}}{\text{mass C}_{12}\text{H}_{22}\text{O}_{11}} \times 100\% = \frac{22.18 \text{ g H}}{342.3 \text{ g C}_{12}\text{H}_{22}\text{O}_{11}} \times 100\% = 6.48\% \ H$$

$$\% \ O = \frac{\text{mass O}}{\text{mass C}_{12}\text{H}_{22}\text{O}_{11}} \times 100\% = \frac{176.0 \text{ g O}}{342.3 \text{ g C}_{12}\text{H}_{22}\text{O}_{11}} \times 100\% = 51.42\% \ O$$

Practice Problems*

7. Barium sulfate, $BaSO_4$, is commonly used to detect gastrointestinal tract abnormalities; it is administered as a water suspension by mouth prior to X-ray imaging. Find the mass percent of each element in barium sulfate.

8. Sodium acetate is a common component in commercial thermal packs. Find the mass percent of each element in sodium acetate, $NaCH_3COO$.

Deriving formulas from percent composition data

An **empirical formula** gives the relative numbers of each element in a substance, using the smallest whole numbers for subscripts. The empirical formula of a compound can be calculated from percent composition data. A formula requires the relative numbers of moles of each element, so percent values must be converted to grams and grams to moles. It is easiest to assume 100 g of compound. Then each percent value is equal to the total grams of that element.

Sample Problem 1: A hydrocarbon consists of 85.7% carbon and 14.3% hydrogen by mass. What is its empirical formula?

    Step 1. Assume 100 g of compound—thus 85.7 g are carbon and 14.3 g are hydrogen.

    Step 2. Use dimensional analysis to find the total moles of each element in the compound.

$$85.7 \ \cancel{\text{g C}} \times \frac{1 \text{ mol C}}{12.01 \ \cancel{\text{g C}}} = 7.14 \text{ mol C}$$

$$14.3 \ \cancel{\text{g H}} \times \frac{1 \text{ mol H}}{1.008 \ \cancel{\text{g H}}} = 14.2 \text{ mol H}$$

    Step 3. Determine the smallest whole-number ratio of moles of elements by dividing all mole values by the smallest value.

$$\frac{7.14 \text{ mol C}}{7.14} = 1 \text{ mol C}$$

$$\frac{14.2 \text{ mol H}}{7.14} = 1.99 \text{ mol H}$$

    The ratio of moles C to moles H = 1:2, so the empirical formula for the compound must be CH_2.

*Answers to odd-numbered problems can be found on page ANS-1.

Sample Problem 2: *The percent composition of one of the oxides of nitrogen is 74.07% oxygen and 25.93% nitrogen. What is the empirical formula of that compound?*

Step 1.   100 g of compound consists of 74.07 g oxygen and 25.93 g nitrogen.

Step 2. $25.93 \; \cancel{g\,N} \times \dfrac{1 \text{ mol N}}{14.01 \cancel{g\,N}} = 1.851 \text{ mol N}$

$74.07 \; \cancel{g\,O} \times \dfrac{1 \text{ mol O}}{16.00 \cancel{g\,O}} = 4.629 \text{ mol O}$

Step 3. $\dfrac{1.851 \text{ mol N}}{1.851} = 1 \text{ mol N}$

$\dfrac{4.629 \text{ mol O}}{1.851} = 2.50 \text{ mol O}$

Step 4.   This ratio (1:2.5) does not consist entirely of whole numbers. Thus all the numbers in the ratio must be multiplied by a number that converts the decimal to a whole number. In this case the number is 2, and the empirical formula becomes N_2O_5.

Practice Problems*

9. The percent composition by mass of an industrially important substance is 2.04% H, 32.72% S, and 65.24% O. What is the formula of this compound?

10. Determine the empirical formula of a compound that contains (by mass) 38.71% C, 9.71% H, and 51.58% O.

Mass Relationships in Chemical Reactions

Since the total number of atoms is conserved in a chemical reaction, their masses must also be conserved, as expected from the law of conservation of mass. In the equation for the formation of water from the elements hydrogen and oxygen, $2H_2(g) + O_2(g) \longrightarrow 2H_2O(l)$, 2 molecules of hydrogen gas and 1 molecule of oxygen gas combine to form 2 molecules of water. One could also interpret the equation this way: 2 mol hydrogen gas react with 1 mol oxygen gas to form 2 mol water. Using the molar mass of each substance, the mass relationships in the table below can be determined. The ratio of moles of hydrogen gas to moles of oxygen gas in forming water will be 2:1. If 10 mol hydrogen gas are available, 5 mol oxygen gas are required.

$2 \, H_2(g)$ +	$O_2(g)$ $\longrightarrow$	$2 \, H_2O(l)$
2 molecules	1 molecule	2 molecules
2 mol	1 mol	2 mol
2 mol × (2.02 g/mol)	1 mol × (32.00 g/mol)	2 mol × (18.02 g/mol)
4.04 g	32.00 g	36.04 g

Solving problems involving the masses of products and/or reactants is conveniently accomplished by dimensional analysis. And remember, all numerical problems involving chemical reactions involve a correctly balanced equation.

*Answers to odd-numbered problems can be found on page ANS-1.

Sample Problem: *Find the mass of water formed when 10.0 g hydrogen gas completely reacts with oxygen gas.*

$$2\ H_2(g)\ +\ O_2(g) \longrightarrow 2\ H_2O(l)$$

Step 1. Find the moles of hydrogen gas represented by 10.0 g, using the molar mass of H_2.

$$10.0\ \cancel{g\ H_2} \times \frac{1\ mol\ H_2}{2.02\ \cancel{g\ H_2}}\ =\ 4.95\ mol\ H_2$$

Step 2. Find the total moles of H_2O produced by 4.95 mol H_2. From the balanced equation, you know that for every 2 mol H_2, 2 mol H_2O are produced.

$$4.95\ \cancel{mol\ H_2} \times \frac{2\ mol\ H_2O}{2\ \cancel{mol\ H_2}}\ =\ 4.95\ mol\ H_2O$$

Step 3. Find the mass of H_2O that contains 4.95 mol H_2O by using the molar mass of water.

$$4.95\ \cancel{mol\ H_2O} \times \frac{18.02\ g\ H_2O}{1\ \cancel{mol\ H_2O}}\ =\ 89.2\ g\ H_2O$$

Most chemistry students find it is more convenient to set up all three steps in one extended calculation. Assure yourself that all units divide and cancel except for grams of water (g H_2O)—an appropriate way to express the answer sought in the problem.

$$10.0\ \cancel{g\ H_2} \times \underbrace{\frac{1\ \cancel{mol\ H_2}}{2.02\ \cancel{g\ H_2}}}_{\substack{\text{Molar mass} \\ \text{of } H_2}} \times \underbrace{\frac{2\ \cancel{mol\ H_2O}}{2\ \cancel{mol\ H_2}}}_{\substack{\text{Coefficients} \\ \text{in equation}}} \times \underbrace{\frac{18.02\ g\ H_2O}{1\ \cancel{mol\ H_2O}}}_{\substack{\text{Molar mass} \\ \text{of } H_2O}}\ =\ 89.2\ g\ H_2O$$

Practice Problems*

11. Find the mass of copper(II) oxide formed if 2.0 g copper metal completely reacts with oxygen gas.

$$2\ Cu(s) + O_2(g) \longrightarrow 2\ CuO(s)$$

12. What mass of water is produced when 25.0 g methane gas reacts completely with oxygen gas?

$$CH_4(g) + 2\ O_2(g) \longrightarrow 2\ H_2O(g) + CO_2(g)$$

Answers to Practice Problems

Appendix B

1. a. 3.69×10^{-5}
   b. 4.52×10^{-2}
   c. 4.52×10^{6}
   d. 3.65×10^{5}
3. 4741 km
5. a. 3
   b. 4
   c. 4
   d. 2
   e. 3

Appendix D

1. 60.05 g/mol
3. 180.16 g/mol
5. 0.1754 mol acetic acid, 0.0362 mol salicylic acid
7. 58.84% Ba, 13.74% S, 27.42% O
9. H_2SO_4
11. 2.5 g CuO

Glossary

A

absolute zero
the lowest temperature theoretically obtainable, which is −273 °C or 0 K (zero kelvin)

absorbance
a measure of the amount of light that is absorbed by a particular substance/solution

accuracy
extent to which a measurement represents its corresponding actual values

acid precipitation
see acid rain

acid rain
fog, sleet, snow, or rain with a pH lower than about 5.6 due to dissolved gases such as SO_2, SO_3, and NO_2

acidic solution
an aqueous solution with a pH less than 7; it turns litmus from blue to red

acid
ion or compound that produces hydrogen ions, H^+ (or hydronium ions, H_3O^+), when dissolved in water

activation energy
the minimum energy required for the successful collision of reactant particles in a chemical reaction

active site
the location on an enzyme where a substrate molecule becomes positioned for a reaction

activity series
the ranking of elements in order of chemical reactivity

addition polymer
a polymer formed by repeated addition reactions at double or triple bonds within monomer units

addition reaction
a reaction at the double or triple bond within an organic molecule

adhesive force
force that causes molecules of different substances to be attracted to one another

adsorb
to take up or hold molecules or particles to the surface of a material

alcohol
a nonaromatic organic compound containing one or more —OH groups

alkali metal family
the group of elements consisting of lithium, sodium, potassium, rubidium, cesium, and francium

alkaline
a basic solution containing an excess of hydroxide ions (OH^-)

alkane
a hydrocarbon containing only single covalent bonds

alkene
a hydrocarbon containing one or more double covalent bonds

alkyne
a hydrocarbon containing one or more triple covalent bonds

alloy
a solid solution consisting of atoms of two or more metals

alpha particle (α)
high-speed, positively charged particle emitted during the decay of some radioactive elements; consists of a helium nucleus, $_2^4He^{2+}$

amino acid
an organic molecule containing a carboxylic acid group and an amine group; serves as a protein building block

anion
a negatively charged ion

anode
an electrode in an electrochemical cell at which oxidation occurs

aqueous solution
a solution in which water is the solvent

aquifer
a structure of porous rock, sand, or gravel that holds water beneath Earth's surface

area
the total space that makes up the surface of an object

aromatic compound
a ring-like compound, such as benzene, that can be represented as having alternating double and single bonds between carbon atoms

arterial plaque
build-up in blood vessel walls that results from fatty material

atherosclerosis
condition commonly known as "hardening of the arteries"

atmosphere
(a) a unit of gas pressure (atm); (b) the gaseous envelope surrounding Earth and composed of four layers: troposphere, stratosphere, mesosphere, and thermosphere

atom
the smallest particle possessing the properties of an element

atomic number
the number of protons in an atom; this value distinguishes atoms of different elements

Avogadro's law
equal volumes of all gases, measured at the same temperature and pressure, contain the same number of gas molecules

B

background radiation
the relatively constant level of natural radioactivity that is always present

balanced
in a chemical equation, when the total number of each type of atom is the same for both the reactants and products

balanced chemical equation
see chemical equation

ball-and-stick model
molecular model where each ball represents an atom, and each stick represents a pair of shared electrons (a single covalent bond) connecting two atoms

base
ion or compound that produces OH^- ions when dissolved in water

base unit
an SI unit that expresses a fundamental physical quantity (such as temperature, length, or mass)

basic solution
an aqueous solution with a pH greater than 7; it turns litmus from red to blue

battery
a device composed of one or more connected voltaic cells that supplies electrical current

beta decay
the radioactive decay of a nucleus accompanied by the emission of a beta particle

beta particle (β)
negatively charged particle emitted during the decay of some radioactive elements; high-speed electron

bias
a tendency to a particular belief or perspective

biodiesel
an alternative fuel or fuel additive for diesel engines made from various materials such as new or recycled vegetable oils and animal fats

biomolecule
large organic molecule found in living systems

blank
a solution or substance known not to contain any ions or molecules of interest

bottoms
components from petroleum found in the lower trays of a fractionating tower after distillation

Boyle's law
the pressure and volume of a gas sample at constant temperature are inversely proportional; $PV = k$

branched-chain alkane
an alkane in which at least one carbon atom is bonded to three or four other carbon atoms

branched polymer
a polymer formed by reactions that create numerous side chains rather than linear chains

brittle
a property of a material that causes it to shatter under pressure

buffer
a substance or combination of dissolved substances capable of resisting changes in pH when limited quantities of either acid or base are added

C

calibration curve
a graph constructed from data collected on solutions of known concentration

Calorie (Cal)
a unit of energy; thermal energy required to raise the temperature of one kilogram of water by one degree Celsius; commonly used to express quantity of food energy; informally called *food calorie*; 1 Cal = 1 000 cal; 1 kJ = 4.184 Cal

calorimeter
measuring device to determine energy released from the burning of a substance or other chemical reactions

calorimetry
procedure to determine the energy released during the combustion of several types of fuel or from other chemical reactions

carbohydrate
substance such as sugar or starch that is composed of carbon, hydrogen, and oxygen atoms; a main source of energy in foods

carbon chain
carbon atoms chemically linked to one another, forming a chainlike molecular structure

carbon cycle
the movement of carbon atoms within Earth's ecosystems, from carbon storage as plant and animal matter, through release as carbon dioxide due to cellular respiration, combustion, and decay, to reacquisition by plants

carbon footprint
the quantity of greenhouse gases emitted based upon individual activities; measured in kilograms of carbon dioxide (CO_2)

carboxylic acid
an organic compound containing the —COOH group

carcinogen
substance known to cause cancer

catalyst
a substance that speeds up a chemical reaction but is itself unchanged

catalytic convertor
the reaction chamber in an auto exhaust system designed to accelerate the conversion of potentially harmful exhaust gases to nitrogen gas, carbon dioxide, and water vapor

cathode
an electrode in an electrochemical cell at which reduction occurs

cathode ray
a beam of electrons emitted from a cathode when electricity is passed through an evacuated tube

cation
a positively charged ion

cellular respiration
the process that is used by organisms to convert complex organic molecules into carbon dioxide and water molecules, with an overall release of energy

chain reaction
in nuclear fission, a reaction that is sustained because it produces enough neutrons to collide with and split additional fissionable nuclei

Charles' law
the volume of a gas sample at constant pressure is directly proportional to its kelvin temperature; $V = kT$

chemical bond
the attractive force that holds atoms or ions together; *see also* covalent bonds; ionic bonds

chemical change
an interaction of matter that results in the formation of one or more new substances

chemical energy
a form of potential energy stored in chemical compounds

chemical equation
a symbolic expression summarizing a chemical reaction, such as $2 H_2(g) + O_2(g) \longrightarrow 2 H_2O(g)$

chemical formula
a symbolic expression representing the elements contained in a substance, together with subscripts that indicate the relative numbers of atoms of each element, such as H_2O

chemical kinetics
study of the rate of chemical reactions

chemical properties
properties only observed or measured by changing the chemical identity of a sample of matter

chemical reaction
the process of forming new substances from reactants that involves the breaking and forming of chemical bonds

chemical symbol
an abbreviation of an element's name, such as N for nitrogen or Fe for iron

cis-trans isomerism
isomers based on arrangement about a double bond in a molecule

claim
one- or two-sentence statement summarizing an important result of an investigation

climate
the average or prevailing weather conditions in a region

cloud chamber
a container filled with supersaturated air that, when cooled and exposed to ionizing radiation, produces visible trails of condensation, tracing paths taken by radioactive emissions

coefficient
a number in a chemical equation that indicates the relative number of units of a reactant or product involved in the reaction

coenzyme
an organic molecule that interacts with an enzyme to facilitate or enhance its activity

cohesive force
attractive force between molecules in a substance, especially in a liquid

collision theory
for a reaction to occur, reactant molecules must collide in proper orientation with sufficient kinetic energy

colloid
a mixture containing solid particles small enough to remain suspended and not settle out

colorimetry
a chemical analysis method that uses color intensity to determine solution concentration

combustion
a chemical reaction with oxygen gas that produces heat and light; burning

complementary proteins
multiple protein sources that provide adequate amounts of all essential amino acids when consumed together

complete protein
a protein source for humans containing adequate amounts of all essential amino acids

complex ion
a single central atom or ion, usually a metal ion, to which other atoms, molecules, or ions are attached

compound
a substance composed of two or more elements bonded together in fixed proportions; a compound cannot be broken down into simpler substances by physical means

compressed natural gas (CNG)
natural gas condensed under high pressure (160–240 atm) and stored in metal cylinders; CNG can serve as a substitute for gasoline or diesel fuel

concentration
see solution concentration

condensation
converting a substance from a gaseous state to a liquid state

condensation polymer
a polymer formed by repeated condensation reactions of one or more monomers

condensation reaction
the chemical combination of two organic molecules, accompanied by the loss of water or other small molecules

condensed formula
a chemical formula that provides additional information about bonding; for example, the condensed formula for propane, C_3H_8, is CH_3—CH_2—CH_3 or $CH_3CH_2CH_3$

conductor
a material that allows electricity (or thermal energy) to flow through it

confirming test
a laboratory test giving a positive result if a particular chemical species is present

control
in an experiment, a trial that duplicates all conditions except for the variable under investigation

covalent bond
a linkage between two atoms involving the sharing of one pair (single bond), two pairs (double bond), or three pairs (triple bond) of electrons

cracking
the process in which hydrocarbon molecules from petroleum are converted to smaller molecules, using thermal energy and a catalyst

criteria pollutant
an Environmental Protection Agency classification for a pollutant commonly found throughout the United States and detrimental to human health or the environment, such as carbon monoxide (CO), sulfur oxides (SO_x), nitrogen oxides (NO_x), ozone (O_3), lead (Pb), and particulate matter (PM)

critical mass
the minimum mass of fissionable material needed to sustain a nuclear chain reaction

cross-linking
polymer chains interconnected by chemical bonds; causes polymer rigidity

crude oil
unrefined liquid petroleum as it is pumped from the ground by oil wells

crystal
A solid 3-D network with a regular arrangement of anions and cations

currency
circulating money, includes both coins and bills (banknotes)

cycloalkane
a saturated hydrocarbon containing carbon atoms joined in a ring

cyclotron
accelerator used by scientists to conduct nuclear reactions, particularly bombardment reactions to produce new elements

D

data
objective pieces of information, such as information gathered in a laboratory investigation

decay series
the decay of a particular radioisotope, which yields a different radioisotope that, in turn, decays; this sequential decay process may continue through several more radioisotopes until a stable nucleus is produced

density
the mass per unit volume of a given material that is often expressed as g/cm^3

dependent variable
measured or observed variable in an experiment that is used to draw conclusions about effects of changes to the independent variable

deposit
naturally occurring collection of ores in the lithosphere

derived unit
an SI unit formed by mathematically combining two or more base units

diagnostic
a test that helps doctors understand what is happening inside the body

diatomic molecule
a molecule made up of two atoms, such as chlorine gas, Cl_2, or carbon monoxide, CO

dilution
process of making a solution less concentrated by adding solvent

dimer
a molecule composed of two monomers

dipeptide
a molecule consisting of two amino acids bonded together

direct water use
water consumed by an end user

disaccharide
a sugar molecule (such as sucrose) composed of two monosaccharide units bonded through a condensation reaction

distillate
the condensed products of distillation

distillation
a process that separates liquid substances based on differences in their boiling points; *see also* fractional distillation

dose
quantity of ionizing radiation that people are exposed to over time

dot structure
see electron-dot structure

double covalent bond
a bond in which four electrons are shared between two adjacent atoms

drinking-water treatment
pre-use purification of water that occurs at a filtration and treatment plant

dry cell
battery in which the electrolyte is in paste form, rather than liquid

ductile (ductility)
a property of a material that permits it to be stretched into a wire without breaking

dynamic equilibrium
see equilibrium

E

elastic collision
molecular collision in which there is no gain or loss in total kinetic energy

electric current
the flow of electrons, as through a wire connecting electrodes in a voltaic cell

electrical potential
the tendency for electrical charge to move through an electrochemical cell (based on an element's relative tendency to lose electrons when in contact with a solution of its ions); it is measured in volts (V)

electrochemistry
the study of chemical changes that produce or are caused by electrical energy

electrode
a strip of metal or other conductor serving as a contact between an ionic solution and the external circuit in an electrochemical cell

electrolysis
the process in which a chemical reaction is caused by passing an electrical current through an ionic solution

electromagnetic radiation
radiation ranging from low-energy radio waves to high-energy X-rays and gamma rays; includes visible light

electromagnetic spectrum
comprising the full range of electromagnetic radiation frequencies; *see also* electromagnetic radiation

electron
a particle possessing a negative electrical charge; electrons surround the nuclei of atoms

electron-dot formula
see Lewis dot structure

electron-dot structure
a structure of a substance or ion in which dots represent the valence electrons in each atom

electronegativity
an expression of the tendency of an atom to attract shared electrons within a chemical bond

electroplating
the deposition of a thin layer of metal on a surface by an electrical process involving oxidation–reduction reactions

elements
the fundamental chemical substances from which all other substances are made

endothermic
a process that requires the addition of energy

endpoint
point where a titration is stopped, usually because an appropriate indicator just changes color

energy efficiency
the use of smaller quantities of energy to achieve the same effect

enzyme
biological catalyst

equilibrium
the point in a reversible reaction where the rate of products forming from reactants is equal to the rate of reactants forming from products; also called *dynamic equilibrium*

essential amino acid
amino acid not synthesized in adequate quantities by the human body and that must be obtained from protein in the diet

ester
an organic compound containing the —COOR group, where R represents any stable arrangement of bonded carbon and hydrogen atoms

evidence
qualitative observations or quantitative data; experimental support for claims; should be used to answer the questions, "How do I know what I know?" and "Why am I making this claim?"; it should be used in presentable format along with an explanation

exothermic
a process that involves the release of energy

experimental design
relying on making measurements on one variable while changing another variable for investigations

extrapolation
the process of estimating a value beyond a known range of data points

F

family (periodic table)
see group

fat
energy-storage molecule composed of carbon, hydrogen, and oxygen

fatty acid
organic compound made up of long hydrocarbon chains with carboxylic acid groups at one end

filtrate
the liquid collected after filtration

filtration
the process of separating solid particles from a liquid by passing the mixture through a material that retains the solid particles

fixed
the process in which nitrogen is combined with other elements to produce nitrogen-containing compounds that plants can use chemically

fluorescence
the emission of visible light when exposed to radiant energy (usually ultraviolet radiation)

food additive
a substance added during food processing that is intended to increase nutritive value or enhance storage life, visual appeal, or ease of production

food group
one of several categories for foods that humans eat, typically grains, vegetables, fruits, milk, meats, and beans

force
a push or pull exerted on an object; expressed by the newton (N), an SI unit

formula unit
a group of atoms or ions represented by a compound's chemical formula; simplest unit of an ionic compound

fossil fuel
a fuel (such as coal, petroleum, or natural gas) believed to be formed from plant or animal remains that were buried under Earth's surface for millions of years

fraction
(a) a mixture of petroleum-based substances with similar boiling points and other properties; (b) one of the substances collected during distillation

fractional distillation
a process of separating a mixture into its components by boiling and condensing the components; *see also* distillation

frequency (υ)
the number of waves that pass a given point each second; in other words, the rate of oscillation; for electromagnetic radiation, the product of frequency and wavelength equals the speed of light

fuel cell
a device for directly converting chemical energy into electrical energy by chemically combining a fuel (such as hydrogen gas) with oxygen gas; does not involve combustion

functional group
an atom or a group of atoms that imparts characteristic properties to an organic compound

G

gamma ray (γ)
high-energy electromagnetic radiation emitted during the decay of some radioactive elements

global warming
the observed and predicted increases in average global surface temperatures

glycogen
the carbohydrate by which animals store glucose

gray (Gy)
the unit expressing the quantity of ionizing radiation delivered to tissue; it equals one joule absorbed per kilogram of body tissue

Green Chemistry
the design of chemical products and processes that require fewer resources and less energy and that reduces or eliminates reliance on and generation of hazardous substances

greenhouse effect
the trapping and returning of infrared radiation to Earth's surface by atmospheric substances such as water and carbon dioxide

greenhouse gas
atmospheric substance that absorbs infrared radiation, such as CO_2, N_2O, and CH_4

groundwater
water from an aquifer or other underground source

group (periodic table)
a vertical column of elements in the periodic table; also called a *family*; group members share similar properties

H

Haber–Bosch process
the industrial synthesis of ammonia from hydrogen and nitrogen gases, which involves high pressure and temperature accompanied by a suitable catalyst

half-cell
a metal (or other electrode material) in contact with a solution of ions to form one half of a voltaic cell

half-life
the time for a radioactive substance to lose half of its radioactivity from decay; at the end of one half-life, 50% of the original radioisotope remains

half-reaction
a chemical equation explicitly showing either a loss of electrons (oxidation) or a gain of electrons (reduction); any oxidation–reduction reaction can be expressed as the sum of two half-reactions

halogen
see halogen family

halogen family
the group of elements consisting of fluorine, chlorine, bromine, iodine, and astatine

hazardous air pollutant
substance in air that is known or suspected to cause cancer or other serious health effects or adverse environmental effects; air toxic

heat
see thermal energy

heat of combustion
the quantity of thermal energy released when a specific quantity of a material burns

heterogeneous mixture
a mixture that is not uniform throughout

high-level radioactive waste
(a) products of nuclear fission, such as those generated in a nuclear reactor; (b) transuranics, products formed when the original uranium-235 fuel absorbs neutrons

histogram
a graph indicating the frequency or number of instances of particular values (or value ranges) within a set of related data

homogeneous mixture
a mixture that is uniform throughout; a solution is a homogeneous mixture

hybrid vehicle
a vehicle that combines two or more power sources; the combination of gasoline and electric power is the most common design

hydrocarbon
a molecular compound composed only of carbon and hydrogen atoms

hydrogen bond
strong intermolecular force in compounds in which a hydrogen atom is bonded directly to an atom of oxygen, nitrogen, or fluorine

hydrogenation
a chemical reaction that adds hydrogen atoms to an organic molecule

hydrologic cycle
see water cycle

hydronium ion
in aqueous solutions, the ion formed when the hydrogen ion released by the acid is bonded to water, commonly represented as $H_3O^+(aq)$

hydrosphere
all parts of Earth where water is found, including oceans, clouds, ice caps, glaciers, lakes, rivers, and underground water supplies

I

ideal gas
a gas that behaves under all conditions as described by the kinetic molecular theory or by the ideal gas law

ideal gas law
a mathematical relationship that describes the behavior of an ideal gas sample, $PV = nRT$, where P = gas pressure; V = gas volume; n = moles of gas; R = gas constant, 0.0821 L • atm/(mol • K); and T = kelvin temperature

independent variable
variable that is manipulated by the investigator during an experiment

indirect water use
water consumed in the preparation, production, or delivery of goods and services

inference
a conclusion based on analysis of data and observations

infrared (IR)
electromagnetic radiation just beyond the red (low-energy) end of the visible spectrum

intermolecular force
force of attraction among molecules

International System of Units (SI)
the modernized metric system

ion
electrically charged atom or group of atoms; negative ions are called *anions*, and positive ions are called *cations*

ionic compound
a substance composed of positive and negative ions

ionize
to convert to ions

ionizing radiation
nuclear radiation and high-energy electromagnetic radiation with sufficient energy to produce ions by ejecting electrons from atoms and molecules

isomer
a molecule that has the same formula as another molecule and differs from it only by the arrangement of atoms or bonds

isomerization
a chemical change involving the rearrangement of atoms or bonds within a molecule without changing its molecular formula

isotope
atom of the same element with differing numbers of neutrons

K

kelvin temperature scale (K)
the absolute temperature scale, where zero kelvins (0 K) represents the theoretical lowest possible temperature; 0 °C = 273 K

kinetic energy
energy associated with the motion of an object

kinetic molecular theory (KMT)
observed gas behavior and gas laws explained by rapidly moving particles (gas molecules) that are relatively far apart and change direction only through collisions with each other or the container walls

kinetics
see chemical kinetics

L

law of conservation of energy
energy can change form but cannot be created or destroyed in any chemical reaction or physical change

law of conservation of matter
matter is neither created nor destroyed in any chemical reaction or physical change

Le Châtelier's principle
the predicted shift in the equilibrium position that partially counteracts the imposed change in conditions

Lewis dot structure
the representation of atoms, ions, and molecules where valence-electron dots surround each atom's symbol; it is useful for indicating covalent bonding; also called *electron-dot structure*

Lewis structure
see Lewis dot structure

life cycle
the sequence of steps that a material or product undergoes from raw materials to product to final disposal

limiting reactant
a starting substance that is used up first in a chemical reaction; sometimes called the *limiting reagent*

liquefied petroleum gas (LPG)
petroleum-based gaseous substance fuel, propane (C_3H_8)

lithosphere
the solid outer layer of Earth, which also includes land areas under oceans and other bodies of water

low-level radioactive waste
nuclear laboratory protective clothing, diagnostic radioisotopes, and air filters from nuclear power plants

luster
the reflection of light from the surface of a material

M

macromineral
mineral essential to human life and occurring in relatively large quantities (at least 5 g) in the body; also called a *major mineral*

macroscopic
large enough to be seen by an unaided eye

magnetic resonance imaging (MRI)
a non-invasive computerized method of imaging soft human tissues by use of a powerful magnet and radio wave

major mineral
see macromineral

malleable
a property of a material that permits it to be flattened without shattering

mass number
the sum of the number of protons and neutrons in the nucleus of an atom of a particular isotope

material's life cycle
several distinct stages, which include aquisition, manufacturing, use/reuse/maintenance, and recycle/waste management

matter
anything that has mass and occupies space

mean
an expression of central tendency obtained by dividing the sum of a set of values by the number of values in the set; also known as the *average value*

median
within an ascending or descending set of values, the number that represents the middle value with an equal number of values above and below it

metabolism
complex series of interrelated chemical reactions that keep organisms alive

metal
a material possessing properties such as luster, ductility, conductivity, and malleability

metalloid
a material with properties intermediate between those of metals and nonmetals

metastable
nuclei in an energetically excited state; designated by the symbol *m*

meter (m)
the SI base unit of length

micromineral
mineral essential to human life and occurring in relatively small quantities (less than 5 g) in the body; also called a *trace mineral*

mineral
(a) a naturally occurring solid substance commonly removed from ores to obtain a particular element of interest or value; (b) an inorganic substance needed by the body to maintain good health

mixture
a combination of materials in which each material retains its separate identity

model
tool to understand and interpret natural phenomena that can be physical, mathematical, or conceptual and represent, explain, or predict observed behavior; models are particularly helpful in chemistry to account for molecular-level interactions

molar concentration (M)
see molarity

molar heat of combustion
the quantity of thermal energy released from burning one mole of a substance

molar mass
the mass (usually in grams) of one mole of a substance

molar volume
the volume occupied by one mole of a substance; at 0 °C and 1 atm, the molar volume of any gas is 22.4 L

molarity (M)
the concentration determined by dividing the total moles of solute by the solution volume (expressed in liters); also known as *molar concentration*

mole (mol)
the SI unit for amount of a substance, equal to 6.02×10^{23} units, where the unit may be any specified entity; it is the chemist's "counting" unit

molecular formula
a chemical expression indicating the total atoms of each element contained in one molecule of a particular substance

molecule
the smallest particle of a substance retaining all the properties of that substance

monatomic ion
ion composed of only one atom, such as chloride (Cl^-)

monomer
a compound whose molecules can react to form the repeating units of a polymer

monosaccharide
a simple sugar (such as glucose or fructose) that cannot be hydrolyzed to produce other sugars

monounsaturated fat
a fat molecule containing one carbon–carbon double bond

mutagen
a material that causes mutations in DNA

mutation
changes in the structure of DNA that may result in production of altered protein material

N

negative oxidation state
the negative number assigned to an atom in a compound when that atom has greater control of bonding electrons than the control exerted by one or more atoms to which it is bonded

net ionic equation
equation written without the spectator ions, such as $Ag^+(aq) + Cl^-(aq) \longrightarrow AgCl(s)$

neutral solution
a water solution in which H^+ (H_3O^+) and OH^- concentrations are equal (pH = 7)

neutralization
combining an acid and a base in amounts that result in the elimination of all excess acid or base

neutron
a particle without electrical charge; found in the nucleus of an atom

newton (N)
the SI unit of force that is roughly equal to the force exerted by a mass of 100 g at Earth's surface

nitrogen cycle
the movement of atmospheric nitrogen atoms through Earth's ecosystems via collection by bacteria, conversion into ammonia or ammonium ions, conversion into nitrate ions, uptake by plants, passage through the food chain, release as ammonia or ammonium ions, and conversion back into atmospheric nitrogen

noble gas family
an unreactive element belonging to the last (right-most) group on the periodic table

nonconductor
a material that does not allow electrical current (or thermal energy) to flow through it

nonionizing radiation
electromagnetic radiation in the visible and lower-energy regions of the electromagnetic spectrum with insufficient energy to form ions when it transfers energy to matter

nonmetal
a material possessing properties such as brittleness, lack of luster, and nonconductivity; nonmetals are often insulators

nonpolar molecule
a molecule that has an even distribution of electrical charge with no regions of partial positive and negative charge

nonrenewable resource
a resource in limited supply that cannot be replenished by natural processes over the time frame of human experience

nuclear fission
splitting an atom into two smaller nuclei

nuclear fusion
the combination of two nuclei to form a new, more massive nucleus

nuclear power plant
a facility where thermal energy generated by the controlled fission of nuclear fuel drives a steam turbine, producing electrical power; in other words, a facility that converts nuclear energy into electricity

nuclear radiation
a form of ionizing radiation that results from changes in the nuclei of atoms

nucleus, atomic
the dense, positively charged central region of an atom that contains protons and neutrons

O

observation
data/information that you can collect with your senses; what you see, hear, feel, or smell

octane rating
a measure of the combustion quality of gasoline compared to the combustion quality of isooctane; the higher the number, the higher the octane rating; also called *octane number*

oil sands
source of petroleum that contains bitumen, a viscous, heavy crude oil

oil shale
sedimentary rock containing a material (kerogen) that can be converted to crude oil

ore
a rock or other solid material from which it is profitable to recover a mineral containing a metal or other useful substances

organic chemistry
a branch of chemistry dealing with hydrocarbons and their derivatives

oxidation
any process in which one or more electrons can be considered as lost by a chemical species

oxidation state
the apparent state of oxidation of an atom; also called *oxidation number*

oxidation–reduction (redox) reaction
a chemical reaction in which oxidation and reduction simultaneously occur

oxidized
see oxidation

oxidizing agent
a species that causes another atom, molecule, or ion to become oxidized; the oxidizing agent becomes reduced in this process

oxygenated fuel
a fuel with oxygen-containing additives, such as methanol, that increase the octane rating and reduce harmful emissions

P

paper chromatography
a method for separating substances that relies on solution components having different attractions to the solvent (mobile phase) and paper (stationary phase)

particulate level
the realm of unseen atoms, molecules, and ions in contrast to the observable macroscopic entities

particulate pollutant
microscopic particle that enters the air from either human activities or natural processes

pascal (Pa)
the SI pressure unit; equal to one newton of force applied per square meter, $1 \text{ Pa} = 1 \text{ N/m}^2$

peptide bond
the chemical bond that links amino acids together in peptides and proteins

percent composition
the percent by mass of each component in a material; or, specifically, the percent by mass of each element within a compound

percent recovery
the proportion of sought material recovered in a process

period (periodic table)
a horizontal row of elements in the periodic table

periodic properties
chemical or physical properties that vary among elements according to trends that repeat as atomic number increases

periodic relationship
regular patterns among chemical and physical properties of elements arrayed on a periodic table

periodic table of the elements
an arrangement of elements in order of increasing atomic number, such that elements with similar properties are located in the same vertical column (group)

petrochemical
any organic compound produced from petroleum or natural gas

pH scale
method used as a convenient way to measure and report acidic, basic, or chemically neutral character of a solution; pH is based on a solution's hydrogen ion (H^+ or H_3O^+) molar concentration

photochemical smog
a potentially hazardous mixture of secondary pollutants formed by solar irradiation of certain primary pollutants in the presence of oxygen

photon
an energy bundle of electromagnetic radiation that travels at the speed of light

photosynthesis
the process by which green plants and some microorganisms use solar energy to convert water and carbon dioxide to carbohydrates (stored chemical energy)

physical change
a change in matter in which the identity of the material involved does not change

physical property
a property that can be observed or measured without changing the identity of the sample of matter

polar molecule
a molecule with regions of partial positive and negative charge resulting from the uneven distribution of electrical charge

pollutant
an undesirable contaminant that adversely affects the chemical, physical, or biological characteristics of the environment

polyatomic ion
an ion composed of two or more atoms, such as the ammonium cation, NH_4^+, or acetate anion, $C_2H_3O_2^-$

polymer
a molecule composed of very large numbers of identical repeating units

polysaccharide
a polymer composed of many monosaccharide units

polyunsaturated fat
a fat molecule containing two or more carbon–carbon double bonds

positive oxidation state
the positive number assigned to an atom in a compound when that atom has less control of its electrons than it has as a free element

positron
a positively charged subatomic particle with the same mass as an electron; the antimatter counterpart to the electron

positron emission tomography (PET)
a technique for examining metabolic activity in tissues (particularly in the brain) by measuring blood flow containing tracers that emit positrons

postulate
an accepted statement used as the basis for developing an argument or explanation; also called an *axiom*

potential energy
energy associated with position

precipitate
an insoluble solid substance that has separated from a solution

precision
describes how closely repeated measurements cluster around the same value

pressure
force applied per unit area; in SI, pressure is expressed in pascals (Pa)

primary air pollutant
a contaminant that directly enters the atmosphere; it is not initially formed by reactions of airborne substances

primary battery
battery designed for a single use and that cannot be recharged

product
a substance formed in a chemical reaction

protein
a major structural component of living tissue made from many linked amino acids

proton
a particle possessing a positive electrical charge that is found in the nuclei of all atoms; the total protons in an element's atom equals its atomic number

Q

qualitative test
a chemical test indicating the presence or absence of an element, ion, or compound in a sample

quantitative test
a chemical test indicating the amount or concentration of an element, ion, or compound in a sample

R

rad
a unit that expresses the quantity of ionizing radiation absorbed by tissue; 1 rad = 0.01 Gy (Gray)

radiation
energy emitted in the form of electromagnetic waves or high-speed particles; refers to both ionizing radiation and nonionizing radiation

radioactive decay
a change in an atom's nucleus due to the spontaneous emission of alpha, beta, or gamma radiation

radioactivity
the spontaneous emission of nuclear radiation

radioisotope
radioactive isotope

range
the difference between the highest and lowest values in a data set

reactant
a starting material in a chemical reaction

reaction rate
an expression of how fast a particular chemical change occurs

redox reaction
see oxidation–reduction reaction

reduced
see reduction

reducing agent
a species that causes another atom, molecule, or ion to become reduced; the reducing agent, in turn, becomes oxidized in this process

reduction
any process in which one or more electrons can be considered as gained by a chemical species

reference solution
a solution of known composition used as a comparison in chemical tests

refined
removal of impurities from a desired material

rem
a unit that expresses the ability of radiation to cause ionization in human tissue; 1 rem = 0.01 Sv (sievert)

renewable resource
a resource that can be replenished by natural processes over the time frame of human experience

reversible reaction
a chemical reaction in which products form reactants at the same time that reactants form products

S

salt bridge
a connection that allows a voltaic cell's two half-cells to be in electrical contact without mixing; specifically, a tube containing an electrolyte (such as potassium chloride solution) that completes the internal circuit of a voltaic cell

saturated fat
a fat molecule containing only single carbon–carbon bonds within its fatty acid components

saturated hydrocarbon
a hydrocarbon consisting of molecules in which each carbon atom is bonded to four other atoms

saturated solution
a solution in which the solvent has dissolved as much solute as it can retain stably at a specified temperature

scientific model
a representation of either a part of the natural world or of a scientific theory

scientific question
question that provides a framework for gathering and analyzing data that will ultimately result in being able to describe, explain, or predict natural phenomena

scientific theory
a coherent set of ideas that explains many related observations or events in the natural world and offer "how"-type explanations of phenomena in the natural world

scintillation counter
a detector of ionizing radiation that measures light emitted by atoms that have been excited by ionizing radiation

secondary air pollutant
a contaminant generated in the atmosphere by chemical reactions between primary air pollutants and natural components of air

shell (electron)
energy level surrounding an atom's nucleus within which one or more electrons reside; outer-shell electrons are commonly called *valence electrons*

sievert (Sv)
an SI unit that expresses the dose equivalent of absorbed radiation that causes the same biological effects as one gray of gamma rays

single covalent bond
a bond in which two electrons are shared by the two bonded atoms

smog
the potentially hazardous combination of smoke and fog; *see also* photochemical smog

solid-state detector
a device used to monitor changes in the movement of electrons through semiconductors as they are exposed to ionizing radiation

solubility
the quantity of a substance that will dissolve in a given quantity of solvent to form a saturated solution at a particular temperature

solubility curve
a graph indicating the solubility of a particular solute at different temperatures

solubility rules
trends that have been identified within known data that make it easier to identify a precipitate

solute
the dissolved species in a solution; the solute is usually the smaller component of a solution

solution
a homogeneous mixture of two or more substances

solution concentration
the quantity of solute dissolved in a specific quantity of solvent or solution

solvent
the dissolving agent in a solution; the solvent is usually the larger component in a solution

space-filling model
model that depicts atoms in contact with each other

species
a general name used in chemistry for atoms, molecules, ions, free radicals, or other well-defined entities

specific heat capacity
the quantity of thermal energy needed to raise the temperature of 1 g of a material by 1 °C; the expression commonly has units of J/(g °C)

spectator ion
ion that does not participate in the reaction

spontaneous
reactions occurring without any input of additional stimuli; the overall energy of products is lower than the overall energy of reactants

standard solution
a solution of known concentration

standard temperature and pressure (STP)
conditions of 0 °C and 1 atm

stoichiometry
the relationships by which quantities of substances involved in a chemical reaction are linked and calculated

straight-chain alkane
an alkane consisting of molecules in which each carbon atom is linked to no more than two other carbon atoms

strong acid
an acid that fully ionizes in solution to liberate H^+ (H_3O^+); no molecular form of the acid remains

strong base
a base that fully liberates OH^- in solution

strong force
the force that holds protons and neutrons together in an atom's nucleus

structural formula
a chemical formula showing the arrangement of atoms and covalent bonds in a molecule, in which each electron pair in a covalent bond is represented by a line between the symbols of two atoms

structural isomers
substances involving rearrangement of atoms or bonds within a molecule but sharing a common molecular formula

subatomic particle
particle smaller than an atom; commonly regarded as electrons, protons, neutrons

subscript
the number printed below the line of type indicating the total atoms of a given element in a chemical formula; in H_2O, for example, the subscript 2 specifies the total H atoms

substance
an element or a compound; that is, a material with a uniform, definite composition and distinct properties

substituted hydrocarbon
carbon-backbone hydrocarbon with other elements substituted for one or more hydrogen atoms

substrate
a molecule that interacts with an enzyme and undergoes a reaction

superconductivity
the ability of a material to conduct an electrical current with zero electrical resistance; with present technology, operating superconductors must be extremely cold

surface water
water found on Earth's surface, such as oceans, rivers, and lakes

suspension
a mixture containing large, dispersed solid particles that can settle out or be separated by filtration

sustainability
present-day activities that preserve the ability of future generations to thrive and meet their resource needs on a habitable Earth

synergistic interaction
an interaction where the combined effect of several factors is greater than the sum of their separate effects

T

temperature inversion
an atmospheric condition where a cool air mass is trapped beneath a less dense warm air mass; it most frequently occurs in a valley or over a city

tetrahedron
a regular triangular pyramid; the four bonds of each carbon atom in an alkane point to the corners of a tetrahedron

therapeutic
treating a medical condition

thermal energy
the energy a material possesses due to its temperature; also known as *heat*

titrant
in a titration, the solution of known concentration that is added until an endpoint is reached

titration
a laboratory procedure for determining the concentrations of dissolved substances

total ionic equation
equation that accurately reflects the form of all substances in solution, such as $Ag^+(aq) + NO_3^-(aq) + Na^+(aq) + Cl^-(aq) \longrightarrow AgCl(s) + Na^+(aq) + NO_3^-(aq)$

trace mineral
see micromineral

tracer
a readily-identifiable material, such as a radioisotope, used to diagnose disease or to determine how the body is responding to treatment

transmutation
the conversion of one element to another either naturally or artificially

transuranium
any element with an atomic number greater than 92 (uranium)

triglyceride
a fat molecule composed of a simple three-carbon alcohol (glycerol) and three fatty acid molecules

trihalomethane (THM)
substance that in sufficiently high concentrations can be harmful to human health; a methane molecule with 3 hydrogen atoms substituted with halogen atoms

troposphere
the layer of the atmosphere closest to Earth's surface where most clouds and weather are located

Tyndall effect
the scattering of a beam of light caused by reflection from suspended particles

U

unsaturated fat
a fat molecule containing one or more carbon–carbon double bonds; it is monounsaturated if each fat molecule has a single double bond and polyunsaturated if it has two or more double bonds

unsaturated hydrocarbon
a hydrocarbon molecule containing one or more double or triple bonds

unsaturated solution
a solution containing a lower concentration of solute than a saturated solution contains at a specified temperature

V

valence electron
electron in the outermost shell of an atom; these relatively loosely held electrons often participate in bonding with other atoms or molecules

vaporization
the phase change that occurs when a substance changes from a liquid state to a gaseous state

viscosity
resistance to flow

vitamin
a biomolecule necessary for growth, reproduction, health, and life

vitrification
the conversion of material into a glassy solid by the application of high temperatures

volatile organic compound (VOC)
reactive carbon-containing substance that readily evaporates into air, such as components of gasoline and organic solvents

voltaic cell
an electrochemical cell in which a spontaneous chemical reaction produces electricity

W

water cycle
repetitive processes of rainfall (or other precipitation), run-off, evaporation, and condensation that circulate water within Earth's crust and atmosphere; also called the *hydrologic cycle*

wavelength (λ)
the distance between corresponding points of two consecutive waves; for electromagnetic radiation, the product of frequency and wavelength equals the speed of light

weak acid
an acid that does not fully ionize in solution to liberate H^+ (H_3O^+) but remains primarily in molecular form

X

X-ray
high-energy electromagnetic radiation that cannot penetrate dense materials such as bone or lead but can penetrate less dense materials

Z

zero oxidation state
the value of the oxidation state of an element's atoms when not chemically combined with any other element

Photo Credits

Project Credits

CHEMCOM Project Credits

ChemCom is the product of teamwork involving individuals from all over the United States over more than twenty-five years. The American Chemical Society is pleased to recognize all who contributed to *ChemCom*. The team responsible for the sixth edition of *ChemCom* is listed on the copyright page. Individuals who contributed to the initial development of *ChemCom*—for the first edition in 1988, the second and third editions in 1993 and 1998, respectively, the fourth edition in 2002, and the fifth edition in 2006—are listed below.

Principal Investigator: W. T. Lippincott

Project Manager: Sylvia Ware

Chief Editor: Henry Heikkinen & Conrad L. Stanitski

Contributing Editor: Mary Castellion

Assistant to Contributing Editor: Arnold Diamond

Editor of Teacher's Guide: Thomas O'Brien & Patricia J. Smith

Revision Team: Diane Bunce, Gregory Crosby, David Holzman, Thomas O'Brien, Joan Senyk, Thomas Wysocki

Editorial Advisory Board: Joseph Breen, Glenn Crosby, James DeRose, I. Dwaine Eubanks, Lucy Pryde Eubanks, Regis Goode, Henry Heikkinen (chair), Mary Kochansky, Ivan Legg, W. T. Lippincott (ex officio), Steven Long, Nina McClelland, Lucy McCorkle, Carlo Parravano, Robert Patrizi, Max Rodel, K. Michael Shea, Patricia Smith, Susan Snyder, Conrad Stanitski, Jeanne Vaughn, Sylvia Ware (ex officio)

Writing Team: Rosa Balaco, James Banks, Joan Beardsley, William Bleam, Kenneth Brody, Ronald Brown, Diane Bunce, Becky Chambers, Alan DeGennaro, Patricia Eckfeldt, Dwaine Eubanks (dir.), Henry Heikkinen (dir.), Bruce Jarvis (dir.), Dan Kallus, Jerry Kent, Grace McGuffie, David Newton (dir.), Thomas O'Brien, Andrew Pogan, David Robson, Amado Sandoval, Joseph Schmuckler (dir.), Richard Shelly, Patricia Smith, Tamar Susskind, Joseph Tarello, Thomas Warren, Robert Wistort, Thomas Wysocki

Steering Committee: Alan Cairncross, William Cook, Derek Davenport, James DeRose, Anna Harrison (ch.), W. T. Lippincott (ex officio), Lucy McCorkle, Donald McCurdy, William Mooney, Moses Passer, Martha Sager, Glenn Seaborg, John Truxall, Jeanne Vaughn

Consultants: Alan Cairncross, Michael Doyle, Donald Fenton, Conrad Fernelius, Victor Fratalli, Peter Girardot, Glen Gordon, Dudley Herron, John Hill, Chester Holmlund, John Holman, Kenneth Kolb, E. N. Kresge, David Lavallee, Charles Lewis, Wayne Marchant, Joseph Moore, Richard Millis, Kenneth Mossman, Herschel Porter, Glenn Seaborg, Victor Viola, William West, John Whitaker

Synthesis Committee: Diane Bunce, Dwaine Eubanks, Anna Harrison, Henry Heikkinen, John Hill, Stanley Kirschner, W. T. Lippincott (ex officio), Lucy McCorkle, Thomas O'Brien, Ronald Perkins, Sylvia Ware (ex officio), Thomas Wysocki

Evaluation Team: Ronald Anderson, Matthew Bruce, Frank Sutman (dir.)

Field Test Coordinator: Sylvia Ware

Field Test Workshops: Dwaine Eubanks

Field Test Directors: Keith Berry, Fitzgerald Bramwell, Mamie Moy, William Nevill, Michael Pavelich, Lucy Pryde, Conrad Stanitski

Pilot Test Teachers: Howard Baldwin, Donald Belanger, Navarro Bharat, Ellen Byrne, Eugene Cashour, Karen Cotter, Joseph Deangelis, Virginia Denney, Diane Doepken, Donald Fritz, Andrew Gettes, Mary Gromko, Robert Haigler, Anna Helms, Allen Hummel, Charlotte Hutton, Elaine Kilbourne, Joseph Linker, Larry Lipton, Grace McGuffie, Nancy Miller, Gloria Mumford, Beverly Nelson, Kathy Nirei, Elliott Nires, Polly Parent, Mary Parker, Dicie Petree, Ellen Pitts, Ruth Rand, Kathy Ravano, Steven Rischling, Charles Ross, Jr., David Roudebush, Joseph Rozaik, Susan Rutherland, George Smeller, Cheryl Snyder, Jade Snyder, Samuel Taylor, Ronald Tempest, Thomas Van Egeren, Gabrielle Vereecke, Howard White, Thomas Wysocki, Joseph Zisk

Field Test Teachers: Vincent Bono, Allison Booth, Naomi Brodsky, Mary D. Brower, Lydia Brown, George Bulovsky, Kay Burrough, Gene Cashour, Frank Cox, Bobbie Craven, Pat Criswell, Jim Davis, Nancy Dickman, Dave W. Gammon, Frank Gibson, Grace Giglio, Theodis Gorre, Margaret Guess, Yvette Hayes, Lu Hensen, Kenn Heydrick, Gary Hurst, Don Holderread, Michael Ironsmith, Lucy Jache, Larry Jerdal, Ed Johnson, Grant Johnson, Robert Kennedy, Anne Kenney, Joyce Knox, Leanne Kogler, Dave Kolquist, Sherman Kopelson, Jon Malmin, Douglas Mandt, Jay Maness, Patricia Martin, Mary Monroe, Mike Morris, Phyllis Murray, Silas Nelson, Larry Nelson, Bill Rademaker, Willie Reed, Jay Rubin, Bill Rudd, David Ruscus, Richard Scheele, Paul Shank, Dawn Smith, John Southworth, Mitzi Swift, Steve Ufer, Bob Van Zant, Daniel Vandercar, Bob Volzer, Terri Wahlberg, Tammy Weatherly, Lee Weaver, Joyce Willis, Belinda Wolfe

Field Test Schools: California: Chula Vista High, Chula Vista; Gompers Secondary School, San Diego; Montgomery High, San Diego; Point Loma High, San Diego; Serra Junior-Senior High, San Diego; Southwest High, San Diego. Colorado: Bear Creek Senior High, Lakewood; Evergreen Senior High, Evergreen; Green Mountain Senior High, Lakewood; Golden Senior High, Golden; Lakewood Senior High, Lakewood; Wheat Ridge Senior High, Wheat Ridge. Hawaii: University of Hawaii Laboratory School, Honolulu. Illinois: Project Individual Education High, Oak Lawn. Iowa: Linn-Mar High, Marion. Louisiana: Booker T. Washington High, Shreveport; Byrd High, Shreveport; Caddo Magnet High, Shreveport; Captain Shreve High, Shreveport; Fair Park High, Shreveport; Green Oaks High, Shreveport; Huntington High, Shreveport; North Caddo High, Vivian; Northwood High, Shreveport. Maryland: Charles Smith Jewish Day School, Rockville; Owings Mills Junior-Senior High, Owings Mills; Parkville High, Baltimore; Sparrows Point Middle-Senior High, Baltimore; Woodlawn High, Baltimore. New Jersey: School No. 10, Patterson. New York: New Dorp High, Staten Island. Texas: Clements High, Sugar Land; Cy-Fair High, Houston.

Virginia: Armstrong High, Richmond; Freeman High, Richmond; Henrico High, Richmond; Highland Springs High, Highland Springs; Marymount School, Richmond; Midlothian High, Midlothian; St. Gertrude's High, Richmond; Thomas Dale High, Chester; Thomas Jefferson High, Richmond; Tucker High, Richmond; Varina High, Richmond. Wisconsin: James Madison High, Madison; Thomas More High, Milwaukee. Washington: Bethel High, Spanaway; Chief Sealth High, Seattle; Clover Park High, Tacoma; Foss Senior High, Tacoma; Hazen High, Renton; Lakes High, Tacoma; Peninsula High, Gig Harbor; Rogers High, Puyallup; Sumner Senior High, Sumner; Washington High, Tacoma; Wilson High, Tacoma

Safety Consultant: Stanley Pine & William H. Breazeale, Jr.

Social Science Consultants: Ross Eshelman, Judith Gillespie

Art: Rabina Fisher, Pat Hoetmer, Alan Kahan (dir.), Kelley Richard, Sharon Wolfgang

Copy Editor: Martha Polkey

Production Consultant: Marcia Vogel

Administrative Assistant: Carolyn Avery

ACS Staff: Rebecca Mason Simmons, Martha K. Turckes

Student Aides: Paul Drago, Stephanie French, Patricia Teleska

Second Edition Revision Team

Project Manager: Keith Michael Shea & Ted Dresie

Chief Editor: Henry Heikkinen

Assistant to Chief Editor: Wilbur Bergquist

Editor of Teacher's Guide: Jon Malmin

Second Edition Editorial Advisory Board: Diane Bunce, Henry Heikkinen (ex officio), S. Allen Heininger, Donald Jones (chair), Jon Malmin, Paul Mazzocchi, Bradley Moore, Carolyn Morse, Keith Michael Shea (ex officio), Sylvia Ware (ex officio)

Teacher Reviewers of First Edition: Vincent Bono, New Dorp High School, New York; Charles Butterfield, Brattle Union High School, Vermont; Regis Goode, Spring Valley High School, South Carolina; George Gross, Union High School, New Jersey; C. Leonard Himes, Edgewater High School, Florida; Gary Hurst, Standley Lake High School, Colorado; Jon Malmin, Peninsula High School, Washington; Maureen Murphy, Essex Junction Educational Center, Vermont; Keith Michael Shea, Hinsdale Central High School, Illinois; Betsy Ross Uhing, Grand Island Senior High School, Nebraska; Jane Voth-Palisi, Concord High School, New Hampshire; Terri Wahlberg, Golden High School, Colorado.

Teacher Reviewers of Second Edition: Michael Clemente, Carlson High School, Gibraltar, MI; Steven Long, Rogers High School, Rogers, AR; William Penker, Neillsville High School, Neillsville, WI; Audrey Mandel, Connetquot High School, Bohemia, NY; Barbara Sitzman, Chatsworth High School, Chatsworth, CA; Kathleen Voorhees, Shore Regional High School, West Long Branch, NJ; Debra Compton, Cy-Fair High School, Houston, TX; Christ Forte, York Community High School, Elmhurst, IL; Gwyneth D. Sharp, Cape Henlopen High School, Lewes, DE; Louis Dittami, Dover-Sherborn High School, Dover, MA; Sandra Mueller, John Burroughs School, St. Louis, MO; Kirk Soule, Sunset High School, Beaverton, OR; Sigrid Wiolkinson, Athens Area High School, Athens,

PA; Millie McDowell, Clayton High School, Clayton, MO; Leslie A. Roughley, Steward School, Richmond, VA; Robert Houle, Bacon Academy, Colchester, CT; Robert Storch, Bishop Ireton High School, Alexandria, VA; Michael Smolarek, Neenah High School, Neenah, WI; Fred Nozawa, Timpview High School, Provo, UT; Michael Sixtus, Mar Vista High School, Imperial Beach, CA

Safety Consultant: Stanley Pine

Editorial: The Stone Cottage

Design: Bonnie Baumann & P.C. & F., Inc.

Art: Additional art for this edition by Seda Sookias Maurer

ACS also offers thanks to the National Science Foundation for its support of the initial development of *ChemCom*, and to NSF project officers Mary Ann Ryan and John Thorpe for their comments, suggestions, and unfailing support.

Fourth Edition Credits

Chief Editor: Henry Heikkinen

Revision Team: Laurie Langdon, Robert Milne, Angela Powers, Christine Gaudinski, Courtney Willis

Revision Assistants: Cassie McClure, Seth Willis

Teacher Edition: Lear Willis, Joseph Zisk

Ancillary Materials: Regis Goode, Mike Clemente, Ruth Leonard

Fourth Edition Editorial Advisory Board: Conrad L. Stanitski (Chair), Boris Berenfeld, Jack Collette, Robert Dayton, Ruth Leonard, Nina I. McClelland, George Miller, Adele Mouakad, Carlo Parravano, Kirk Soulé, Maria Walsh, Sylvia A. Ware (ex officio), Henry Heikkinen (ex officio)

ACS: Sylvia Ware, Janet Boese, Michael Tinnesand, Guy Belleman, Patti Galvan, Helen Herlocker, Beverly DeAngelo

ACS Safety Committee: Henry Clayton Ramsey (Chair), Wayne Wolsey, Kevin Joseph Edgar, Herbert Bryce

Technical Reviewers: Steve Cawthron, Kenneth Hughes, Susan C. Karr, Mary Kirchhoff, David Miller, Charles Poukish, Mary Ann Ryan, Tracy Williamson

Teacher Edition Reviewers: Drew Lanthrum, Karen Morris

Safety Consultants: Stanley Pine, Herbert Bryce

Fifth Edition Credits

Chief Editor: Henry Heikkinen

Project Manager: Angela Powers

Revision Team: Laurie Langdon, Robert Milne, Wendy Naughton

ACS: Sylvia Ware, Michael Tinnesand, Terri Taylor, Helen Herlocker, Jodi Wesemann

ACS Project Editor: Rebecca Strehlow

Technical Reviewer: Conrad Stanitski

Technical Contributors: Janet Cohen, Susan Heikkinen, Jack A. Ladson, COLOR Science Consultancy

Fourth Edition Reviewers: Kirsten Almo, Mark Boehlor, Susan Berrend, Martin Besant, Regis Goode, Gary Jackson, Jane Meadows, Walt Shacklett, Terri Taylor

Teacher Contributors: Pat Chriswell, Robert Dayton, Regis Goode, Drew Lanthrum, Joelle Lastica, Steve Long, Cece Schwennsen, Barbara Sitzman, Terri Taylor

Index

Chart of the Elements

Element	Symbol	Atomic Number	Average Atomic	Element	Symbol	Atomic Number	Average Atomic
Actinium	Ac	89	[227]	Erbium	Er	68	167.26
Aluminum	Al	13	26.98	Europium	Eu	63	151.96
Americium	Am	95	[243]	Fermium	Fm	100	[257]
Antimony	Sb	51	121.76	Fluorine	F	9	19.00
Argon	Ar	18	39.95	Francium	Fr	87	[223]
Arsenic	As	33	74.92	Gadolinium	Gd	64	157.25
Astatine	At	85	[210]	Gallium	Ga	31	69.72
Barium	Ba	56	137.33	Germanium	Ge	32	72.64
Berkelium	Bk	97	[247]	Gold	Au	79	196.97
Beryllium	Be	4	9.01	Hafnium	Hf	72	178.49
Bismuth	Bi	83	208.98	Hassium	Hs	108	[277]
Bohrium	Bh	107	[272]	Helium	He	2	4.003
Boron	B	5	10.81	Holmium	Ho	67	164.93
Bromine	Br	35	79.90	Hydrogen	H	1	1.008
Cadmium	Cd	48	112.41	Indium	In	49	114.82
Calcium	Ca	20	40.08	Iodine	I	53	126.90
Californium	Cf	98	[251]	Iridium	Ir	77	192.22
Carbon	C	6	12.01	Iron	Fe	26	55.85
Cerium	Ce	58	140.12	Krypton	Kr	36	83.80
Cesium	Cs	55	132.91	Lanthanum	La	57	138.91
Chlorine	Cl	17	35.45	Lawrencium	Lr	103	[262]
Chromium	Cr	24	52.00	Lead	Pb	82	207.2
Cobalt	Co	27	58.93	Lithium	Li	3	6.94
Copernicium	Cn	112	[285]	Lutetium	Lu	71	174.97
Copper	Cu	29	63.55	Magnesium	Mg	12	24.31
Curium	Cm	96	[247]	Manganese	Mn	25	54.94
Darmstadtium	Ds	110	[281]	Meitnerium	Mt	109	[276]
Dubnium	Db	105	[268]	Mendelevium	Md	101	[258]
Dysprosium	Dy	66	162.50	Mercury	Hg	80	200.59
Einsteinium	Es	99	[252]	Molybdenum	Mo	42	95.96

Chart of the Elements

Element	Symbol	Atomic Number	Average Atomic	Element	Symbol	Atomic Number	Average Atomic
Neodymium	Nd	60	144.24	Silicon	Si	14	28.09
Neon	Ne	10	20.18	Silver	Ag	47	107.87
Neptunium	Np	93	[237]	Sodium	Na	11	22.99
Nickel	Ni	28	58.69	Strontium	Sr	38	87.62
Niobium	Nb	41	92.91	Sulfur	S	16	32.07
Nitrogen	N	7	14.01	Tantalum	Ta	73	180.95
Nobelium	No	102	[259]	Technetium	Tc	43	[98]
Osmium	Os	76	190.23	Tellurium	Te	52	127.60
Oxygen	O	8	16.00	Terbium	Tb	65	158.93
Palladium	Pd	46	106.42	Thallium	Tl	81	204.38
Phosphorus	P	15	30.97	Thorium	Th	90	232.04
Platinum	Pt	78	195.08	Thulium	Tm	69	168.93
Plutonium	Pu	94	[244]	Tin	Sn	50	118.71
Polonium	Po	84	[209]	Titanium	Ti	22	47.87
Potassium	K	19	39.10	Tungsten	W	74	183.84
Praseodymium	Pr	59	140.91	Ununhexium	Uuh	116	[293]
Promethium	Pm	61	[145]	Ununoctium	Uuo	118	[294]
Protactinium	Pa	91	231.04	Ununpentium	Uup	115	[288]
Radium	Ra	88	[226]	Ununquadium	Uuq	114	[289]
Radon	Rn	86	[222]	Ununtrium	Uut	113	[284]
Rhenium	Re	75	186.21	Uranium	U	92	238.03
Rhodium	Rh	45	102.91	Vanadium	V	23	50.94
Roentgenium	Rg	111	[280]	Xenon	Xe	54	131.29
Rubidium	Rb	37	85.47	Ytterbium	Yb	70	173.05
Ruthenium	Ru	44	101.07	Yttrium	Y	39	88.91
Rutherfordium	Rf	104	[265]	Zinc	Zn	30	65.38
Samarium	Sm	62	150.36	Zirconium	Zr	40	91.22
Scandium	Sc	21	44.96				
Seaborgium	Sg	106	[271]				
Selenium	Se	34	78.96